Beginning with Linear Algebra

by

Eric Carlen and Conceição Carvalho

preliminary version

Table of Contents

Overview of Chapter One

This chapter introduces the mathematical objects that we'll be working with in the rest of the book: vectors, matrices and linear transformations.

We start right away in the first section by considering examples of functions that take lists of data as input, and return lists of data as output. Such functions come up in all branches of science and engineering, and you are surely familiar with many examples. For example, a weather service may give you a list of numerical data: air pressure, temperature, humidity, etc., as a function of your position as specified by a list of coordinates – say lattitude and longitude.

These can be considered as *multivariable* functions, with several variables on both the input and output sides, but if we think of the whole lists of data on the input and output as new kind of variable – a *vector variable*, then these functions can be considered as vector valued functions of a *single* vector variable. The later point of view has significant advantages, as we shall see.

The functions of a single numerical variable that are probably most familiar to you are those built up out of algebraic operations like addition, multiplication and taking roots. For example, the function $f(x) = \sqrt{1 + x^2}$ is of this type. Algebra will be involved in the vector functions we consider here – after all, the subject is called linear algebra. In the second section, we introduce the basic algebraic operations on vectors, and in particular the notion of a linear combination of vectors. We then introduce the important class of vector valued functions of a vector variable that we will be working with in the rest of the book: *linear transformations*. We explain how every linear transformation can be conveniently represented by a *matrix*, and define a *matrix–vector* multiplication so that multiplying a vector by a matrix gives the result of the corrsponding linear transformation applied to that vector. We give criteria for deciding when a transformation is linear. We apply these to show that rotations in the plane are linear transformations, and we determine the corresponding rotation matrices as an example.

In the third section we show that the composition of two linear transformations is another linear transformation, and then introduce *matrix–matrix* multiplication so that the product of two matrices represents the composition product of the corresponding linear transformations. We introduce left inverses, right inverses, and inverses, with examples.

The fourth section introduces the geometry of n–dimensional Euclidean space, and in particular the *dot product*. (We do not mention the cross product here, which has no role to play in solving linear system of equations, which will be our focus in the first chapters. We postpone that until Chapter Four, on determinants).

In the fifth section we return to matrix multiplication, and look at it in terms of the columns and rows of the matrices involved, and the dot product as well. The formulas derived here are very useful! We introduce the transpose of a matrix, and investigate the important class of matrices that preserve the lengths of vectors.

Section six is the longest in the chapter, but it is full of pictures. It is only about linear transformations in $I\!R^2$, so it is rather specialized, and could be dispensed with. However, the intuition that can be gained from these pictures will help most students.

Section 1: Vectors and Multivariable Transformations

Many functions considered in mathematics and science take *lists of data* as input, and return *lists of data* as output. Here we will study an important class of such functions, and solve equations in which they figure. However, before we say what this "important class" is, here are some examples of functions that take *lists of data* as input, and return *lists of data* as output.

Example 1 (Fluid Velocity) Consider a region in the plane with points specified by Cartesian coordinates (x, y). Suppose this planar region represents a surface over which some fluid is flowing. Fix a point (x_0, y_0) in it, and at time $t = 0$, drop a small Styrofoam bead into the fluid there. The bead will be carried "downstream" by the fluid, and by observing its motion, we can study the flow pattern of the fluid.

Fix a small time step $\Delta t > 0$, and let $(\tilde{x}_0, \tilde{y}_0)$ be the point to which the bead is transported at time t. The quantities

$$\frac{\tilde{x}_0 - x_0}{\Delta t} \qquad \text{and} \qquad \frac{\tilde{y}_0 - y_0}{\Delta t}$$

will approximately measure the x and y components of the fluid velocity at (x_0, y_0) and $t = 0$. Experiments of this type are often performed. Usually lots of little beads would go in at once, and their displacements would be measured using strobe photography. Here is a graph showing the displacements of a number of such beads dropped into a fluid at $t = 0$ at points on a rectangular grid:

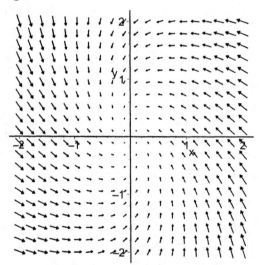

The "tail" of each arrow in the graph marks the spot at which a bead went in, and the "head" marks the spot to which that bead was carried in time Δt. You can get a pretty good idea of what the fluid is doing from the graph.

Now, even if we didn't drop a bead in at the point (x, y), we could have. If we did, it would have been carried to a well defined point $(\tilde{x}, \tilde{y})$ at time Δt, giving us

$$u(x, y) = \frac{\tilde{x} - x}{\Delta t} \qquad \text{and} \qquad v(x, y) = \frac{\tilde{y} - y}{\Delta t} \ .$$

We can interpret $u(x,y)$ and $v(x,y)$ as approximations for the horizontal and vertical velocities of the fluid at the point (x,y) at time $t = 0$. Then we have:

Input list: (x,y), the list of coordinates specifying the location.

Output list: (u,v), the list of velocity components specifying the fluid velocity at that location.

It often helps to indicate that the data in the output list depends on the data in the input list, in which case we write $u(x,y)$ and $v(x,y)$ to indicate this dependence. In any case, the rule that assigns the "output list" of data $(u(x,y), v(x,y))$ to the "input list" (x,y) is a function assigning one two–item list to another.

The graph shown above was actually drawn using

$$
\begin{aligned}
u(x,y) &= -a(2x + y) \\
v(x,y) &= a(2x - y)
\end{aligned}
\tag{1.1}
$$

for some value of a that gives arrows of reasonable length.

Example 2 (Voltages and Currents) Here is a diagram of a simple electric circuit with two voltage sources and three resistors.

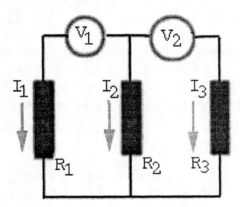

When the voltage sources are activated (for example, when batteries are put in place), electric current will flow. Let I_1, I_2 and I_3 denote the currents, (measured in *amps*) flowing through the three resistors (whose resistance R is measures in *Ohms*). We call the current positive if it flows in the direction of the corresponding arrow, and negative otherwise. Let the values of the three resistances, measured in Ohms be R_1, R_2 and R_3. To make our mathematical points in the simplest possible setting, let's suppose all three resistances are equal: $R_1 = R_2 = R_3 = R$. Finally, let the two voltages be V_1 and V_2 as indicated.

Regard the resistances as a fixed attribute of the circuit, while the voltages can be varied with voltage regulators or a change of batteries. Then there are two lists of variables to consider:

Voltage list: (V_1, V_2), the list of voltages.

Current list: (I_1, I_2, I_3), the list of currents.

Which is the input list and which is the output list? That depends on the question you ask. If you are given one list of data, you can figure out the other one using certain physical laws, namely *Ohms Law* and *Kirchhoff's rules*. Let's suppose that we are given the list of currents, and want to figure out what the voltages are. The function that tells us which voltage list corresponds to our given current list has the current list as input, and the voltage list as output.

What is this function? As we've said, it can be deduced using Ohm's law and Kirchoff's rules, but we won't explain these physical laws, because that isn't our point. We simply jump to the conclusion, and give the rule for obtaining the voltage list from the current list, as an *example of a list to list function*, which is our point right now. Here is what the physics gives us:

$$V_1 = I_2 R - I_1 R$$
$$V_2 = I_3 R - I_2 R \tag{1.2}$$

The equations (1.2) specify the functional dependence of the voltages V_1 and V_2 on the currents, I_1, I_2 and I_3. Let's write this as

$$(V_1, V_2) = f(I_1, I_2, I_3) , \tag{1.3}$$

which is just a shorthand notation for the computational rule expressed in (1.2). This is the function specifying how the voltage list depends on the current list.

Now let's consider another question. Suppose this time that the voltages are given, and you want to figure out what the currents are. Just to be concrete, let's ask:

What currents will flow in our circuit if we plug a 6 volt battery in place so that $V_1 = 6$, and a 9 battery in place so that $V_2 = 9$?

To answer this question, we can try to solve the equations (1.2) for I_1, I_2 and I_3. But that won't work: we have only two equations, and there are three unknowns to determine. Why is this? Surely when we plug in the 6 and 9 volt batteries, the three currents start to flow, and thier values must be determined.

This is indeed the case. Physics provides one more piece of information. Another of Kirchhoff's rules states that

$$I_1 + I_2 + I_3 = 0 . \tag{1.4}$$

The equation (1.4) expresses the fact that electrical charge cannot be steadily flowing into or out of the node at the bottom of the circuit. It is a mathematical expression of the physical law of *electrical charge conservation*.

Regardless of where it comes from or what it means physically, this equation tells us that not just any list of three numbers can be a list of the currents flowing in this circuit. No matter how the voltages are set, (1.4) must hold. So if we know any two values on the list of currents, we know the third. Therefore, we can *eliminate* any one of these current variables from the list as redundant information.

Let's eliminate I_2 from (1.2) using $I_2 = -(I_1 + I_3)$:

$$V_1 = -(I_3 - 2I_1)R$$
$$V_2 = (2I_3 + I_1)R$$

(1.5)

This gives us a second functional relationship

$$(V_1, V_2) = g(I_1, I_3) \, ,$$

(1.6)

where g is a shorthand notation for the computational rules expressed in (1.5). This time there is no physical restriction on the input list. Any pair of values for I_1 and I_3 is physically admissible. (Except of course that in a real circuit, if either one gets *too* large, something will melt! But let's ignore this and consider an "ideal" circuit for which Ohm's law and Kirchhoff's rules hold no matter what.)

We can now *solve the equations* (1.5) to determine the currents as a function of the voltages: Multiplying the top equation in (1.5) through by 2, and adding it to the bottom equation, we obtain

$$2V_1 + V_2 = -3RI_1 \, .$$

Multiplying the bottom equation in (1.5) through by 2, and adding it to the top equation, we obtain

$$V_1 + 2V_2 = 3RI_3 \, .$$

Hence

$$I_1 = -\frac{2}{3R}V_1 - \frac{1}{3R}V_2$$
$$I_3 = \frac{1}{3R}V_1 + \frac{2}{3R}V_2 \, .$$

(1.7)

In view of (1.4),

$$I_2 = \frac{1}{3R}(V_1 - V_2) \, .$$

(1.8)

Together, (1.7) and (1.8) give us a rule for determining the list of currents (I_1, I_2, I_3) given the list of voltages (V_1, V_2). As such, it is a function from the set of "two–entry lists of numbers" to the set of "three–entry lists of numbers". Let h denote this functional dependence so that

$$(I_1, I_2, I_3) = h(V_1, V_2) \, .$$

For example, (1.7) and (1.8) tell us that if V_1 is 6 volts, and V_2 is 9 volts, and R is measured in Ohms, then from (1.7) and (1.8), we have $I_1 = -7/R$ Amps, $I_2 = -1/R$ Amps, and $I_3 = 8/R$ Amps. That is

$$\left(-\frac{7}{R}, \frac{-1}{R}, \frac{8}{R}\right) = h(6, 9) \, .$$

This example has many important features to which we return later. What we want to focus at this point are simply the several functional relationships between the "list of voltages", and the "list of currents" as examples of functions taking lists to lists. Here is our next example:

Example 3 (Sorting Lists of Numbers) Let $(x_1, x_2, \ldots, x_n)$ be any list of n numbers. For example, consider the list

$$(-3, 5, 0, 1, 2) .$$

Very often, it is useful to "sort" such lists so that the largest number is listed first, then the second largest, and so on. Such a sorting operation would rearrange the list above into

$$(5, 2, 1, 0, -3) .$$

Again, what we have here is a mathematically well defined transformation of one list into another. We define the *sorting function* g_{sort} on all finite lists of numbers to be the function whose output value at a given list is the corresponding sorted list. For example,

$$(5, 2, 1, 0, -3) = g_{\text{sort}}(5, 2, 1, 0, -3) . \tag{1.9}$$

We will see many more examples of such functions in what follows, but these are enough to get going. Each of these examples has involved a function, or in other words, a transformation of one list into another. There are two ways to think about such functions:

(1) *As a list of functions of several numerical variables, namely the numerical variables constituting the entries of the input list, as in (1.1)*

(2) *As a single "list valued" function of a single "list" variable, as in (1.9).*

In many instances, the second way is more natural, and computationally more convenient as well. If you have done some programming, you have most likely run into "list" or "array" variables. If so, you've run into "methods" or "procedures" for working on array type variables, transforming one such array into another. The sorting transformation in Example 3 is a case in point.

Before going any further, it will be well to fix some terminology and notations once and for all. The term "list" has a clear meaning which is why we've used it up to here. But it is not the standard mathematical term, which is "vector":

Definition (Vectors in $\mathbb{R}^n$) A *vector* is an ordered list of n numbers x_j, $j = 1, 2, \ldots, n$, for some positive integer n, which is called the *dimension* of the vector. The integers $j = 1, 2, \ldots, n$ that order the list are called the *indices*, and the corresponding numbers x_j are called the *entries*. That is, for each $j = 1, 2, \ldots, n$, x_j is the jth entry on the list. The set of all n dimensional vectors is denoted $\mathbb{R}^n$.

The term *vector* comes from the Latin for "to carry". The diagram in Example 1 is what people were thinking of when the term was coined. The little arrows in that diagram show the displacements across which the little Styrofoam spheres get carried. Because of this, vectors are sometimes defined as "mathematical quantities having direction as well as magnitude". For the vectors arising in the fluid flow problem of Example 1, it is clear what "direction" and "carrying" mean.

It is not so obvious that the "carrying" idea is relevant for a vector whose entries are the three electrical currents I_1, I_2 and I_3 discussed in Example 2. What does "direction" even mean for a vector whose entries are the magnitudes of electrical currents? And what if our circuit had been more complicated, involving many more than three currents?

Perhaps surprisingly, there is a useful* notion of length and direction in any number of dimensions. We'll explain all that soon enough, but at the beginning, vectors are just lists.

We will generally denote vectors by bold face lower case roman letters with, for example, $\mathbf{x}$ denoting the vector whose jth entry is x_j. Sometimes it will help to be more explicit, in which case we specify the vector $\mathbf{x}$ by listing its entries in *column form*:

$$\mathbf{x} = \begin{bmatrix} x_1 \\ x_2 \\ \vdots \\ x_n \end{bmatrix} .$$

The next definition provides the preferred terminology for discussing the functional dependence of one vector on another.

Definition: (Transformation from $I\!\!R^n$ to $I\!\!R^m$) Let m and n be positive integers. A *transformation* from $I\!\!R^n$ to $I\!\!R^m$ is simply a function f assigning a uniquely determined output vector $\mathbf{y}$ in $I\!\!R^m$ to each input vector $\mathbf{x}$ in $I\!\!R^n$, in which case we write

$$\mathbf{y} = f(\mathbf{x}) .$$

In this book, the terms *transformation* and *function* are synonymous.

We have seen examples of transformations already. Indeed, the "position to velocity" transformation in Example 1, specified in (1.1), can be written as a transformation from $I\!\!R^2$ to $I\!\!R^2$ if we just introduce the vectors $\begin{bmatrix} x \\ y \end{bmatrix}$ and $\begin{bmatrix} u \\ v \end{bmatrix}$, and then we can express (1.1) as $\begin{bmatrix} u \\ v \end{bmatrix} = \begin{bmatrix} u(x,y) \\ v(x,y) \end{bmatrix} = f\left(\begin{bmatrix} x \\ y \end{bmatrix} \right)$ where

$$f\left(\begin{bmatrix} x \\ y \end{bmatrix} \right) = \begin{bmatrix} -a(2x+y) \\ a(2x-y) \end{bmatrix} . \tag{1.10}$$

When we write $\begin{bmatrix} u \\ v \end{bmatrix} = \begin{bmatrix} u(x,y) \\ v(x,y) \end{bmatrix}$, we are simply emphasizing that the output vector of our transformation, $\begin{bmatrix} u \\ v \end{bmatrix}$, depends on x and y. When this is clear from the context, we'll just use the short form $\begin{bmatrix} u \\ v \end{bmatrix}$.

---

* By "useful", we mean useful for solving equations, among other things In other words, useful in a practical sense, even in, say, eight dimensions.

Likewise, (1.2) can be written as a transformation from $I\!R^3$ to $I\!R^2$ if we just introduce the vectors $\begin{bmatrix} I_1 \\ I_2 \\ I_3 \end{bmatrix}$ and $\begin{bmatrix} V_1 \\ V_2 \end{bmatrix}$, while (1.7) and (1.8) together give us a linear transformation form $I\!R^2$ to $I\!R^3$ in terms of these same vectors. Do this now, writing down formulas for the transformations like the one in (1.10).

Before going further, we briefly recall some terminology and a number of definitions concerning *functions from one set to another*. For the most part, these are probably familiar. But because these things are so fundamental, a careful review is worthwhile.

Functions: A Brief Review

Let X and Y be two sets. They could be sets of numbers, sets of lists, sets of triangles – it doesn't matter. Just some pair of sets of mathematical objects. A function f from X to Y is a rule associating exactly one member of the set Y to each member of the set X.

The set X of inputs for which f is defined is usually called its *domain*, and the set Y of its possible outputs is called its *range*. Here are some kinds of functions:

(i) Algebraic functions of a real variable. In this case both sets X and Y are $I\!R$, or some subset of $I\!R$, and the functions is given by an algeraic formula like $f(x) = \sqrt{1 + x^2}$.

(ii) Transcendental functions of a real variable. Again in this case both sets X and Y are $I\!R$, or some subset of $I\!R$, but the functions is not given by an algeraic formula. For example, $\sin(\theta)$ isn't given by any algebraic formula. Instead, for $0 \le \theta \le \pi/2$, $\sin(\theta)$ is the length of the side of a right triangle opposite a vertex with angle θ divided by the length of the hypotenuse. This is a well defined function specifying, by geometric means instead of algebraic means, an output number given an input number.

(iii) Functions on finite sets. Consider for example, $X = \{1, 2, 3, 4, 5\}$ and $Y = \{2, 4, 6, 8\}$. That is, X is the set consisting of the first 5 natural numbers, and Y is the set consisting of the first four even natural numbers. In this case, since there are only finitely many inputs to consider, we can specify the action of the function by a *table*. A convenient way to do this is to list the outputs right beneath the inputs. For example, the table

$$
\begin{array}{ccccc}
1 & 2 & 3 & 4 & 5 \\
\downarrow & \downarrow & \downarrow & \downarrow & \downarrow \\
8 & 4 & 8 & 2 & 2
\end{array}
\qquad (1.11)
$$

descibes a function f from X to Y for which $f(1) = 8$, $f(2) = 2$, and so forth. The arrows are there to emphasize the input–output relation.

The important thing to bear in mind is that functions don't need to be given by a *formula*, just some rule for determining the output given the input. This might be given by a formula, but it might be given by a table as well. A table can only specify a function f if the domain X is finite set. (Or else the table woud be infinitely long, which isn't much use).

Also, while the "table" type of function may look artificial, the vector functions that we will work with are most conveniently studied in table terms*

Given another function g from Y to some third set Z, we can form the *composition product* $g \circ f$ which is a function from X to Z through the rule

$$g \circ f(x) = g(f(x))$$

for all x in the domain of f. For example, if $f(x) = 1 + x^2$ and $g(y) = 1/y$ for $y \neq 0$, then

$$g \circ f(x) = \frac{1}{1 + x^2} .$$

Let's do another example where the functions are given by tables. Let f be the function given by the table (1.11). Let $Z = \{2, 3, 5, 7\}$ be the set consisting of the first 4 prime numbers. Then with $Y = \{2, 4, 6, 8\}$ as above, define a function g from Y to Z by

$$
\begin{array}{cccc}
2 & 4 & 6 & 8 \\
\downarrow & \downarrow & \downarrow & \downarrow \\
5 & 3 & 2 & 7
\end{array}
\tag{1.12}
$$

Let's figure out what the table is for $g \circ f$. Since $f(1) = 8$, and $g(8) = 7$, $g \circ f(1) = g(f(1)) = g(8) = 7$. Since $f(2) = 4$, and $g(4) = 3$, $g \circ f(2) = g(f(2)) = g(4) = 3$. Continuing in this way, we find the table for $g \circ f$:

$$
\begin{array}{ccccc}
1 & 2 & 3 & 4 & 5 \\
\downarrow & \downarrow & \downarrow & \downarrow & \downarrow \\
7 & 3 & 7 & 5 & 5
\end{array}
\tag{1.13}
$$

The notation $g \circ f$ suggests a product, and indeed, $g \circ f$ is sometimes called the "composition product" of f and g. Note however, that it is not a commutative product. Indeed, $g \circ f$ is only defined if the the range of f is contained in the domain of g. So it can be the case that $f \circ g$ isn't even defined though $g \circ f$ is. As a simple example, let $g(y) = 0$ for all y, and let $f(x) = 1/x$ for $x \neq 0$. Then $g \circ f$ is defined, and $f \circ g$ is not.

However, the composition product is associative: If f, g and h are any three functions with g defined on the range of f, and h defined on the range of g, then

$$(h \circ g) \circ f = h \circ (g \circ f) .$$

In fact, by the definition, for any x in the domain of f,

$$((h \circ g) \circ f)(x) = (h \circ g)(f(x)) = h(g(f(x)))$$

---

* In fact, as we will see in the next section, "matrices" are just a way of writing certain vector functions in a sort of table form.

and likewise
$$(h \circ (g \circ f))(x) = h((g \circ f)(x)) = h(g(f(x))) \ .$$
Either way, you get the same thing.

We are almost done with this review. There is one more operation on functions, besides composition, that we will be concerned with: *Inverting functions*. Just as not every pair of functions can be composed, not every function can be inverted. Let's see what we require of a function if we are going to try to "run it backwards" and deduce the inputs from the outputs:

If every element in Y is an actual output value of f; that is if for every y in Y there is *at least one* x in X so that $f(x) = y$, then we say f transforms X *onto* Y. For example, the function f defined in (1.11) does not transform X onto Y because 6, which belongs to Y is not an output value of f. There is no answer to the question: "Which input from X produces the output 6 in Y?". On the other hand, the function g defined in (1.12) does transform Y onto Z. Every element of $Z = \{2, 3, 5, 7\}$ is an output value of g.

One more property is required for a function f to be invertible: If f has the property that
$$f(x_1) = f(x_2) \Rightarrow x_1 = x_2$$
so that no two different inputs produce the same output, then we say that f is *one–to–one*. for example, the function f defined in (1.11) is not one–to–one since it assigns the same output value, namely 8, to two different inputs, namely 1 and 3. (It also assigns the output 2 to both of the inputs 4 and 5, but once we've found one problem, that's it: the function isn't one–to–one). There are two answers to the question: "Which input from X produces the output 8 in Y?" On the other hand the function g defined in (1.12) is one–to–one: Every inputs gets assigned its own unique output value. There is just one answer, namely 4, to the question: "Which input from Y produces the output 3 in Z?"

A function f from X to Y is *invertible* exactly when it is a one–to–one function from X onto Y. Since every y in Y is an actual output value of f; i.e., $y = f(x)$ for some x in X, and since it is possible to determine (in principle at least) what the input x was if we are given the output y, we can define a new function f^{-1} from Y back to X by defining $f^{-1}(y)$ to be the unique x such that $f(x) = y$. Clearly then, $f^{-1} \circ f$ is the *identity function* on X: $f^{-1} \circ f(x) = x$ for all x in X. The function f^{-1} is called the *inverse* of f. Invertible functions set up a one–to–one correspondence between their domains and ranges.

For example, the function g defined in (1.12) is invertible. (We've already checked that it is onto and one–to–one). To specify g^{-1}, we just give a table which has the elements of Z as input values, and underneath we list the elements of Y that are associated to them by g. The result is:

$$
\begin{array}{cccc}
2 & 3 & 5 & 7 \\
\downarrow & \downarrow & \downarrow & \downarrow \\
6 & 4 & 2 & 8
\end{array}
\qquad (1.14)
$$

There are two "bullet points" to bear in mind as a summary of all this:

• *The composition product of functions is not commutative in general, but it is associative, which is to say that $h \circ (g \circ f) = (h \circ g) \circ f$ whenever the domain of g contains the range of f and the domain of h contains the range of g*

- *A function is invertible if and only if it is both onto and one–to–one.*

Back to vectors in $I\!R^n$

Now let's get back to functions whose domains and ranges are in $I\!R^m$ and $I\!R^n$ for some m and n.

Example 4 Consider the transformation f from $I\!R^2$ to $I\!R^2$ defined by

$$f\left(\begin{bmatrix} x \\ y \end{bmatrix}\right) = \begin{bmatrix} x + 2y \\ y \end{bmatrix}.$$

Here we have a mathematical formula, "built" out of algebraic operations applied to the individual variables x and y, for the output vector in $I\!R^2$ that is assigned to each input vector in $I\!R^2$. If we let $\mathbf{u} = \begin{bmatrix} u \\ v \end{bmatrix}$ denote the output vector we have

$$\mathbf{u} = f(\mathbf{x}) \tag{1.15}$$

or in long form,

$$\begin{bmatrix} u \\ v \end{bmatrix} = f\left(\begin{bmatrix} x \\ y \end{bmatrix}\right). \tag{1.16}$$

This single vector equation, written in either short form (1.15) or long form (1.16), is equivalent to the pair of equations

$$\begin{aligned} u &= x + 2y \\ v &= y \end{aligned} \tag{1.17}$$

relating the four numerical valued variables x, y, u and v.

Now is this transformation from $I\!R^2$ to $I\!R^2$ invertible? Yes, it is. To see this, let's fix a vector $\mathbf{u}$ in $I\!R^2$, and try to solve (1.15). The transformation f is onto if and only if (1.15) has a solution for *every* $\mathbf{u}$ in $I\!R^2$, and it is one to one if and only if (1.15) never has more than one solutions for *any* $\mathbf{u}$ in $I\!R^2$. *Finding inverses is intimately connected with solving equations, which is why we are discussing the topic.*

Since presumably you are more familiar with numerical variables, let approach solving (1.15) through (1.17), and solve for x and y given u and v.

Solving (1.17) is easy: The second equation tells us $y = v$, and substituting this into the first equation, we have $u = x + 2v$, or $x = u - 2v$. Notice that x and y are uniquely determined by u and v, so that for every $\mathbf{u}$ there is exactly one $\mathbf{x}$ satisfying (1.15). In other words, f is both onto and one–to–one, and f^{-1} is given by

$$f^{-1}\left(\begin{bmatrix} u \\ v \end{bmatrix}\right) = \begin{bmatrix} u - 2v \\ v \end{bmatrix}.$$

In this example we analyzed f by considering u and v separately. They can be though of as algebraic functions of the numerical variables x and y. This is the "multivariable" point of view. There is another way to look at this function in terms of algebraic operations on vector variables. This way of looking at things, which we introduce in the next section, turns out to be very advantageous*

---

* There wouldn't be much of an advantage in the context of Example 4; it is so simple that *any* approach deals with it pretty quickly. But in more realistic problems with more numerical variables, the vector approach shows its strength.

Section 2: Vector Operations and Linear Transformations

The most familiar sort of functions of a single real variable x, say $f(x) = x^2 + 1$, are "built up" out of certain algebraic operations that can be applied to a real variable. An important class of vector functions arises this way too. Before giving examples, we must first introduce some algebraic operations on vectors.

Definition (Scalar Multiplication) Given a number a in $I\!R$ and a vector $\mathbf{x} = \begin{bmatrix} x_1 \\ x_2 \\ \vdots \\ x_n \end{bmatrix}$,

we define the product of a and $\mathbf{x}$, namely $a\mathbf{x}$, by

$$\left\{ \quad a\mathbf{x} = \begin{bmatrix} ax_1 \\ ax_2 \\ \vdots \\ ax_n \end{bmatrix} \right. .$$

For any vector $\mathbf{x}$, $-\mathbf{x}$ denotes the product of -1 and $\mathbf{x}$.

Example 1 Here are several examples, which should be clear without much discussion:

$$2 \begin{bmatrix} 1 \\ -1 \\ 0 \end{bmatrix} = \begin{bmatrix} 2 \\ -2 \\ 0 \end{bmatrix}$$

$$\pi \begin{bmatrix} 1/2 \\ -1/2 \end{bmatrix} = \begin{bmatrix} \pi/2 \\ -\pi/2 \end{bmatrix} = \frac{1}{2} \begin{bmatrix} \pi \\ -\pi \end{bmatrix}$$

$$0 \begin{bmatrix} 1/2 \\ -1/2 \end{bmatrix} = \begin{bmatrix} 0 \\ 0 \end{bmatrix}$$

Next, we define *addition of vectors*:

Definition (Vector Addition) Given two vectors $\mathbf{x}$ and $\mathbf{y}$ in $I\!R^n$ for some n, we define their *vector sum*, $\mathbf{x} + \mathbf{y}$, by summing the corresponding entries:

$$\left\{ \quad \mathbf{x} + \mathbf{y} = \begin{bmatrix} x_1 + y_1 \\ x_2 + y_2 \\ \vdots \\ x_n + y_n \end{bmatrix} \right. .$$

We define the *vector difference* of $\mathbf{x}$ and $\mathbf{y}$, $\mathbf{x} - \mathbf{y}$ by $\mathbf{x} - \mathbf{y} = \mathbf{x} + (-\mathbf{y})$.

Notice that these operation do not mix up the entries of the vectors involved at all: The third entry, say, of the sum depends only on the third entries of the summands. For this reason, vector addition inherits the commutative, associative properties of addition in the real numbers. It is just the addition of real numbers "done in parallel".

Example 2 Here are some examples:

$$\begin{bmatrix} 1 \\ -2 \\ -1 \end{bmatrix} + \begin{bmatrix} 1 \\ 1 \\ 1 \end{bmatrix} = \begin{bmatrix} 2 \\ -1 \\ 0 \end{bmatrix}$$

$$\begin{bmatrix} 1 \\ -2 \\ -1 \end{bmatrix} + \begin{bmatrix} -1 \\ 2 \\ 1 \end{bmatrix} = \begin{bmatrix} 0 \\ 0 \\ 0 \end{bmatrix}$$

$$\begin{bmatrix} 0 \\ 0 \\ 0 \end{bmatrix} + \begin{bmatrix} 1 \\ -2 \\ -1 \end{bmatrix} = \begin{bmatrix} 1 \\ -2 \\ -1 \end{bmatrix}$$

There is a geometric way to think about vector addition in $I\!R^2$. This comes about through identifying the vector $\begin{bmatrix} x \\ y \end{bmatrix}$ in $I\!R^2$ with the point (x, y) in the Euclidean plane. We can then represent this vector geometrically by drawing an arrow with its tail at the origin and its head at (x, y). The following diagram shows three vectors represented this way: $\mathbf{x} = \begin{bmatrix} 3/2 \\ 1 \end{bmatrix}$, $\mathbf{y} = \begin{bmatrix} -1/2 \\ 1/2 \end{bmatrix}$ and their sum, $\mathbf{x} + \mathbf{y} = \begin{bmatrix} 1 \\ 3/2 \end{bmatrix}$.

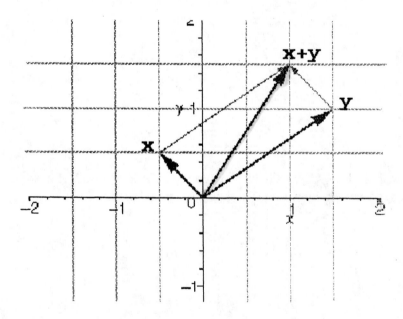

The vectors $\mathbf{x}$, $\mathbf{y}$ and $\mathbf{x} + \mathbf{y}$ themselves are drawn in bold. There are also two arrows

1-14

drawn more lightly: one is a parallel copy of **x** "transported" so its tail is at the head of **y**, while the other is a parallel copy of **y** "transported" so its tail is at the head of **x**. These four arrows run along the sides of the parallelogram whose vertices are the origin, and the points corresponding to **x**, **y** and **x** + **y**. As you see, the arrow representing **x** + **y** is the diagonal of this parallelogram that has its "tail end" at the origin.

A similar diagram could be drawn for any pair of vectors and their sum, and you see that we can think of vector addition in the plane as corresponding to the following operation: Represent the vectors by arrows as in the diagram. Transport one arrow without turning it – that is, in a parrallel motion – to bring its tail to the other arrow's head. The head of the transported arrow is now at the point corresponding to the sum of the vectors.

• *This geometrical way of thinking about vector addition is useful for many purposes as we shall soon see.*

In fact, it is even more plain when it comes to subtraction of vectors.

Example 3 Let **x** and **y** be two vectors in the plane $I\!\!R^2$, and let **w** = **x** − **y**. Then

$$\mathbf{x} = \mathbf{y} + (\mathbf{x} - \mathbf{y}) = \mathbf{y} + \mathbf{w} .$$

Using the same diagram, with the arrow labeled a bit differently, we see that **w** = **x** − **y** is the arrow running from the head of **y** to the head of **x**, "parrallel transported" so that its tail is at the origin.

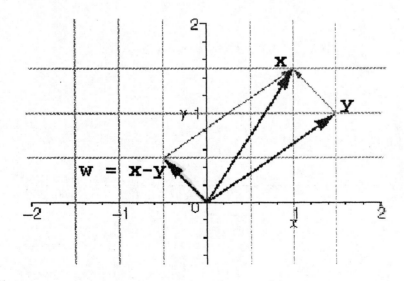

Having defined addition and subtraction of vectors, it might seem natural at this point to go on and define a vector product $\mathbf{x} \times \mathbf{y}$ by

$$\mathbf{x} \times \mathbf{y} = \begin{bmatrix} x_1 \times y_1 \\ x_2 \times y_2 \\ \vdots \\ x_n \times y_n \end{bmatrix} .$$

We *do not* do this. The reason is that there are precious few examples of interesting vector transformations f for which, in the sense of this "definition",

$$f(\mathbf{x} \times \mathbf{y}) = f(\mathbf{x}) \times f(\mathbf{y}) .$$

On the other hand, the world abounds with interesting examples of vector transformations f from $I\!R^n$ to $I\!R^m$ with the following property:

Definition: (Linear Transformation) A transformation from $I\!R^n$ to $I\!R^m$; i.e., a function from $I\!R^n$ to $I\!R^m$, is a *linear transformation* in case

$$f(a\mathbf{x} + b\mathbf{y}) = af(\mathbf{x}) + bf(\mathbf{y}) \tag{2.1}$$

for all real numbers a and b, and all vectors $\mathbf{x}$ and $\mathbf{y}$ in $I\!R^n$.

Example 4 We have seen many examples of transformations from $I\!R^n$ to $I\!R^m$ for various values of m and n in the previous section. Which of these are linear? Let's look at two examples. Rewrite (1.5) and (1.6) in vector form:

$$\begin{bmatrix} V_1 \\ V_2 \end{bmatrix} = g\left(\begin{bmatrix} I_1 \\ I_3 \end{bmatrix}\right) = \begin{bmatrix} -(2I_1 - I_3)R \\ (2I_3 + I_1)R \end{bmatrix} .$$

It is now easy to check that

$$g\left(a\begin{bmatrix} I_1 \\ I_3 \end{bmatrix} + b\begin{bmatrix} \tilde{I}_1 \\ \tilde{I}_3 \end{bmatrix}\right) = g\left(\begin{bmatrix} aI_1 + b\tilde{I}_1 \\ aI_3 + b\tilde{I}_3 \end{bmatrix}\right)$$

$$= \begin{bmatrix} -(2(aI_1 + b\tilde{I}_1) - (aI_3 + b\tilde{I}_3))R \\ (2(aI_3 + b\tilde{I}_3) + (aI_1 + b\tilde{I}_1))R \end{bmatrix}$$

$$= a\begin{bmatrix} -(2I_1 - I_3)R \\ (2I_3 + I_1)R \end{bmatrix} + b\begin{bmatrix} -(2\tilde{I}_1 - \tilde{I}_3)R \\ (2\tilde{I}_3 + \tilde{I}_1)R \end{bmatrix}$$

$$= ag\left(\begin{bmatrix} I_1 \\ I_3 \end{bmatrix}\right) + bg\left(\begin{bmatrix} \tilde{I}_1 \\ \tilde{I}_3 \end{bmatrix}\right) .$$

Therefore, this transformation is linear.

On the other hand, consider the sorting function g_{sort}. Writing it as a vector function, we have

$$g_{\text{sort}}\left(\begin{bmatrix} 1 \\ 2 \\ 3 \\ 4 \\ 5 \end{bmatrix} + \begin{bmatrix} 5 \\ 4 \\ 3 \\ 2 \\ 1 \end{bmatrix}\right) = g_{\text{sort}}\left(\begin{bmatrix} 6 \\ 6 \\ 6 \\ 6 \\ 6 \end{bmatrix}\right) = \begin{bmatrix} 6 \\ 6 \\ 6 \\ 6 \\ 6 \end{bmatrix}$$

while

$$g_{\text{sort}}\left(\begin{bmatrix} 1 \\ 2 \\ 3 \\ 4 \\ 5 \end{bmatrix}\right) + g_{\text{sort}}\left(\begin{bmatrix} 5 \\ 4 \\ 3 \\ 2 \\ 1 \end{bmatrix}\right) = \begin{bmatrix} 5 \\ 4 \\ 3 \\ 2 \\ 1 \end{bmatrix} + \begin{bmatrix} 5 \\ 4 \\ 3 \\ 2 \\ 1 \end{bmatrix} = \begin{bmatrix} 10 \\ 8 \\ 6 \\ 4 \\ 2 \end{bmatrix} .$$

(Here we have taken $a = b = 1$.) Evidently our sorting transformation is not linear.

Another pair of definitions will help us "divide and conquer" questions about linearity:

Definition: (Homogeneous and Additive) A transformation from $I\!R^n$ to $I\!R^m$ is *homogenous* in case for all $\mathbf{x}$ in $I\!R^n$ and all numbers a,

$$f(a\mathbf{x}) = af(\mathbf{x}) . \tag{2.2}$$

A transformation from $I\!R^n$ to $I\!R^m$ is *additive* in case for all $\mathbf{x}$ and $\mathbf{y}$ in $I\!R^n$

$$f(\mathbf{x} + \mathbf{y}) = f(\mathbf{x}) + f(\mathbf{y}) . \tag{2.3}$$

If you think about it a bit, you'll see that *a transformation from $I\!R^n$ to $I\!R^m$ is linear if and only if it is both homogenous and additive.* In fact, in Example 3 we showed that g_{sort} was not linear by showing that it wasn't even additive.

We can make some more progress by bringing geometry into the game:

Example 5 Consider the transformation f from $I\!R^2$ to $I\!R^2$ corresponding to rotation through the angle $\pi/2$ in the counterclockwise direction. That is, we identify a vector $\mathbf{x} = \begin{bmatrix} x \\ y \end{bmatrix}$ with the point (x, y) in the plane, and then rotate it to produce a new point (x', y'), and finally put $f\left(\begin{bmatrix} x \\ y \end{bmatrix}\right) = \begin{bmatrix} x' \\ y' \end{bmatrix}$. The following diagram shows the three vectors $\mathbf{x} = \begin{bmatrix} 3/2 \\ 1 \end{bmatrix}$, $\mathbf{y} = \begin{bmatrix} -1/2 \\ 1/2 \end{bmatrix}$ and their sum, $\mathbf{x} + \mathbf{y} = \begin{bmatrix} 1 \\ 3/2 \end{bmatrix}$, together with the three vectors obtained by rotating them counterclockwise through $\pi/2$ radians:

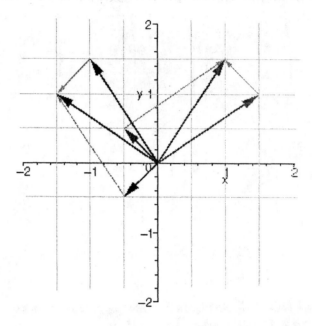

We have also drawn in the ancillary arrows used in the geometric construction of the vector sum. Now since the whole parallelogram used in forming the vectors sum of **x** and **y** is rotated *as a unit*, you see that

$$f(\mathbf{x} + \mathbf{y}) = f(\mathbf{x}) + f(\mathbf{y}) \ .$$

That is, this transformation is additive. It is also geometrically clear, if you ponder a bit, that for any **x** in $I\!R^2$ and any number a, $f(a\mathbf{x}) = af(\mathbf{x})$. Hence this transformation is homogenous. Since it is both additive and homogeneous, it is linear.

Notice two things: First, we proved that this transformation was linear just using its geometric definition; we didn't use, or need to use, any algebraic formula for $f(\mathbf{x})$. This is an important difference between Example 5 and Example 4. In fact, we will soon use the knowledge that f is linear to write down a formula for it.

Second, it didn't matter at all that the angle was $\pi/4$, or that the direction of rotation was counterclockwise. We just picked a specific angle and direction for the sake of drawing a picture. But think about it and you'll see the following:

•*The linear transformation of $I\!R^2$ induced by rotation in the plane through any angle, clockwise or counterclockwise, is always linear.*

Here is another example of this type:

Example 6 Consider the transformation from $I\!R^2$ to $I\!R^2$ given geometrically by reflection about the line $y = x$ the plane. Again, this reflection transforms the whole parallelogram representing the addition of **x** and **y** without any distortion so that reflection is clearly additive. Make sure you draw the picture; look back at hte diagram in the previous example if you are not sure what to draw.

Also, if you ponder it a bit, you will see that the reflection of $a\mathbf{x}$ is the same as a times the reflection of $\mathbf{x}$, so reflection is homogenous. Therefore, reflection about the line $y = x$ corresponds to a linear transformation of $I\!R^2$.

For the most part, in the rest of these pages, we shall restrict ourselves to the consideration of transformations from $I\!R^n$ to $I\!R^m$ with this very special property.

Aside on the significance of linear transformations

Aren't we giving up a lot by focusing so much on this special class of transformations? Yes and no, but mostly no.

True, not all naturally occurring transformations are linear, but nature does abound with vector valued functions that are linear (or at least pretty much so). Electric circuits are but one example; we will see more.

Second, and most important, the main strategy in the differential calculus of functions of a single real variable is to find the equation of the tangent line to the graph of a given (differentiable) function $f(x)$ at any given point x_0. Now the equation for the tangent line has the form

$$y = m(x - x_0) + f(x_0)$$

where the slope m is the derivative of f at x_0. The equation $f(x) = 0$ can be difficult to solve by algebra alone – and this may even be impossible. For example, if $f(x)$ is a generic fifth degree polynomial in x, there is no closed form algebraic solution of this equation. On the other hand, let $f_{\text{lin}}(x)$ be the *linearization* of f at x_0, which is to say, the function whose graph is the tangent line to the graph of f at x_0; namely

$$f_{\text{lin}}(x) = f'(x_0)(x - x_0) + f(x_0) \ .$$

It is trivial to solve $f_{\text{lin}}(x) = 0$; the unique solution is $x = x_0 - f(x_0)/f'(x_0)$. This formula is the basis of Newton's method for the iterative solution of $f(x) = 0$.

It turns out that that this strategy of *approximating* non-linear functions by linear functions, determined through the differential calculus, is applicable to an amazing variety of problems. However, the utility of this strategy in higher dimensions depends on ease with which we can deal with linear equations of vector variables.

With one variable, we simply didn't need a theory of linear functions and equations: Its all just too simple for that. The solution of $mx + b = 0$ is simply $x = -b/m$ if $m \neq 0$. If $m = 0$ there is no solution unless $b = 0$ too, in which case *every real number solves the equation.* That's it – the whole theory of linear equations in one variable.

Already in two variables, things are more complex, and the situation simply cannot be summarized in a few short lines. However, what does emerge, in the rest of these pages, is a beautiful theory of *linear* vector functions and equations, which provides, through the sort of "linear approximation strategy" employed in Newton's method, a powerful handle on a much wider class of nonlinear vector functions and equations. But all of this comes later. For now we focus on the study of linear vector functions and equations, which after all, have many, many applications independent of their use in approximating nonlinear functions and equations.

End of aside on the significance of linearity

The reason linear transformations are so nice has to do with the nice way they treat "linear combinations", a notion that we now define. You've already seen an example though: the "combination" $a\mathbf{x} + b\mathbf{y}$ appearing in the definition of a linear transformation is a particular example of a linear combination:

Definition (Linear Combination) Given k vectors $\mathbf{x}_1, \mathbf{x}_2 \ldots, \mathbf{x}_k$ in $I\!\!R^n$, any vector of the form

$$a_1\mathbf{x}_1 + a_2\mathbf{x}_2 + \cdots + a_k\mathbf{x}_k$$

for any k numbers $a_1, a_2, \ldots, a_k$ is called a *linear combination* of $\mathbf{x}_1, \mathbf{x}_2 \ldots, \mathbf{x}_k$.

Theorem 1 (Before and After Linear Combinations) *Fix any pair of positive integers n and m, and any linear transformation f from $I\!\!R^n$ to $I\!\!R^m$. Then for any k vectors $\mathbf{x}_1, \mathbf{x}_2 \ldots, \mathbf{x}_k$ in $I\!\!R^n$ and any k numbers $a_1, a_2, \ldots, a_k$,*

$$f(a_1\mathbf{x}_1 + a_2\mathbf{x}_2 + \cdots + a_k\mathbf{x}_k) = a_1 f(\mathbf{x}_1) + a_2 f(\mathbf{x}_2) + \cdots + a_k f(\mathbf{x}_k). \qquad (2.4)$$

Proof: When $k = 2$, (2.4) is just the defining property of a linear transformation. The general case is proved by induction. Suppose (2.4) has been established for any linear combination of $k - 1$ vectors. Then with $\mathbf{w}$ defined by

$$\mathbf{w} = a_1\mathbf{x}_1 + a_2\mathbf{x}_2 + \cdots + a_{k-1}\mathbf{x}_{k-1},$$

we have

$$a_1\mathbf{x}_1 + a_2\mathbf{x}_2 + \cdots + a_k\mathbf{x}_k = \mathbf{w} + a_k\mathbf{x}_k.$$

By this, (2.1), and then the inductive hypothesis,

$$f(a_1\mathbf{x}_1 + a_2\mathbf{x}_2 + \cdots + a_k\mathbf{x}_k) = f(\mathbf{w} + a_k\mathbf{x}_k)$$
$$= f(\mathbf{w}) + a_k f(\mathbf{x}_k)$$
$$= (a_1 f(\mathbf{x}_1) + a_2 f(\mathbf{x}_2) + \cdots + a_{k-1} f(\mathbf{x}_{k-1})) + a_k f(\mathbf{x}_k).$$

But this is the same as (2.4) ∎

We are now ready to introduce one really nice property of linear transformations from $I\!\!R^n$ to $I\!\!R^n$, which is that they can be represented by *tables*. Recall from the discussion of functions in the first section that if the domain of a function f is a finite set, say $\{1, 2, 3, \ldots, n\}$, one can specify the function in table form by listing the output values of the function in a "table" directly below the input values:

$$
\begin{array}{ccccc}
1 & 2 & 3 & \cdots & n \\
\downarrow & \downarrow & \downarrow & & \downarrow \\
f(1) & f(2) & f(3) & \cdots & f(n)
\end{array}
$$

1-20

Generally, when there are infinitely many input values, the table gets infinitely long, and ceases to be of use. However:

• *Despite the fact that there are infinitely many vectors in $I\!\!R^n$, we can still use a finite table to specify a linear transformation from $I\!\!R^n$ to $I\!\!R^m$.*

To explain this absolutely fundamental fact, we introduce the *standard basis vectors* in $I\!\!R^n$:

Definition (Standard Basis Vectors) For any fixed positive integer n, and any j with $j = 1, 2, 3, \ldots, n$, let $\mathbf{e}_j$ denote the vector in $I\!\!R^n$ which has 1 for its jth entry, and 0 for all of its other entries.

For example, the three standard basis vectors in $I\!\!R^3$ are

$$\mathbf{e}_1 = \begin{bmatrix} 1 \\ 0 \\ 0 \end{bmatrix} \quad \mathbf{e}_2 = \begin{bmatrix} 0 \\ 1 \\ 0 \end{bmatrix} \quad \text{and} \quad \mathbf{e}_3 = \begin{bmatrix} 0 \\ 0 \\ 1 \end{bmatrix} .$$

Now it is obvious that any vector $\begin{bmatrix} x \\ y \\ z \end{bmatrix}$ in $I\!\!R^3$ can be written as a linear combination of the standard basis vectors:

$$\begin{bmatrix} x \\ y \\ z \end{bmatrix} = x\mathbf{e}_1 + y\mathbf{e}_2 + z\mathbf{e}_3 .$$

By Theorem 1, if f is a linear transformation from $I\!\!R^3$ to $I\!\!R^m$, any m, then

$$f\left(\begin{bmatrix} x \\ y \\ z \end{bmatrix} \right) = f(x\mathbf{e}_1 + y\mathbf{e}_2 + z\mathbf{e}_3) = xf(\mathbf{e}_1) + yf(\mathbf{e}_2) + zf(\mathbf{e}_3) . \tag{2.5}$$

So if we have a table that lists the three output vectors $f(\mathbf{e}_1)$, $f(\mathbf{e}_2)$ and $f(\mathbf{e}_3)$, we can use it and (2.5) to compute $f\left(\begin{bmatrix} x \\ y \\ z \end{bmatrix} \right)$ for arbitrary values of x, y and z.

Suppose $\mathbf{v}_1$, $\mathbf{v}_2$ and $\mathbf{v}_3$ are three given vectors in $I\!\!R^m$ and

$$f(\mathbf{e}_1) = \mathbf{v}_1 \quad , \quad f(\mathbf{e}_2) = \mathbf{v}_2 \quad \text{and} \quad f(\mathbf{e}_3) = \mathbf{v}_3 . \tag{2.6}$$

We can write this in table form as

$$\begin{array}{ccc} \mathbf{e}_1 & \mathbf{e}_2 & \mathbf{e}_3 \\ \downarrow & \downarrow & \downarrow \\ \mathbf{v}_1 & \mathbf{v}_2 & \mathbf{v}_3 \end{array} . \tag{2.7}$$

1-21

Then the information provided in this table (2.7) suffices to compute $f(\mathbf{x})$ for *all of the infinitely many* vectors $\mathbf{x}$ in $\mathbb{R}^3$ when f is linear, since then, by (2.5) and (2.6)

$$f\left(\begin{bmatrix} x \\ y \\ z \end{bmatrix}\right) = x\mathbf{v}_1 + y\mathbf{v}_2 + z\mathbf{v}_3 \ . \tag{2.8}$$

acturally, there are more symbols in (2.7) than we really need. we just need the list $[\mathbf{v}_1, \mathbf{v}_2, \mathbf{v}_3]$ if we remember that $b\mathbf{v}_1$ comes from $\mathbf{e}_1$, $\mathbf{v}_2$ comes from $\mathbf{e}_2$, and so forth.

- *Thus the list of vectors $[\mathbf{v}_1, \mathbf{v}_2, \mathbf{v}_3]$ suffices to specify the linear transformation f.*

Example 7

$$\mathbf{v}_1 = \begin{bmatrix} 1 \\ 2 \end{bmatrix} \quad \mathbf{v}_2 = \begin{bmatrix} 2 \\ 3 \end{bmatrix} \quad \text{and} \quad \mathbf{v}_3 = \begin{bmatrix} 1 \\ 1 \end{bmatrix} \ ,$$

and suppose f is a linear transformation from $\mathbb{R}^3$ to $\mathbb{R}^2$ such that (2.6) holds for these vectors $\mathbf{v}_1$, $\mathbf{v}_2$ and $\mathbf{v}_3$. We can then compute, say, $f\left(\begin{bmatrix} 1 \\ 2 \\ 3 \end{bmatrix}\right)$ using (2.8). The result is

$$f\left(\begin{bmatrix} 1 \\ 2 \\ 3 \end{bmatrix}\right) = \mathbf{v}_1 + 2\mathbf{v}_2 + 3\mathbf{v}_3 = \begin{bmatrix} 8 \\ 11 \end{bmatrix} \ .$$

Everything we needed to know to do this computation is given in the list

$$\left[\begin{bmatrix} 1 \\ 2 \end{bmatrix}, \begin{bmatrix} 2 \\ 3 \end{bmatrix}, \begin{bmatrix} 1 \\ 1 \end{bmatrix}\right] \ . \tag{2.9}$$

There are quite a few braces in this expression, and they don't really tell us much. Let's suppress them, and form the 2×3 rectangular array

$$\begin{bmatrix} 1 & 2 & 1 \\ 2 & 3 & 1 \end{bmatrix} \ . \tag{2.10}$$

(Whenever we refer to an $m \times n$ rectangular array we will *always* use the convention the that m is the number of rows and n is the number of columns.) You can clearly recover (2.9), and hence f itself from (2.10). The jth column of this array is $\mathbf{v}_j$ for each $j = 1, 2, 3$, so it contains all that you need to compute $f\left(\begin{bmatrix} x \\ y \\ z \end{bmatrix}\right)$, through (2.8).

1-22

- *The rectangular array of numbers (2.10) is just an efficient way of recording the information required to compute $f(\mathbf{x})$ for any $\mathbf{x}$ in $\mathbb{R}^3$. Though we write it as a rectangular array of numbers, it is often helpful to think of it as a list of vectors $[\mathbf{v}_1, \mathbf{v}_2, \mathbf{v}_3]$ as in (2.9).*

Definition (Matrix) For positive integers m and n, an $m \times n$ *matrix* A is the $m \times n$ array of numbers $A_{i,j}$, $1 \leq i \leq m$, $1 \leq j \leq n$ with $A_{i,j}$ being the ith entry in the jth column, or, what is the same, the jth entry in the ith row. Given n vectors $\{\mathbf{v}_1, \mathbf{v}_2, \ldots, \mathbf{v}_n\}$, in $\mathbb{R}^m$, we let $[\mathbf{v}_1, \mathbf{v}_2, \ldots, \mathbf{v}_n]$ denote the $m \times n$ matrix whose jth column is $\mathbf{v}_j$ for each $j = 1, 2, \ldots, n$. We will generally denote matrices by upper–case Roman letters.

Definition (Matrix Corresponding to a Linear Transformation) If f is a linear transformation from $\mathbb{R}^n$ to $\mathbb{R}^m$, the matrix A_f *corresponding to* f is the $m \times n$ matrix

$$A_f = [f(\mathbf{e}_1), f(\mathbf{e}_2), \ldots, f(\mathbf{e}_n)] . \tag{2.11}$$

Example 8 Let f be that linear transformation of $\mathbb{R}^2$ induced by rotation through the angle $\pi/2$ radians in the counterclockwise direction. We saw in Example 5 that this is indeed a linear transformation. If you draw a picture representing $\mathbf{e}_1$ and $\mathbf{e}_2$ as arrows, and their rotations, you can see that

$$f(\mathbf{e}_1) = \mathbf{e}_2 \quad \text{and} \quad f(\mathbf{e}_2) = -\mathbf{e}_1 .$$

Therefore, the matrix A_f corresponding to this linear transformation is

$$A_f = [f(\mathbf{e}_1), f(\mathbf{e}_2)] = [\mathbf{e}_2, -\mathbf{e}_1] = \begin{bmatrix} 0 & -1 \\ 1 & 0 \end{bmatrix} .$$

More generally, let f be that linear transformation of $\mathbb{R}^2$ induced by rotation through the angle θ radians in the counterclockwise direction. Then drawing the same sort of diagram (make sure you draw it!) we find

$$f(\mathbf{e}_1) = \begin{bmatrix} \cos(\theta) \\ \sin(\theta) \end{bmatrix} \quad \text{and} \quad f(\mathbf{e}_2) = \begin{bmatrix} -\sin(\theta) \\ \cos(\theta) \end{bmatrix} .$$

Therefore, the matrix A_f corresponding to this linear transformation is

$$A_f = [f(\mathbf{e}_1), f(\mathbf{e}_2)] = \begin{bmatrix} \cos(\theta) & -\sin(\theta) \\ \sin(\theta) & \cos(\theta) \end{bmatrix} . \tag{2.12}$$

The matrices in (2.12) are called 2×2 rotation matrices, and we'll be seeing a lot of them.

We now define *matrix–vector* multiplication. We do this with a purpose in mind. The object is to set things up so that if f is a linear transformation from $\mathbb{R}^n$ to $\mathbb{R}^m$, and A_f is the corresponding matrix, then for every vector $\mathbf{x}$ in $\mathbb{R}^n$, we have

$$f(\mathbf{x}) = A_f \mathbf{x}$$

where the right hand side is the matrix–vector product that we are about to define. Note that we indicate the product by writing the vector *just to the right of the matrix*, without any special symbol for the multiplication. Note also that the left hand side is a vector in $I\!R^m$, so the product of an $m \times n$ matrix and a vector in $I\!R^n$ is another vector in $I\!R^m$. The definition is being "cooked up" so that multiplication by A_f "does" the transformation f.

Definition (Matrix–Vector Multiplication) Let A be an $m \times n$ matrix whose jth column is the m dimensional vector $\mathbf{v}_j$, and let $\mathbf{x} = \begin{bmatrix} x_1 \\ x_2 \\ \vdots \\ x_n \end{bmatrix}$ be an n–dimensional vector.

Then the *matrix–vector product* of $A = [\mathbf{v}_1, \mathbf{v}_2, \ldots, \mathbf{v}_n]$ and $\mathbf{x}$ is the m dimensional vector

$$A\mathbf{x} = [\mathbf{v}_1, \mathbf{v}_2, \ldots, \mathbf{v}_n] \begin{bmatrix} x_1 \\ x_2 \\ \vdots \\ x_n \end{bmatrix} = x_1\mathbf{v}_1 + x_2\mathbf{v}_2 + \ldots + x_n\mathbf{v}_n \ . \tag{2.13}$$

Example 9 Consider the matrix $\begin{bmatrix} 1 & 2 & 1 \\ 2 & 3 & 1 \end{bmatrix}$ and the vector $\begin{bmatrix} 1 \\ 2 \\ 3 \end{bmatrix}$. Then

$$\begin{bmatrix} 1 & 2 & 1 \\ 2 & 3 & 1 \end{bmatrix} \begin{bmatrix} 1 \\ 2 \\ 3 \end{bmatrix} = 1\begin{bmatrix} 1 \\ 2 \end{bmatrix} + 2\begin{bmatrix} 2 \\ 3 \end{bmatrix} + 3\begin{bmatrix} 1 \\ 1 \end{bmatrix} = \begin{bmatrix} 1 \\ 2 \end{bmatrix} + \begin{bmatrix} 4 \\ 6 \end{bmatrix} + \begin{bmatrix} 3 \\ 3 \end{bmatrix} = \begin{bmatrix} 8 \\ 11 \end{bmatrix} \ .$$

Likewise,

$$\begin{bmatrix} 1 & 2 \\ 2 & -3 \\ 1 & -2 \end{bmatrix} \begin{bmatrix} 3 \\ 2 \end{bmatrix} = 3\begin{bmatrix} 1 \\ 2 \\ 1 \end{bmatrix} + 2\begin{bmatrix} 2 \\ -3 \\ -2 \end{bmatrix} = \begin{bmatrix} 3 \\ 6 \\ 3 \end{bmatrix} + \begin{bmatrix} 4 \\ -6 \\ -4 \end{bmatrix} = \begin{bmatrix} 7 \\ 0 \\ -1 \end{bmatrix} \ .$$

However,

$$\begin{bmatrix} 1 & 2 \\ 2 & -3 \\ 1 & -2 \end{bmatrix} \begin{bmatrix} 1 \\ 2 \\ 3 \end{bmatrix}$$

is simply not defined. There is no definition for the product of a 3×2 matrix and a 3 dimensional vector. (One *could* make one up. But *useful* mathematical definitions are made for a reason. The purpose behind the definition we did make was to get our hands on a convenient way to write down and work with linear transformations. The next theorem says we got it right.)

1-24

Theorem 2 (Correspondence between Matrices and Linear Transformations)
For any linear transformation f from $\mathbb{R}^n$ to $\mathbb{R}^m$, let $A_f = [f(\mathbf{e}_1), f(\mathbf{e}_2), \ldots, f(\mathbf{e}_n)]$ be the corresponding $m \times n$ matrix. Then for all all $\mathbf{x}$ in $\mathbb{R}^n$,

$$f(\mathbf{x}) = A_f \mathbf{x} \ . \tag{2.14}$$

Conversely, given an $m \times n$ matrix A, define a transformation f_A from $\mathbb{R}^n$ to $\mathbb{R}^m$ by

$$f_A(\mathbf{x}) = A\mathbf{x} \ . \tag{2.15}$$

Then this transformation is linear, and A itself is the matrix corresponding to f_A. In particular, the jth column of A is $A\mathbf{e}_j$.

This theorem says that there is a one–to–one correspondence between matrices and linear transformations, with the matrix–vector product giving the "action" of the transformation.

Proof: First, by definition,

$$f\left(\begin{bmatrix} x_1 \\ x_2 \\ \vdots \\ x_n \end{bmatrix}\right) = x_1 f(\mathbf{e}_1) + x_2 f(\mathbf{e}_2) + \ldots + x_n f(\mathbf{e}_n) \ . \tag{2.16}$$

Also by definition, $A_f = [f(\mathbf{e}_1), f(\mathbf{e}_2), \ldots, f(\mathbf{e}_n)]$. Comparing (2.13) and (2.16), we have

$$f\left(\begin{bmatrix} x_1 \\ x_2 \\ \vdots \\ x_n \end{bmatrix}\right) = [f(\mathbf{e}_1), f(\mathbf{e}_2), \ldots, f(\mathbf{e}_n)] \begin{bmatrix} x_1 \\ x_2 \\ \vdots \\ x_n \end{bmatrix}$$

and this proves (2.14).

For the second part, let a and b be any two numbers, and let $\mathbf{x}$ and $\mathbf{y}$ be any two vectors in $\mathbb{R}^n$. Then writing $A = [\mathbf{v}_1, \mathbf{v}_2, \ldots, \mathbf{v}_n]$,

$$A(a\mathbf{x} + b\mathbf{y}) = [\mathbf{v}_1, \mathbf{v}_2, \ldots, \mathbf{v}_n] \begin{bmatrix} ax_1 + by_1 \\ ax_2 + by_2 \\ \vdots \\ ax_n + by_n \end{bmatrix}$$

$$= (ax_1 + by_1)\mathbf{v}_1 + (ax_2 + by_2)\mathbf{v}_2 + \cdots + (ax_n + by_n)\mathbf{v}_n$$

$$= a\left(x_1\mathbf{v}_1 + x_2\mathbf{v}_2 + \cdots x_n\mathbf{v}_n\right) + b\left(y_1\mathbf{v}_1 + y_x\mathbf{v}_2 + \cdots y_n\mathbf{v}_n\right)$$

$$= a[\mathbf{v}_1, \mathbf{v}_2, \ldots, \mathbf{v}_n] \begin{bmatrix} x_1 \\ x_2 \\ \vdots \\ x_n \end{bmatrix} + b[\mathbf{v}_1, \mathbf{v}_2, \ldots, \mathbf{v}_n] \begin{bmatrix} y_1 \\ y_2 \\ \vdots \\ y_n \end{bmatrix}$$

$$= aA\mathbf{x} + bA\mathbf{y} \ .$$

Now writing $f_A(\mathbf{x}) = A\mathbf{x}$ this says that $f_A(a\mathbf{x} + b\mathbf{y}) = af(\mathbf{x}) + bf(\mathbf{y})$ so that f_A is indeed linear. We also have that

$$f_A(\mathbf{e}_j) = [\mathbf{v}_1, \mathbf{v}_2, \ldots, \mathbf{v}_n] \begin{bmatrix} 0 \\ 0 \\ \vdots \\ 1 \\ \vdots \\ 0 \end{bmatrix} = \mathbf{v}_j$$

where the 1 is in the jth place by the definition of $\mathbf{e}_j$. This says that the jth column of the matrix corresponding to f_A, is $\mathbf{v}_j$, the jth column of A itself, and this proves the last part. ∎

Example 10 Let $\mathbf{x}$ be the vector $\mathbf{x} = \begin{bmatrix} 2 \\ 6 \end{bmatrix}$. Let $\mathbf{x}'$ be the vector obtained from $\mathbf{x}$ by rotating it counterclockwise through an angle $\pi/3$. What is $\mathbf{x}'$?

Let f be the linear transformation of $I\!\!R^2$ given by rotation in a counterclockwise direction through an angle $\pi/3$ so that $\mathbf{x}' = f(\mathbf{x})$. We know from Example 8 that the corresponding matrix is

$$A_f = \begin{bmatrix} \cos(\pi/3) & -\sin(\pi/3) \\ \sin(\pi/3) & \cos(\pi/3) \end{bmatrix} = \frac{1}{2} \begin{bmatrix} 1 & -\sqrt{3} \\ \sqrt{3} & 1 \end{bmatrix} .$$

Therefore, by Theorem 2,

$$\mathbf{x}' = f(\mathbf{x}) = \frac{1}{2} \begin{bmatrix} 1 & -\sqrt{3} \\ \sqrt{3} & 1 \end{bmatrix} \begin{bmatrix} 2 \\ 6 \end{bmatrix} = \begin{bmatrix} 1 - 3\sqrt{3} \\ \sqrt{3} + 1 \end{bmatrix}$$

There is another way to express matrix–vector product of A and $\mathbf{x}$ simply in terms of the numerical entries of A and $\mathbf{x}$:

Theorem 3 (Entry Form of Matrix–Vector Multiplication) *Let A be an $m \times n$ matrix, and let $\mathbf{x}$ be a vector in $I\!\!R^n$. Then the ith entry of the vector $A\mathbf{x}$ is*

$$\sum_{j=1}^{n} A_{i,j} x_j . \tag{2.17}$$

Proof: The ith entry of $x_1\mathbf{v}_1 + x_2\mathbf{v}_2 + \ldots + x_n\mathbf{v}_n$ is

$$x_1(\mathbf{v}_1)_i + x_2(\mathbf{v}_2)_i + \ldots + x_n(\mathbf{v}_n)_i .$$

But by definition $(\mathbf{v}_j)_i = A_{i,j}$, and so

$$x_1(\mathbf{v}_1)_i + x_2(\mathbf{v}_2)_i + \ldots + x_n(\mathbf{v}_n)_i = \sum_{j=1}^{n} A_{i,j} x_j \ .$$

■

Example 11 Consider the first product from Example 8, namely of the matrix $A = \begin{bmatrix} 1 & 2 & 1 \\ 2 & 3 & 1 \end{bmatrix}$ and the vector $\mathbf{x} = \begin{bmatrix} 1 \\ 2 \\ 3 \end{bmatrix}$. The first entry of the product is

$$A_{1,1}x_1 + A_{1,2}x_2 + A_{1,3}x_3 = 1 + 4 + 3 = 8 \ .$$

The second entry of the product is

$$A_{2,1}x_1 + A_{2,2}x_2 + A_{2,3}x_3 = 2 + 6 + 3 = 11 \ .$$

Therefore, the product is $\begin{bmatrix} 8 \\ 11 \end{bmatrix}$, as we found before.

There is at least one more useful way to write the matrix–vector product, but it requires the "dot product", of two vectors. We'll get there soon in this chapter. In the meantime, try Theorem 3 out on the rest of the matrix–vector products in Example 8. This way you can calculate the entries of the products one at a time intead of "in parallel". This can be easier, and you may only want to know some of the entries of the product anyway. Most important of all though is to try computing a number of examples both ways.

Section 3: The Matrix Product

Consider one linear transformation f from $I\!R^n$ to $I\!R^m$, and another linear transformation g from $I\!R^m$ to $I\!R^p$. Since the output of f is a vector in $I\!R^m$, which is what g takes as input, we can form the composition

$$h = g \circ f$$

which means that for all $\mathbf{x}$ in $I\!R^n$, $h(\mathbf{x}) = g(f(\mathbf{x}))$. Now for any two vectors $\mathbf{x}_1$, $\mathbf{x}_2$ in $I\!R^n$, and any two numbers a_1, a_2,

$$h(a_1\mathbf{x}_1 + a_2\mathbf{x}_2) = g(f(a_1\mathbf{x}_1 + a_2\mathbf{x}_2))$$
$$= g(a_1 f(\mathbf{x}_1) + a_2 f(\mathbf{x}_2))$$
$$= a_1 g(f(\mathbf{x}_1)) + a_2 g(f(\mathbf{x}_2))$$
$$= a_1 h(\mathbf{x}_1) + a_2 h(\mathbf{x}_2) \ .$$

The four equalities here are: The definition of h, the linearity of f, the linearity of g, and then the definition of h again. This simple analysis leads to the following fundamental fact:

Theorem 1 (The Composition of Linear Transformations is Linear) *Suppose that f is a linear transformation from $I\!R^n$ to $I\!R^m$, and g is a linear transformation from $I\!R^m$ to $I\!R^p$. Then $g \circ f$ is a linear transformation from $I\!R^n$ to $I\!R^p$*

Now there is a one–to–one correspondence between matrices and linear transformations. With f, g and $g \circ f$ as in the theorem above, let A_f, A_g and $A_{g \circ f}$ be the corresponding matrices. What is the relation between these matrices? We now define *matrix–matrix* multiplication so that

$$A_{g \circ f} = A_g A_f \ .$$

That is, the matrix product will be defined so that it "does" composition of linear transformations. To do this, note first of all that A_f is an $m \times n$ matrix, A_g is a $p \times m$ matrix and $A_{g \circ f}$ is a $p \times n$ matrix.

Definition (Matrix–Matrix Multiplication) Let A be an $m \times n$ matrix and B be a $p \times m$ matrix. Let $\{\mathbf{v}_1, \mathbf{v}_2, \ldots, \mathbf{v}_n\}$ be the columns of A so that $A = [\mathbf{v}_1, \mathbf{v}_2, \ldots, \mathbf{v}_n]$. Then BA is the $p \times n$ matrix given by

$$BA = [B\mathbf{v}_1, B\mathbf{v}_2, \ldots, B\mathbf{v}_n] \ . \tag{3.1}$$

Notice that we are using matrix–vector multiplication to define matrix–matrix multiplication.

• *Matrix–matrix multiplication is just matrix–vector multiplication done in parallel.*

Example 1 Let's compute the product BA where

$$B = \begin{bmatrix} 1 & 2 \\ -2 & 1 \end{bmatrix} \quad \text{and} \quad A = \begin{bmatrix} 0 & -1 \\ 1 & 2 \end{bmatrix} .$$

Since $A = [\mathbf{v}_1, \mathbf{v}_2]$ where $\mathbf{v}_1 = \begin{bmatrix} 0 \\ 1 \end{bmatrix}$ and $\begin{bmatrix} -1 \\ 2 \end{bmatrix}$, we just have to compute $B\mathbf{v}_1$ and $B\mathbf{v}_2$ since $BA = [B\mathbf{v}_1, B\mathbf{v}_2]$. Let's do this:

$$B\mathbf{v}_1 = \begin{bmatrix} 1 & 2 \\ -2 & 1 \end{bmatrix} \begin{bmatrix} 0 \\ 1 \end{bmatrix} = 0 \begin{bmatrix} 1 \\ -2 \end{bmatrix} + 1 \begin{bmatrix} 2 \\ 1 \end{bmatrix} = \begin{bmatrix} 2 \\ 1 \end{bmatrix} \tag{3.2}$$

and

$$B\mathbf{v}_2 = \begin{bmatrix} 1 & 2 \\ -2 & 1 \end{bmatrix} \begin{bmatrix} -1 \\ 2 \end{bmatrix} = -1 \begin{bmatrix} 1 \\ -2 \end{bmatrix} + 2 \begin{bmatrix} 2 \\ 1 \end{bmatrix} = \begin{bmatrix} 3 \\ 4 \end{bmatrix} \tag{3.3}$$

Now just put it all together:

$$BA = [B\mathbf{v}_1, B\mathbf{v}_2] = \begin{bmatrix} 2 & 3 \\ 1 & 4 \end{bmatrix} .$$

The fact that you can compute $B\mathbf{v}_1$ and $B\mathbf{v}_2$ in parallel is useful if you are writing a computer program to compute matrix products. Even if you are computing matrix products with pencil and paper, it is helpful that the computations of the columns of the product are independent of one another. *There are a lot of numbers involved in a matrix product, but you only need concern yourself with a few at a time.*

Now that we've seen how to compute matrix–matrix products, let's think back to where they came from: We said that our definition was cooked up so that the matrix–matrix product would "do" composition. The following theorem says that we got it right:

Theorem 2 (Matrix Multiplication Does Composition) *Let f be a linear transformation from $\mathbb{R}^n$ to $\mathbb{R}^m$, and g be a linear transformation from $\mathbb{R}^m$ to $\mathbb{R}^p$ so that the composition $h = g \circ f$ is defined. Let A_f, A_g and $A_{g \circ f}$ be the corresponding matrices. Then*

$$A_{g \circ f} = A_g A_f . \tag{3.4}$$

Proof: The jth column of $A_{g \circ f}$ is, by definition, $g(f(\mathbf{e}_j))$, so that

$$A_{g \circ f} = [g(f(\mathbf{e}_1)), g(f(\mathbf{e}_2)), \dots, g(f(\mathbf{e}_n))] .$$

Now, we defined matrix–vector multiplication so that for all $\mathbf{x}$, $g(\mathbf{x}) = A_g \mathbf{x}$. In particular, taking $\mathbf{x} = \mathbf{e}_j$, $g(f(\mathbf{e}_j)) = A_g f(\mathbf{e}_j)$, and therefore,

$$A_{g \circ f} = [A_g f(\mathbf{e}_1), A_g f(\mathbf{e}_2), \dots, A_g f(\mathbf{e}_n)] . \tag{3.5}$$

On the other hand $A_f = [f(\mathbf{e}_1), f(\mathbf{e}_2), \ldots, f(\mathbf{e}_n)]$, and so by (3.1),

$$A_g A_f = [A_g f(\mathbf{e}_1), A_g f(\mathbf{e}_2), \ldots, A_g f(\mathbf{e}_n)] . \tag{3.6}$$

Comparing (3.5) and (3.6), we see that indeed $A_{g \circ f} = A_g A_f$. ∎

It is useful to have a formula for matrix–matrix multiplication that is expressed directly in terms of the numerical entries of the matrices involved. Here is how to deduce this. Let A be an $m \times n$ matrix. Then by Theorem 1.2.2, * the jth column of A is $A\mathbf{e}_j$, and we have

$$A\mathbf{e}_j = \begin{bmatrix} A_{1,j} \\ A_{2,j} \\ \vdots \\ A_{n,j} \end{bmatrix} .$$

Now for any $p \times m$ matrix B, by the definition of matrix–vector multiplication, the ith entry of

$$B A\mathbf{e}_j = B \begin{bmatrix} A_{1,j} \\ A_{2,j} \\ \vdots \\ A_{n,j} \end{bmatrix}$$

is

$$\sum_{k=1}^{m} B_{i,k} A_{k,j} .$$

This proves the following result:

Theorem 3 (Matrix–Matrix Multiplication in Entry Terms) *Let A be an $m \times n$ matrix and B be a $p \times m$ matrix. Then BA is the $p \times n$ matrix whose i, jth entry with*

$$(BA)_{i,j} = \sum_{k=1}^{m} B_{i,k} A_{k,j} . \tag{3.7}$$

Example 2 Compute the matrix product from Example 1 using Theorem 3. To get the $1, 1$ entry in the product of 2×2 matrices A and B, we add up $B_{1,1}A_{1,1} + B_{1,2}A_{2,1}$, and so forth. In this example,

$$B_{1,1}A_{1,1} + B_{1,2}A_{2,1} = 1 \cdot 0 + 2 \cdot 1 = 2 .$$

---

* Theorem 1.2.2 denotes Theorem 2 of Section 2 of Chapter 1, with the theorem number last. When we refer to a theorem in the same section, we just use the theorem number.

That gives us the upper left entry. Doing the same thing for the other three we get

$$\begin{bmatrix} 1 & 2 \\ -2 & 1 \end{bmatrix} \begin{bmatrix} 0 & -1 \\ 1 & 2 \end{bmatrix} = \begin{bmatrix} 1 \cdot 0 + 2 \cdot 1 & 1 \cdot (-1) + 2 \cdot 2 \\ -2 \cdot 0 + 1 \cdot 1 & (-2) \cdot (-1) + 1 \cdot 2 \end{bmatrix} = \begin{bmatrix} 2 & 3 \\ 1 & 4 \end{bmatrix} ,$$

just as before. Notice the pattern here: $BA_{i,j}$ is the sum of products of the corresponding entries of the ith row of B and the jth column of A. We'll follow up on this "pairing" of the rows and columns in the sections to come.

Properties of Matrix Multiplication

Given any two specific matrices A and B where B is $m \times n$ and A is $n \times p$, we know how to compute the product BA. The first two examples in this section dealt with that. But what can we say about matrix multiplication *in general*? Here is one such thing:

• *Matrix multiplication is associative. That is, whenever C is a $r \times m$ matrix, B is an $m \times n$ matrix, and A is an $n \times p$ matrix,*

$$(CB)A = C(BA) . \tag{3.8}$$

Why is this true? You could check it by using (3.7) and slogging through the sums. But Theorem 2 makes it easy. Let f_C, f_B and f_A be the linear transformations corresponding to C, B and A respectively. Then by Theorem 2, $(CB)A$ corrresponds to $(f_C \circ f_B) \circ f_A$ and $C(BA)$ corresponds to $f_C \circ (f_B \circ f_A)$. But we know from the review of functions in the first section of this chapter that $(f_C \circ f_B) \circ f_A = f_C \circ (f_B \circ f_A)$. Hence (3.8) holds. In other words, *the matrix product is associative because the composition product is associative.*

The exact same reasoning shows that for any $\mathbf{x}$ in $I\!R^p$, and A and B as above,

$$(AB)\mathbf{x} = A(B\mathbf{x}) . \tag{3.9}$$

In fact, we can think of $\mathbf{x}$ as a $p \times 1$ matrix, and then the matrix–vector product is just a special case of the matrix–matrix product. In some contexts this is a bit backwards, but here it enables us to think of (3.9) as a special case of (3.8).

Next, just as the composition product is not commutative in general , the matrix product is not commutative in general. Indeed, consider the matrices A and B from Example 1. We computed there that $BA = \begin{bmatrix} 2 & 3 \\ 1 & 4 \end{bmatrix}$. Try computing AB now. You will find that

$AB = \begin{bmatrix} 2 & -1 \\ -3 & 4 \end{bmatrix}$. Notice that $AB \neq BA$. In this case, at least both AB and BA were

defined. But if B is $m \times n$ and A is $n \times p$ and $m \neq p$, then while BA is defined, AB is not.

What do we learn from this? If you are multiplying out a string of matrices, you don't need to worry about where any parentheses are; the result doesn't depend on that. You can just ignore any parentheses. But the product may well depend on the left to right order of the matrices in the product, so don't go changing that!

The Identity Matrix

Now we have a one–to–one correspondence between matrices and linear transformations. What matrix corresponds to the identity transformation $f_I(\mathbf{x}) = \mathbf{x}$? (This transformation is linear – think about it – so it does correspond to a matrix). By Theorem 1.2.2, the matrix is just $[\mathbf{e}_1, \mathbf{e}_2, \ldots, \mathbf{e}_n]$.

Definition The $n \times n$ identity matrix I, or $I_{n \times n}$ when we need to indicate the size, is the matrix given by $I = [\mathbf{e}_1, \mathbf{e}_2, \ldots, \mathbf{e}_n]$. In terms of entries, the i, jth entry is

$$I_{i,j} = \begin{cases} 1 & \text{if } i = j \\ 0 & \text{if } i \neq j \end{cases} . \tag{3.10}$$

For example, the 3×3 identity matrix is

$$\begin{bmatrix} 1 & 0 & 0 \\ 0 & 1 & 0 \\ 0 & 0 & 1 \end{bmatrix} .$$

The pattern is the same in every dimension: every diagonal entry (entries with equal indices) is one, and every off–diagonal entry (entries with unequal indices) is zero.

Now if B is any $m \times n$ matrix, and I is the $n \times n$ identity matrix, by (3.10) and Theorem 3,

$$(BI)_{i,j} = \sum_{k=1}^{n} B_{i,k} I_{k,j} = B_{i,j}$$

since $I_{k,j} = 0$ for $k \neq j$, and $I_{j,j} = 1$. That is, $BI = B$. But this is also clear from the fact that I represents the identity transformation. If f_B is the linear transformation corresponding to B (as in Theorem 1.2.2), then BI corresponds to $f_B \circ f_I = f_B$. That is, $BI = B$.

Now in the same way, if A is any $n \times p$ matrix, $f_A = f_I \circ f_A$ and so by Thoerem 1.2.2 again, $A = IA$. We can check this again using (3.10) and Theorem 3, but this already shows that $IA = A$ for any $n \times p$ matrix A.

Inverse Transformations and Inverse Matrices

Now if A is an $n \times m$ matrix and B is an $m \times n$ matrix, then we can form the matrix product BA. Suppose this product is the $m \times m$ identity matrix. Then $BA\mathbf{x} = \mathbf{x}$ for all $\mathbf{x}$ in $\mathbb{R}^m$. Since matrix multiplication corresponds to composition of linear functions, this says that f_B, the linear transformation corresponding to B *undoes* the effects of f_A, the linear transformation corresponding to A. This leads to the following definition

Definition (Left and Right Inverses) Let A be an $n \times m$ matrix. Then an $m \times n$ matrix B is a *left inverse* of A in case $BA = I_{m \times m}$, and a $m \times n$ matrix C is a *right inverse* of A in case $AC = I_{n \times n}$.

Notice that both left and right inverses of A, if either exist, have "the same shape"; that is, they both must be $m \times n$ for the multiplications to even make sense. But something even more is true when both a left and right inverse exists: Suppose that an $n \times m$ matrix A has both a left inverse B and a right inverse C. Then by (3.8), the associativity of matrix multiplication,

$$C = IC = (BA)C = B(AC) = BI = B .$$

(This is the first of many uses of (3.8)!) This not only says that $B = C$, but that there is exactly one left inverse, and exactly one right inverse, and they are the same. Indeed suppose $\tilde{B}$ is any other left inverse. Then the same argument applies that $\tilde{B} = C$. Since $B = C$, we have $\tilde{B} = B$, and there is only one left inverse. In the same way, we see there is only one right inverse. This leads to the following definition

Definition (Inverses and Invertible Matrices) An $n \times m$ matrix A is *invertible* in case it has both a left and a right inverse. As we've seen above, in this case it has just one left inverse, and just one right inverse, and they are the same. This matrix is called the *inverse of A*, and it is usually denoted by A^{-1}.

Example 3 Let A and B be invertible $n \times n$ matrics. Is AB ivertible? Yes, and in fact

$$(AB)^{-1} = B^{-1}A^{-1} . \tag{3.11}$$

To see this, note that by the associativity of matrix multiplication,

$$(B^{-1}A^{-1})(AB) = A(B^{-1}B)A^{-1} = AA^{-1} = I .$$

The fact that the order of multiplication is switched is natural: To get $AB\mathbf{x}$, you first apply B to $\mathbf{x}$, and then you apply A to the result, $B\mathbf{x}$. To recover $\mathbf{x}$, you first undo A, and then undo B.

Example 4 Let A be the general 2×2 matrix

$$A = \begin{bmatrix} a & b \\ c & d \end{bmatrix} .$$

Suppose that A has a right inverse B. Can we find a formula for it? Yes, and here's how. Let $\mathbf{v}_1$ and $\mathbf{v}_2$ be the columns of B, so that

$$B = [\mathbf{v}_1, \mathbf{v}_2] .$$

Then by the definition of the matrix–matrix product,

$$AB = A[\mathbf{v}_1, \mathbf{v}_2] = [A\mathbf{v}_1, A\mathbf{v}_2] .$$

Since $I_{2\times 2} = [\mathbf{e}_1, \mathbf{e}_2]$, we have that $AB = I_{2\times 2}$ if and only if

$$[A\mathbf{v}_1, A\mathbf{v}_2] = [\mathbf{e}_1, \mathbf{e}_2] .$$

That is, finding the matrix B we are looking for amounts to finding a pair of vectors $\mathbf{v}_1$ and $\mathbf{v}_2$ such that

$$A\mathbf{v}_1 = \mathbf{e}_1 \qquad \text{and} \qquad A\mathbf{v}_2 = \mathbf{e}_2 \,,$$

Let's see if we can find $\mathbf{v}_1$. Assuming for the moment that it does exist, let's write $\mathbf{v}_1 = \begin{bmatrix} x \\ y \end{bmatrix}$. Then $A\mathbf{v}_1 = \begin{bmatrix} ax + by \\ cx + dy \end{bmatrix}$ so that $A\mathbf{v}_1 = \mathbf{e}_1$ is equivalent to the system of equations

$$\begin{aligned} ax + by &= 1 \\ cx + dy &= 0 \end{aligned} \tag{3.12}$$

Let's solve this for x and y. If we can, we've found $\mathbf{v}_1$!

Multiply the top equation in (3.12) through by d and the bottom equation through by b to obtain

$$\begin{aligned} dax + dby &= d \\ bcx + bdy &= 0 \end{aligned}$$

Subtracting, we find

$$(ad - bc)x = d \,. \tag{3.13}$$

In the same way, if we multiply the top equation in (3.12) through by c and the bottom equation through by a and subtract, we obtain

$$(ad - bc)y = -c \,. \tag{3.14}$$

Combining (3.13) and (3.14), we see that as long as $ad - bc \neq 0$, there is a unique solution to $A\mathbf{x} = \mathbf{e}_1$, and $\mathbf{v}_1$ must be this solution, so we must have

$$\mathbf{v}_1 = \frac{1}{ad - bc} \begin{bmatrix} d \\ -c \end{bmatrix} \,.$$

We're halfway there, and we'll leave the other half as an exercise. In the exact same way, you'll see that as long as $ad - bc \neq 0$, $A\mathbf{x} = \mathbf{e}_2$ also has a unique solution which is

$$\mathbf{v}_2 = \frac{1}{ad - bc} \begin{bmatrix} -b \\ a \end{bmatrix} \,.$$

The conclusion is that when $A = \begin{bmatrix} a & b \\ c & d \end{bmatrix}$ satisfies $ad - bc \neq 0$, A has a unique right inverse B which is given by the explicit formula

$$B = \frac{1}{ad - bc} \begin{bmatrix} d & -b \\ -c & a \end{bmatrix} \,. \tag{3.15}$$

At this point you should multiply out AB and check that indeed, $AB = I$. You should also check that $BA = I$. That is, B is not just a right inverse, but it is an inverse of A. so now we know that whenever $A = \begin{bmatrix} a & b \\ c & d \end{bmatrix}$ satisfies $ad - bc \neq 0$, it is invertible, and the inverse is given by the explicit formula (3.15).

To wrap this example up, let's apply it with some actual numbers. Let $A = \begin{bmatrix} 0 & 2 \\ -1 & 3 \end{bmatrix}$. Then since (3.15) gives the inverse of A,

$$A^{-1} = \frac{1}{2} \begin{bmatrix} 3 & -2 \\ 1 & 0 \end{bmatrix} .$$

(Check this by doing the multiplications explicitly!)

If a matrix A is invertible, and you happen to know the inverse A^{-1}, then you can easily solve the vector equation

$$Ax = b$$

for any b in $\mathbb{R}^m$. Indeed,

$$A(A^{-1}b) = (AA^{-1})b = Ib = b$$

so that $x_0 = A^{-1}b$ is a solution. *Here we just used the fact that A had a right inverse, so whenever A has even just a right inverse B , $Ax = b$ has at least one solution, namely Bb.*

Are there any others? Suppose that x_1 also satisfies $A\tilde{x} = b$. Then

$$A(x_1 - x_0) = Ax_1 - Ax_0 = b - b = 0 .$$

Now multiplying both sides of $A(x_1 - x_0) = 0$ on the left by A^{-1}, we get

$$x_1 - x_0 = A^{-1}A(x_1 - x_0) = A^{-1}0 = 0$$

so that $x_1 = x_0$ after all, and there is no other solution. *In this argument, we used only the fact that A had a left inverse, so whenever A has even just a left inverse, $Ax = b$ never has more than one solution.*

Putting the two together, we see that *matrix inverses, when they exist, give us the unique solutions to vector equations such as $Ax = b$ through $x = A^{-1}b$.*

Example 4 Let's solve the equation $Ax = b$ for $A = \begin{bmatrix} 0 & 2 \\ -1 & 3 \end{bmatrix}$ and $b = \begin{bmatrix} 1 \\ 2 \end{bmatrix}$. As we've seen in Example 3, A is invertible, and by what we've learned, the unique solution is

$$A^{-1}b = \frac{1}{2} \begin{bmatrix} 3 & -2 \\ 1 & 0 \end{bmatrix} \begin{bmatrix} 1 \\ 2 \end{bmatrix} = \begin{bmatrix} -1/2 \\ 1 \end{bmatrix} .$$

How do you know when there is no inverse?

How can you tell when you'd be wasting your time looking for an inverse? There is a usefull criterion for this:

• *If A is an $m \times n$ matrix and there is a non-zero vector $\mathbf{w}$ in $\mathbb{R}^n$ with $A\mathbf{w} = 0$, then A does not have a left inverse (and therefore is not invertible).*

The reason is simple: Since we always have $A0 = 0$, if we have $A\mathbf{w} = 0$ for some $\mathbf{w} \neq 0$, then the linear transformation corresponding to A is not one–to–one. More directly, suppose B were a left inverse. Then

$$\mathbf{w} = I\mathbf{w} = (BA)\mathbf{w} = B(A\mathbf{w}) = B0 = 0 \ ,$$

which contradicts the assumption that $\mathbf{w} \neq 0$.

Example 5 Let's apply this to show that $A = \begin{bmatrix} 1 & 2 & 3 \\ 2 & 4 & 2 \\ 3 & 6 & 0 \end{bmatrix}$ does not have a left inverse.

Since the second column is twice the first, $-2\mathbf{v}_1 + \mathbf{v}_2 = 0$, where $\mathbf{v}_j$ is the jth column of A. Therefore, let $\mathbf{w} = \begin{bmatrix} -2 \\ 1 \\ 0 \end{bmatrix}$ so that

$$A\mathbf{w} = -2\mathbf{v}_1 + \mathbf{v}_2 = 0 \ .$$

Since $\mathbf{w}$ is non–zero, A is not invertible.

This example may look artificial, but we found $\mathbf{x}$ by the strategy of "looking for patterns in A", and that very often is worth doing.

We will see in the next chapter how to decide when matrix inverses exist, and how to compute them when they do, for matrices that are larger than 2×2. However, it turns out that

•*Only square matrices have a chance to be invertible.*

This can be explained now, using the criterion above if we "borrow" one simple result on vector equations.

One thing we will see in the next chapter, where we systematically study equations of the type $A\mathbf{x} = \mathbf{b}$ is that *if A has more columns than rows, there is always a non-zero vector $\mathbf{w}$ satisfying*

$$A\mathbf{w} = 0 \ . \tag{3.16}$$

Roughly speaking, this is because if A is $m \times n$ with $n > m$ there are more input variables than there are output variables: $\mathbf{w}$ is in $\mathbb{R}^n$, and $\mathbf{b}$ is in $\mathbb{R}^m$. As we will see, this means that we can always find a non-zero solution of (3.16). Thus, according to the criterion of Example 5, *matrices with left inverses cannot have more columns than rows*, since in this case there is a non–zero solution to (3.16).

This clinches it: If A is an invertible $m \times n$ matrix, then we must have $n \leq m$ since A has a left inverse. But A^{-1} is an $n \times m$ matrix, and it too has a left inverse, namely A since $AA^{-1} = I$. But then A^{-1} cannot have more columns than rows, so $m \leq n$. Putting the two inequalities together, we have $m = n$, and so A must be square if it is invertible. We will state this as a theorem in the next chapter after we've proved that (3.16) always has non–zero solutions when A has more columns than rows. But it makes sense to discuss the facts here, so you don't wonder why you only run into examples of inverses of square matrices in the exercises for this section!

Now let's return to Example 3. There we found a formula for the inverse of $A = \begin{bmatrix} a & b \\ c & d \end{bmatrix}$ assuming that $ad - bc \neq 0$. What about the case $ad - bc = 0$? We can again apply our criterion to show that in this case A doesn't even have a left inverse.

In case $ad - bc = 0$,

$$\begin{bmatrix} a & b \\ c & d \end{bmatrix} \begin{bmatrix} -b \\ a \end{bmatrix} = \begin{bmatrix} 0 \\ ad - bc \end{bmatrix} = 0 \tag{3.17}$$

and

$$\begin{bmatrix} a & b \\ c & d \end{bmatrix} \begin{bmatrix} d \\ -c \end{bmatrix} = \begin{bmatrix} ad - bc \\ 0 \end{bmatrix} = 0 \ . \tag{3.18}$$

Now as long as A is not the zero matrix; i.e., as long as at least one of a, b, c or d is non–zero, at least one of (3.17) or (3.18) gives us a non–zero solution of $A\mathbf{x} = 0$, so in this case A has no left inverse, and is not invertible. But if A is the zero matrix, *every* non–zero vector is a non–zero solution to $A\mathbf{x} = 0$, so A certainly doesn't have a left inverse in this case. The conclusion is that since $ad - bc = 0$ leads to (3.17) and (3.18), A is never invertible, and in fact never even has a left inverse, when $ad - bc = 0$. Combining this with what we learned in Example 3 gives us the following result.

Thoerem 4 (Inverting 2×2 Matrices) *Let A be the 2×2 matrix $A = \begin{bmatrix} a & b \\ c & d \end{bmatrix}$. Then A is invertible if and only if $ad - bc \neq 0$, in which case*

$$A^{-1} = \frac{1}{ad - bc} \begin{bmatrix} d & -b \\ -c & a \end{bmatrix} \ . \tag{3.19}$$

There is one more thing we should say about matrix inverses; a word of caution: Multiplication by an inverse matrix A^{-1} undoes multiplication by A. That is,

$$A^{-1}(AB) = (A^{-1}A)B = B$$

as we have observed above. In this sense, multiplication by A^{-1} is a bit like "dividing by A". However, if C is an $n \times n$ matrix, and A is an invertible $n \times n$ matrix, we have to be careful about the fact that often

$$A^{-1}C \neq CA^{-1} \ . \tag{3.20}$$

For this reason *we do not write*

$$\frac{C}{A}$$

to denote matrix division: It is not clear which of the matrices in (3.20) this might mean.

Addition of matrices

We close this section with one more operation on matrices – much simpler than the matrix product. Let A and B be two $m \times n$ matrices, and let f_A and f_B be the corresponding linear transformations from $I\!R^n$ to $I\!R^m$. We can define a new transformation g from $I\!R^n$ to $I\!R^m$ by adding the outputs of these two: That is, put

$$g(\mathbf{x}) = f_A(\mathbf{x}) + \phi_B(\mathbf{x}) .$$

It is easy to check that g is additive and homogenous since, after all, it is obtained by addition. (Still, make sure to check the details now). Therefore $(f_A + f_B)$ corresponds to a matrix. How is this matrix related to A and B?

Let C be the matrix corresponding to g. Then the jth column of C is

$$g(\mathbf{e}_j) = f_A(\mathbf{e}_j) + f_B(\mathbf{e}_j) = A\mathbf{e}_j + B\mathbf{e}_j .$$

Therefore, $C_{i,j}$ is the ith entry of this vector, namely $(A\mathbf{e}_j)_i + (B\mathbf{e}_j)_i = A_{i,j} + B_{i,j}$. Therefore, $C_{i,j} = A_{i,j} + B_{i,j}$. This brings us to the following definition:

Definition (Addition of Matrices) Let A and B be two $m \times n$ matrices. Then $A + B$ is the $m \times n$ matrix given by

$$(A + B)_{i,j} = A_{i,j} + B_{i,j} . \tag{3.21}$$

Notice that $A + B$ is only defined if both A and B have the same "size", $m \times n$. Also notice that $A + B$ represents the linear transformation obtained by adding the outputs of the linear transformations represented by A and B. That is,

$$(A + B)\mathbf{x} = A\mathbf{x} + B\mathbf{x} . \tag{3.22}$$

Section 4: The Dot Product and the Geometry of $\mathbb{R}^n$

So far we have emphasized vectors as lists of data. However, we can identify vectors in $\mathbb{R}^2$ or $\mathbb{R}^3$ with the corresponding points in Euclidean space, and this provides a useful geometric perspective on vectors and vector equations, at least in these dimensions. In fact, the geometric point of view provides powerful insight in all dimensions, and in a surprisingly straightforward way. But let's begin the discussion in 2 dimensions, where we can easily visualize the content of geometric statements.

Vectors were originally considered in $\mathbb{R}^2$ and $\mathbb{R}^3$, and were defined as quantities having both *magnitude* and *direction*. For example, consider a vector $\mathbf{x} = \begin{bmatrix} x_1 \\ x_2 \end{bmatrix}$ in $\mathbb{R}^2$, and identify it with the corresponding point (x_1, x_2) in the plane. We are being a bit pedantic in using one notation for vectors in $\mathbb{R}^2$, and another for points in the plane. But here is the purpose of this distinction: We have already defined certain *algebraic operations*, namely scalar multiplication and vector addition on $\mathbb{R}^2$, and we are already familiar with the *geometry* of the plane. Now we are going to *put the algebra and the geometry together*, through the identification of

$$\begin{bmatrix} x_1 \\ x_2 \end{bmatrix} \qquad \text{with} \qquad (x_1, x_2) \ . \tag{4.1}$$

Through this identification, we might, for example, think of $\mathbf{x} = \begin{bmatrix} x_1 \\ x_2 \end{bmatrix}$ as representing the *displacement* from the origin $(0,0)$ of a particle in two dimensional Euclidean space.

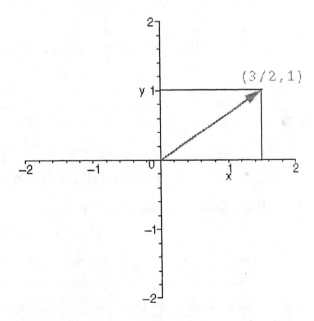

Regarding $\mathbf{x}$ as representing a displacement, we can call it a "displacement vector". (This is not a mathematical definition of a new type of vector, but rather a convenient terminology that recalls what this particular list represents. The same will be true of "velocity vectors", "acceleration vectors", etc.) The magnitude of this *displacement vector* is the *Euclidean distance* from the origin $(0,0)$ to the point (x_1, x_2). This distance is given by the Pythagorean formula

$$\sqrt{x_1^2 + x_2^2} \, ,$$

and it tells us the "magnitude", or "size" of the displacement. The plainest term for discussing the size of a displacement is "length", as in "length of travel". Therefore, for a vector $\mathbf{x} = \begin{bmatrix} x_1 \\ x_2 \end{bmatrix}$ in $I\!R^2$, we define the length of $\mathbf{x}$, which is denoted by $|\mathbf{x}|$, to be

$$|\mathbf{x}| = \sqrt{x_1^2 + x_2^2} \, . \tag{4.2}$$

Now let's put algebra and geometry together. Consider two vectors $\mathbf{x} = \begin{bmatrix} x_1 \\ x_2 \end{bmatrix}$ and $\mathbf{y} = \begin{bmatrix} y_1 \\ y_2 \end{bmatrix}$ in $I\!R^2$. What is $|\mathbf{x} - \mathbf{y}|$, the length of their difference? That's not hard to work out. By the definition,

$$\mathbf{x} - \mathbf{y} = \begin{bmatrix} x_1 - y_1 \\ x_2 - y_2 \end{bmatrix}$$

so that by (4.2)

$$|\mathbf{x} - \mathbf{y}|^2 = (x_1 - y_1)^2 + (x_2 - y_2)^2 \, .$$

The right hand side is the square of the Euclidean distance from (x_1, x_2) to (y_1, y_2), and so *the Euclidean distance from (x_1, x_2) to (y_1, y_2) is just $|\mathbf{x} - \mathbf{y}|$, the length of the difference between $\mathbf{x}$ and $\mathbf{y}$.*

Let's take the algebra a bit further:

$$
\begin{aligned}
|\mathbf{x} - \mathbf{y}|^2 &= (x_1 - y_1)^2 + (x_2 - y_2)^2 \\
&= x_1^2 - 2x_1 y_1 + y_1^2 + x_2^2 - 2x_2 y_2 + y_2^2 \\
&= (x_1^2 + x_2^2) + (y_1^2 + y_2^2) - 2(x_1 y_1 + x_2 y_2) \\
&= |\mathbf{x}|^2 + |\mathbf{y}|^2 - 2(x_1 y_1 + x_2 y_2) \, ,
\end{aligned}
\tag{4.3}
$$

where we used (4.2) in the first and last lines.

This leads us to an important definition, that we give directly for $I\!R^n$ since the definition is purely algebraic, and doesn't require any notion of n dimensional geometry.

Definition (Dot Product in $I\!R^n$) Given two vectors $\mathbf{x}$ and $\mathbf{y}$ in $I\!R^n$, their *dot product* is the number $\mathbf{x} \cdot \mathbf{y}$ given by

$$\mathbf{x} \cdot \mathbf{y} = x_1 y_1 + x_2 y_2 + \cdots + x_n y_n \, . \tag{4.4}$$

Example 1 Computing dot products is very simple. Here are some examples

$$\begin{bmatrix} 1 \\ 2 \end{bmatrix} \cdot \begin{bmatrix} 3 \\ 4 \end{bmatrix} = 1 \times 3 + 2 \times 4 = 11 \ .$$

The same thing is perhaps even clearer if we just do the products on the right before writing them down:

$$\begin{bmatrix} 1 \\ 2 \end{bmatrix} \cdot \begin{bmatrix} 3 \\ 4 \end{bmatrix} = 3 + 8 = 11 \ .$$

In the same way,

$$\begin{bmatrix} 0 \\ 4 \\ 2 \end{bmatrix} \cdot \begin{bmatrix} 2 \\ -1 \\ 2 \end{bmatrix} = 0 - 4 + 4 = 0 \ . \tag{4.5}$$

Notice that the dot product of two vectors is a number, and the dot product is only defined when the two vectors have the same dimension.

The next theorem summarizes some important algebraic properties of the dot product:

Theorem 1 (Algebraic Properties of the Dot Product) *Let* $\mathbf{x}$ *and* $\mathbf{y}$ *be any two vectors in* $\mathbb{R}^n$. *Then the dot product is commutative, which is to say*

$$\mathbf{x} \cdot \mathbf{y} = \mathbf{y} \cdot \mathbf{x} \ , \tag{4.6}$$

and moreover,

$$\mathbf{x} \cdot \mathbf{x} = 0 \iff \mathbf{x} = 0 \ . \tag{4.7}$$

Also the dot product is distributive, which is to say that for any k *vectors* $\mathbf{x}_1, \mathbf{x}_2, \ldots, \mathbf{x}_k$ *in* $\mathbb{R}^n$, *any other vector* $\mathbf{y}$ *in* $\mathbb{R}^n$, *and any* k *numbers* $a_1, a_2, \ldots, a_n$,

$$\mathbf{y} \cdot (a_1\mathbf{x}_1 + a_2\mathbf{x}_2 + \cdots + a_k\mathbf{x}_k) = a_1\mathbf{y} \cdot bx_1 + a_2\mathbf{y} \cdot bx_2 + \cdots + a_k\mathbf{y} \cdot bx_k \ . \tag{4.8}$$

Proof: To see the identities (4.6) and (4.8), just write out both sides in terms of the entries. Similarly, writing $\mathbf{x} \cdot \mathbf{x}$ in terms of the entries of $\mathbf{x}$,

$$\mathbf{x} \cdot \mathbf{x} = x_1^2 + x_2^2 + \cdots + x_n^2 \ .$$

Each term on the right is non–negative, and so $\mathbf{x} \cdot \mathbf{x} = 0$ if and only if $x_j = 0$ for each j, which is what $\mathbf{x} = 0$ means. ∎

Now let's return to our algebraic formula (4.3) for the distance between two vectors in $\mathbb{R}^2$, and rewrite it in terms of the dot product:

$$|\mathbf{x} - \mathbf{y}|^2 = |\mathbf{x}|^2 + |\mathbf{y}|^2 - 2\mathbf{x} \cdot \mathbf{y} \tag{4.9}$$

This formula has a simple geometric interpretation in $I\!R^2$: Consider the triangle with vertices $(0,0)$, (x_1, x_2) and (y_1, y_2). Let a, b and c denote the lengths of the three sides of this triangle, with a being the length of the side opposite the vertex (x_1, x_2), with b being the length of the side opposite the vertex (y_1, y_2), and with c being the length of the side opposite the vertex $(0,0)$.

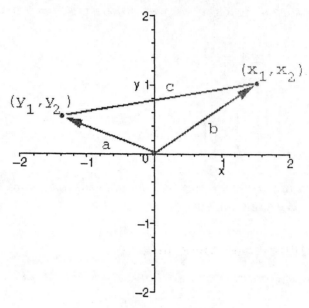

The law of cosines says that

$$c^2 = a^2 + b^2 - 2ab\cos(\theta) \tag{4.10}$$

where θ is the angle at the vertex $(0,0)$. Now, as is clear from the diagram, $a = |\mathbf{x}|$, $b = |\mathbf{y}|$, and $c = |\mathbf{x} - \mathbf{y}|$. Hence (4.10) is the same as

$$|\mathbf{x} - \mathbf{y}|^2 = |\mathbf{x}|^2 + |\mathbf{y}|^2 - 2|\mathbf{x}||\mathbf{y}|\cos(\theta) . \tag{4.11}$$

Comparing (4.9) and (4.11) we see

$$|\mathbf{x}||\mathbf{y}|\cos(\theta) = \mathbf{x} \cdot \mathbf{y} . \tag{4.12}$$

Considering the diagram, we see that θ is the angle between the vectors $\mathbf{x}$ and $\mathbf{y}$. We see from (4.12) that we can express this angle in terms of the dot product through

$$\cos(\theta) = \frac{\mathbf{x} \cdot \mathbf{y}}{|\mathbf{x}||\mathbf{y}|} ,$$

and hence

$$\theta = \cos^{-1}\left(\frac{\mathbf{x} \cdot \mathbf{y}}{|\mathbf{x}||\mathbf{y}|}\right) . \tag{4.13}$$

1-42

By the definition of the inverse cosine function, we always have $0 \leq \theta \leq \pi$, which is correct for an interior angle of a triangle.

Some special values of θ are especially interesting. Notice that $\cos(\theta) = 0$ if and only if $\mathbf{x} \cdot \mathbf{y} = 0$, and of course $\cos(\theta) = 0$ if and only if $\theta = \pi/2$. That is, *our triangle is a right triangle with* $\mathbf{x}$ *and* $\mathbf{y}$ *running along the orthogonal sides exactly when* $\mathbf{x} \cdot \mathbf{y} = 0$. Computing the dot product therefore gives us a test for when two vectors in $I\!\!R^2$ "point in perpendicular directions": This is the case if and only if their dot product is zero.

The formula is also interesting when $\theta = 0$. Then $\cos(\theta) = 1$ and $\mathbf{x} \cdot \mathbf{y} = |\mathbf{x}||\mathbf{y}|$. Of course $\theta = 0$ when $\mathbf{y} = \mathbf{x}$, and so we have

$$|\mathbf{x}|^2 = \mathbf{x} \cdot \mathbf{x} \ . \tag{4.14}$$

In fact, we don't need (4.13) to deduce this formula: From the definition of the length, and from the definition of the dot product, we see that both sides of (4.14) equal $x_1^2 + x_2^2$. *Either way, we see that the dot product is useful for calculating lengths, angles and distances.*

Now let's generalize our formulas for computing angles and distance so that they can be applied in n dimensions. Since our formulas for distance and angle in the plane are expressed in terms of the dot product which we have already defined in n dimensions, it is natural to proceed as follows:

Definition (Length, Distance and Angle in $I\!\!R^n$) Let $\mathbf{x}$ be any vector in $I\!\!R^n$. The *length* of $\mathbf{x}$, denoted by $|\mathbf{x}|$, is defined by

$$|\mathbf{x}|^2 = \mathbf{x} \cdot \mathbf{x} \ . \tag{4.15}$$

The *distance* between two vectors $\mathbf{x}$ and $\mathbf{y}$ in $I\!\!R^n$ is the length of their difference, namely $|\mathbf{x} - \mathbf{y}|$. Finally, the *angle* θ between any two non-zero vectors $\mathbf{x}$ and $\mathbf{y}$ in $I\!\!R^n$ is given by

$$\theta = \cos^{-1}\left(\frac{\mathbf{x} \cdot \mathbf{y}}{|\mathbf{x}||\mathbf{y}|}\right) \ . \tag{4.16}$$

By the properties of the inverse cosine function, as above, $0 \leq \theta \leq \pi$.

That was quick, but does it really make sense, and does it do anything for us computationally? The two questions are closely related, and of course the answer is "yes" in both cases. Let's start with one possible cause for concern: Since $|\cos(\theta)| \leq 1$ no matter what the value of θ may be, we need to have

$$|\mathbf{x} \cdot \mathbf{y}| \leq |\mathbf{x}||\mathbf{y}| \ ,$$

otherwise (4.16) just doesn't define the cosine of any angle at all. Now we know this is true in $I\!\!R^2$, where we derived (4.16) using the law of cosines. Is it true in higher dimension as well?

We are now in a position to address this question using algebra alone since we have algebraic formulas for $|\mathbf{x} \cdot \mathbf{y}|$, $|\mathbf{x}|$ and $|\mathbf{y}|$. Here is how to do this. Consider any two non-zero vectors $\mathbf{x}$ and $\mathbf{y}$ in $I\!\!R^n$. Divide by their length to make them *unit vectors*; i.e., vectors of unit length. In this way we define

$$\mathbf{u} = \frac{1}{|\mathbf{x}|}\mathbf{x} \quad \text{and} \quad \mathbf{v} = \frac{1}{|\mathbf{y}|}\mathbf{y} \ .$$

Notice that

$$\mathbf{u} \cdot \mathbf{v} = \frac{\mathbf{x} \cdot \mathbf{y}}{|\mathbf{x}||\mathbf{y}|} \tag{4.17}$$

so to show that (4.16) makes sense we just have to show that $-1 \leq \mathbf{u} \cdot \mathbf{v} \leq 1$.

To get a handle on $\mathbf{u} \cdot \mathbf{v}$, let's compute $|\mathbf{u} - \mathbf{v}|^2$. (This is probably not an obvious thing to do at this stage*, but you don't need to invent the idea – you only need to understand it). We find, using Theorem 1 several times,

$$|\mathbf{u} - \mathbf{v}|^2 = (\mathbf{u} - \mathbf{v}) \cdot (\mathbf{u} - \mathbf{v})$$

$$= \mathbf{u} \cdot \mathbf{u} + \mathbf{v} \cdot \mathbf{v} - 2\mathbf{u} \cdot \mathbf{v}$$

$$= 2 - 2\mathbf{u} \cdot \mathbf{v} \ .$$

This says that

$$\mathbf{u} \cdot \mathbf{v} = 1 - \frac{|\mathbf{u} - \mathbf{v}|^2}{2} \ ,$$

and in particular that $\mathbf{u} \cdot \mathbf{v} \leq 1$.

In the exact same way, computing $|\mathbf{u} + \mathbf{v}|^2$ we find

$$\mathbf{u} \cdot \mathbf{v} = -1 + \frac{|\mathbf{u} + \mathbf{v}|^2}{2} \ ,$$

and in particular that $\mathbf{u} \cdot \mathbf{v} \geq -1$. Putting it all together, we have $-1 \leq \mathbf{u} \cdot \mathbf{v} \leq 1$, and hence by (4.17),

$$-1 \leq \frac{\mathbf{x} \cdot \mathbf{y}}{|\mathbf{x}||\mathbf{y}|} \leq 1 \ , \tag{4.18}$$

and (4.16) does define θ.

The result we just deduced, (4.18), has many other uses besides showing that (4.16) makes sense. We can rewrite it in a more standard form if we multiply through by $|\mathbf{x}||\mathbf{y}|$. Then (4.18) becomes

$$|\mathbf{x} \cdot \mathbf{y}| \leq |\mathbf{x}||\mathbf{y}| \ . \tag{4.19}$$

Better, yet, we see from (4.17) that $|\mathbf{x} \cdot \mathbf{y}| = |\mathbf{x}||\mathbf{y}|$ if an only if $|\mathbf{u} \cdot \mathbf{v}| = 1$, which in turn happens only if either $|\mathbf{u} - \mathbf{v}| = 0$ or $|\mathbf{u} + \mathbf{v}| = 0$. This happens if and only if

$$\frac{1}{|\mathbf{x}|}\mathbf{x} = \pm\frac{1}{|\mathbf{y}|}\mathbf{y} \qquad \text{or equivalently} \qquad |\mathbf{y}|\mathbf{x} = \pm|\mathbf{x}|\mathbf{y} \ .$$

In geometric terms, there is equality in (4.19) if and only if $\mathbf{x}$ and $\mathbf{y}$ are proportional. This agrees with our definition of θ since when equality holds in (4.19), we have $\cos(\theta) = \pm 1$,

---

* In fact, we are indebted to Michael Loss for suggesting this argument to us.

which means either $\theta = 0$ or $\theta = \pi$. If $\mathbf{x}$ and $\mathbf{y}$ are multiples of one another, then these vectors either "point in the same direction", which corresponds to $\theta = 0$, or they "point in opposite directions", which corresponds to $\theta = \pi$.

The inequality (4.19) that we just derived is known as the *Schwarz inequality*. It is also the key to seeing that the length of $\mathbf{x} - \mathbf{y}$ is in fact a good measure of the distance between $\mathbf{x}$ and $\mathbf{y}$.

The point here is that anything that deserves to be called a measure of "the distance between two vectors" *had better act like a distance*. In mathematics, this has a precise meaning, so that it is only proper to call something a measure of "distance" between vectors provided the following three properties hold:

(i) For any two vectors $\mathbf{x}$ and $\mathbf{y}$, the *distance* from $\mathbf{x}$ to $\mathbf{y}$ is zero if and only if $\mathbf{x} = \mathbf{y}$.

(ii) For any two vectors $\mathbf{x}$ and $\mathbf{y}$, the *distance* from $\mathbf{x}$ to $\mathbf{y}$ is the same as the *distance* from $\mathbf{y}$ to $\mathbf{x}$.

(iii) For any three vectors $\mathbf{x}$, $\mathbf{y}$ and $\mathbf{z}$, the *distance* from $\mathbf{x}$ to $\mathbf{z}$ is no more than the sum of the distances from $\mathbf{x}$ to $\mathbf{y}$ and from $\mathbf{y}$ to $\mathbf{z}$.

This list of three properties may seem somewhat arbitrary. But we will see that these properties are the ones that "come up all the time" in analysis of vector problems, so sometime down the road it will become clear that this is the right list of properties to require.

Also, all of these properties hold for the Euclidean distance in the plane. The first two are quite evident. The third one is called the *triangle inequality*. Given three vectors $\mathbf{x}$, $\mathbf{y}$ and $\mathbf{z}$ in $I\!\!R^2$, consider them as points in the plane. The distance from $\mathbf{x}$ to $\mathbf{y}$ is the length of one side of the triangle, the distance from $\mathbf{y}$ to $\mathbf{z}$ is another, and the distance from $\mathbf{x}$ to $\mathbf{z}$ is the third. Since the length of any side of a triangle in the plane is no greater than the sum of the lengths of the other two sides, the Euclidean distance in the plane does satisfy the triangle inequality.

The intuitive meaning of the triangle inequality comes from thinking of the distance between two points as representing the "length of the shortest path between them". If you insist on going from $\mathbf{x}$ to $\mathbf{z}$ by passing though $\mathbf{y}$ on the way, that can only lengthen your path.

We've seen by a geometrical argument that the distance we've defined for vectors has this property in $I\!\!R^2$. We can use the Schwarz inequality to give an algebraic demonstration of the fact that the triangle inequality holds in every dimension.

Consider three vectors $\mathbf{x}$, $\mathbf{y}$ and $\mathbf{z}$ in $I\!\!R^n$. We want to show that

$$|\mathbf{x} - \mathbf{z}| \leq |\mathbf{x} - \mathbf{y}| + |\mathbf{y} - \mathbf{z}| \tag{4.20}$$

which is to say that the distance from $\mathbf{x}$ to $\mathbf{z}$ is no more than the sum of the distances from $\mathbf{x}$ to $\mathbf{y}$ and then from $\mathbf{y}$ on to $\mathbf{z}$. Let $\mathbf{v} = \mathbf{x} - \mathbf{y}$ and $\mathbf{w} = \mathbf{y} - \mathbf{z}$ so that

$$\mathbf{x} - \mathbf{z} = \mathbf{v} + \mathbf{w} \ .$$

Then

$$|\mathbf{v} + \mathbf{w}|^2 = (\mathbf{v} + \mathbf{w}) \cdot (\mathbf{v} + \mathbf{w})$$

$$= \mathbf{v} \cdot \mathbf{v} + \mathbf{w} \cdot \mathbf{w} + 2\mathbf{v} \cdot \mathbf{w}$$

$$= |\mathbf{v}|^2 + |\mathbf{w}|^2 + 2\mathbf{v} \cdot \mathbf{w}$$

Now apply the Schwarz inequality (4.19) to conclude

$$|\mathbf{v} + \mathbf{w}|^2 \le |\mathbf{v}|^2 + |\mathbf{w}|^2 + 2|\mathbf{w}||\mathbf{v}| = (|\mathbf{v}| + |\mathbf{w}|)^2 \,,$$

and hence, taking the square root of both sides,

$$|\mathbf{v} + \mathbf{w}| \le |\mathbf{v}| + |\mathbf{w}| \,. \tag{4.21}$$

Recalling the definitions of $\mathbf{v}$ and $\mathbf{w}$, we see that the triangle inequality (4.20) does indeed hold. Thus we have seen that our n dimensional distance function has the properties (i), (ii) and (iii) required of a measure of distance. The inequality (4.21) that we deduced along the way is also interesting: It says that the length of a sum of vectors is no more than the sum of their lengths. Again, this is easy to visualize in the plane. This inequality is useul too, and so it has a name: the Minkowski inequality. We summarize these results in the following theorem:

Theorem 2 (The Schwarz, Minkowski and Triangle Inequalities) *For any two vectors $\mathbf{x}$ and $\mathbf{y}$ in $I\!R^n$, (4.19) holds, and there is equality if and only if $\mathbf{x}$ is a multiple of $\mathbf{y}$. For any three vectors $\mathbf{x}$, $\mathbf{y}$ and $\mathbf{z}$ in $I\!R^n$, (4.20) holds, and for any two vectors $\mathbf{v}$ and $\mathbf{w}$ in $I\!R^n$, (4.21) holds.*

Now let's return to angles, this time in $I\!R^n$. We've already seen that $|\mathbf{x} \cdot \mathbf{y}| = |\mathbf{x}||\mathbf{y}|$ if and only if $\mathbf{x}$ and $\mathbf{y}$ are multiples of one another, which corresponds to $\cos(\theta) = \pm 1$ in (4.16), which in turn corresponds to $\theta = 0$ or $\theta = \pi$. Another interesting case, and indeed the most important computationally is $\theta = \pi/2$, in which case $\cos(\theta) = 0$, and hence $\mathbf{x} \cdot \mathbf{y} = 0$. That is, with our definition of angles in n dimensions, two vectors "point at right angles to one another", or more briefly, are "orthogonal", if and only if their dot product is zero. We will make so much use of this fact that it deserves a separate definition:

Definition (Orthogonal Vectors) Two vectors in $I\!R^n$, $\mathbf{x}$ and $\mathbf{y}$, are *orthogonal* to one another in case $\mathbf{x} \cdot \mathbf{y} = 0$. In the same way, we say that a collection $\{\mathbf{x}_1, \mathbf{x}_2, \ldots, \mathbf{x}_k\}$ of k vectors in $I\!R^n$ is *orthogonal* in case

$$\mathbf{x}_i \cdot \mathbf{x}_j = 0 \tag{4.22}$$

for all $i \ne j$ in $1, 2, \ldots, k$.

One reason the orthogonality property is important is that it is easy to compute lengths (and hence distances) in the presence of orthogonality:

Theorem 3 (Lengths of Orthogonal Sums) *Let $\{\mathbf{x}_1, \mathbf{x}_2, \ldots, \mathbf{x}_k\}$ be a collection of k orthogonal vectors in $I\!R^n$, and let $\mathbf{x} = \mathbf{x}_1 + \mathbf{x}_2 + \cdots + \mathbf{x}_k$ be their sum. Then*

$$|\mathbf{x}|^2 = |\mathbf{x}_1|^2 + |\mathbf{x}_2|^2 + \cdots + |\mathbf{x}_k|^2 \,. \tag{4.23}$$

Proof Since $\mathbf{x} = \sum_{i=1}^{k} \mathbf{x}_i$,

$$|\mathbf{x}|^2 = \left(\sum_{i=1}^{k} \mathbf{x}_i\right) \cdot \left(\sum_{j=1}^{k} \mathbf{x}_j\right)$$

$$= \sum_{i,j=1}^{k} \mathbf{x}_i \cdot \mathbf{x}_j = \sum_{i=1}^{k} |\mathbf{x}_i|^2 + \sum_{i,j=1, i\neq j}^{k} \mathbf{x}_i \cdot \mathbf{x}_j$$

$$= \sum_{i=1}^{k} |\mathbf{x}_i|^2 .$$

The orthogonality was used in the last equality: Because of (4.22), there is no contribution form the terms $\mathbf{x}_i \cdot \mathbf{x}_j$ for $i \neq j$.

This theorem can be viewed as a generalization of the Pythagorean Theorem to higher dimensions.

Section 5: Matrix Multiplication Revisited

We have seen that it is natural to consider an $m \times n$ matrix A, not only as a rectangular array of the mn numbers $A_{i,j}$, but as an array of vectors: Thinking of A as representing a linear transformation f_A from R^n to R^m the jth column of A, $\mathbf{c}_j$, is $\mathbf{c}_j = f_A(\mathbf{e}_j)$. Using this notation, we can write A as an horizontal array of n vectors in R^m:

$$A = [\mathbf{c}_1, \mathbf{c}_2, \ldots, \mathbf{c}_n] \,. \tag{5.1}$$

This is the *column representation* of the matrix A.

There is another way to write A as an array of vectors: We can use the rows instead. Let $\{\mathbf{r}_1, \mathbf{r}_2, \ldots, \mathbf{r}_m\}$ be the m rows of A. Consider A as a vertical array of m row vectors in R^n:

$$A = \begin{bmatrix} \mathbf{r}_1 \\ \mathbf{r}_2 \\ \vdots \\ \mathbf{r}_m \end{bmatrix} \,. \tag{5.2}$$

This is the *row representation* of the matrix A.

These two representations are important in theory as well as practice. Indeed, those of you who have done programming with arrays know that programming languages generally don't *directly* support rectangular arrays. To work with rectangular arrays in a program, you have to store and operate on them as arrays of arrays, as in (5.1) and (5.2). We've already had some practice in looking at matrix–vector and matirx–matrix multiplication in terms of the column representation. In this section we shall gain a useful new perspective by bringing in the row representation and the dot product.

Let's begin with matrix–vector multiplication.

Theorem 1 (Matrix–Vector Multiplication in Row and Column Terms) *Let A be an $m \times n$ matrix with the column and row representations (5.1) and (5.2). Let* $\mathbf{v} = \begin{bmatrix} v_1 \\ v_2 \\ \vdots \\ v_n \end{bmatrix}$

be any vector in R^n. Then $A\mathbf{v}$ is the linear combination of the columns of A given by

$$A\mathbf{v} = v_1\mathbf{c}_1 + v_2\mathbf{c}_2 + \cdots + v_n\mathbf{c}_n \,, \tag{5.3}$$

and in terms of the rows of A we have

$$A\mathbf{v} = \begin{bmatrix} \mathbf{r}_1 \cdot \mathbf{v} \\ \mathbf{r}_2 \cdot \mathbf{v} \\ \vdots \\ \mathbf{r}_m \cdot \mathbf{v} \end{bmatrix} \,. \tag{5.4}$$

Proof: We don't need to prove (5.3); it is the definition of $A\mathbf{v}$! We are restating it only for the sake of symmetry betwen the rows and columns.

The new formula (5.4) follows directly from (2.17), namely,

$$(A\mathbf{v})_i = \sum_{j=1}^{n} A_{i,j} v_j , \qquad (5.5)$$

together with (5.2) which says that

$$A_{i,j} = (\mathbf{r}_i)_j . \qquad (5.6)$$

Substituting (5.6) into (5.5), and using (4.4), the definition of the dot product, we deduce

$$(A\mathbf{v})_i = \sum_{j=1}^{n} (\mathbf{r}_i)_j v_j = \mathbf{r}_i \cdot \mathbf{v} ,$$

which is the same as (5.4). ■

Example 1 Let $A = \begin{bmatrix} 1 & 2 & 3 \\ 3 & 2 & 1 \end{bmatrix}$ and let $\mathbf{v} = \begin{bmatrix} 1 \\ -1 \\ 2 \end{bmatrix}$. The rows of A are

$$\mathbf{r}_1 = \begin{bmatrix} 1 \\ 2 \\ 3 \end{bmatrix} \quad \text{and} \quad \mathbf{r}_2 = \begin{bmatrix} 3 \\ 2 \\ 1 \end{bmatrix} .$$

Important convention: We are writing the rows vertically since that is our convention for writing vectors. We will always do this: *When we pull a row out of a matrix and think of it as a vector, we will write it vertically with its left entry on top and its right entry on bottom.*

Now we can easily compute that $\mathbf{r}_1 \cdot \mathbf{v} = 5$ and $\mathbf{r}_2 \cdot \mathbf{v} = 4$ with the result that

$$A\mathbf{v} = \begin{bmatrix} 5 \\ 4 \end{bmatrix} .$$

This is also what we get using (5.3) which gives us

$$A\mathbf{v} = \mathbf{c}_1 - \mathbf{c}_2 + 2\mathbf{c}_3 = \begin{bmatrix} 1 \\ 3 \end{bmatrix} - \begin{bmatrix} 2 \\ 2 \end{bmatrix} + 2\begin{bmatrix} 3 \\ 1 \end{bmatrix} = \begin{bmatrix} 5 \\ 4 \end{bmatrix} .$$

Why do we need two formulas for matrix–vector multiplication? Each has its particular advantages and uses. For example, the formula (5.4) brings *geometry* into the business of solving vector equations. Indeed, you see right away from (5.4) that $A\mathbf{v} = 0$ if and only if $\mathbf{v}$ is orthogonal to each row of A.*

Now let's go on to the matrix–matrix product. Since this is just the matrix–vector product done "in parallel", we can apply Theorem 1.

Let B be an $n \times p$ matrix, so that the product AB, with A as before, is defined. Let $\mathbf{d}_j$ denote the jth column of B, so that

$$B = [\mathbf{d}_1, \mathbf{d}_2, \ldots, \mathbf{d}_p] , \tag{5.7}$$

and hence $AB = [A\mathbf{d}_1, A\mathbf{d}_2, \ldots, A\mathbf{d}_p]$. Theorem 1 gives us two ways to express each $A\mathbf{d}_j$, and this in turn gives us the following:

Theorem 2 (Matrix–Matrix Multiplication in Row and Column Terms) *Let A be an $m \times n$ matrix with the column and row representations (5.1) and (5.2). Let B be an $n \times p$ matrix with the column representation (5.7). Then:*

(1) The i,jth entry of AB is the dot product of the ith row of A with the jth column of B. Using the notation of (5.2) and (5.7),

$$(AB)_{i,j} = \mathbf{r}_i \cdot \mathbf{d}_j . \tag{5.8}$$

(2) Each column of AB is a linear combination of the columns of A, and in particular, the jth column of AB is a linear combination of the columns of A with multiples taken from the jth column of B.

Proof: Since $(AB)_{i,j}$ is the i component of the jth column of AB, namely $A\mathbf{d}_j$, (5.8) follows directly from (5.4). This proves part *(1)*.

For the second part, we use (5.3). Since the jth column of AB is $A\mathbf{d}_j$, and since

$$\mathbf{d}_j = \begin{bmatrix} B_{1,j} \\ B_{2,j} \\ \vdots \\ B_{n,j} \end{bmatrix} ,$$

(5.3) with $\mathbf{d}_j$ in place of $\mathbf{v}$ gives us

$$A\mathbf{d}_j = B_{1,j}\mathbf{c}_1 + B_{2,j}\mathbf{c}_2 + \cdots + B_{n,j}\mathbf{c}_n . \tag{5.9}$$

this proves the second part, and even gives us a formula for the linear combination. Most people find the formula itself hard to remember, but can reconstruct it from the words in the theorem, and that is why we left it out of the statement of Theorem 2. ∎

---

* This may well seem like a mere curiosity right now, but it is our first hint of a very significant connection between the geometric ideas introduced in the last section, and the problem of solving vector equations.

Example 2 Let $A = \begin{bmatrix} 1 & 2 & 3 \\ 3 & 2 & 1 \end{bmatrix}$ and let $B = \begin{bmatrix} 1 & 0 & 3 \\ 0 & 1 & 4 \\ 2 & 0 & 1 \end{bmatrix}$. Let's use (5.9) to compute

the third column of AB. This should clarify the meaning of (5.9), if nothing else. The columns of A are

$$\mathbf{c}_1 = \begin{bmatrix} 1 \\ 3 \end{bmatrix} \quad , \quad \mathbf{c}_2 = \begin{bmatrix} 2 \\ 2 \end{bmatrix} \quad \text{and} \quad \mathbf{c}_3 = \begin{bmatrix} 3 \\ 1 \end{bmatrix} .$$

The third column of B is $\begin{bmatrix} 3 \\ 4 \\ 1 \end{bmatrix}$. Hence, according to (5.9), the third column of AB is

$$3\mathbf{c}_1 + 4\mathbf{c}_2 + \mathbf{c}_3 = 3 \begin{bmatrix} 1 \\ 3 \end{bmatrix} + 4 \begin{bmatrix} 2 \\ 2 \end{bmatrix} + \begin{bmatrix} 3 \\ 1 \end{bmatrix} = \begin{bmatrix} 14 \\ 18 \end{bmatrix} .$$

Before our next example, we give a definition.

Definition (Diagonal Matrices) An $m \times n$ matrix is a *diagonal matrix* in case it is a square matrix; i.e., $m = n$, such that $A_{i,j} = 0$ for all $i \neq j$. We write $\text{diag}(a_1, a_2, \ldots, a_n)$ to denote the $n \times n$ diagonal matrix A with $A_{i,i} = a_i$ for $i = 1, 2 \ldots, n$.

Example 3 Let $A = \text{diag}(a_1, a_2, \ldots, a_n)$. Notice that A has the row and column representations

$$A = \begin{bmatrix} a_1 \mathbf{e}_1 \\ a_1 \mathbf{e}_2 \\ \vdots \\ a_n \mathbf{e}_n \end{bmatrix} \quad \text{and} \quad A = [a_1 \mathbf{e}_1, a_2 \mathbf{e}_2, \ldots, a_n \mathbf{e}_n] .$$

Since similar formulas hold for any other $n \times n$ diagonal matrix $B = \text{diag}(b_1, b_2, \ldots, b_n)$, we have from (5.8) that

$$(AB)_{i,j} = a_i b_j \mathbf{e}_i \cdot \mathbf{e}_j = \begin{cases} a_i b_i & \text{if } i = j \\ 0 & \text{if } i \neq j \end{cases} . \tag{5.10}$$

Therefore

$$AB = \text{diag}(a_1 b_1, a_2 b_2, \ldots, a_n b_n) = \text{diag}(b_1 a_1, b_2 a_2, \ldots, b_n b_n) = BA .$$

In other words, the product of any two diagonal matrices A and B is again diagonal, $AB = BA$, and the diagonal entries of the product are just products of the corresponding diagonal entries of A and B.

Here is another simple observation we can make right now using (5.8): Suppose that A is invertible, and that B is its inverse. Then by (3.10),

$$(AB)_{i,j} = \mathbf{r}_i \cdot \mathbf{d}_j = I_{i,j} = \begin{cases} 1 & \text{if } i = j \\ 0 & \text{if } i \neq j \end{cases} . \tag{5.11}$$

Therefore the ith column of the inverse of A is orthogonal to the jth row of A for all $j \neq i$. Again, this is a hint of the important role that geometry will play in solving vector equations.

The transpose of a matirx

There is another relation between the row and column representations of A and the dot product. To explain this, we need a definition:

Definition (Transpose of a Matrix) Let A be an $m \times n$ matrix. Then the *transpose of A is the* $n \times m$ matrix A^t whose jth column is the jth row of A. In other words the columns of A are the rows of A^t and *vice–versa*. That is, the i, jth entry of A^t, $(A^t)_{i,j}$, is given by

$$(A^t)_{i,j} = A_{j,i} . \tag{5.12}$$

Example 4 Here are some examples of matrices and their transposes:

$$A = \begin{bmatrix} 1 & 3 & 2 \\ 0 & 2 & 1 \end{bmatrix} \quad \text{and} \quad A^t = \begin{bmatrix} 1 & 0 \\ 3 & 2 \\ 2 & 1 \end{bmatrix} .$$

$$B = \begin{bmatrix} 1 & 0 \\ 3 & 2 \\ 2 & 1 \end{bmatrix} \quad \text{and} \quad B^t = \begin{bmatrix} 1 & 3 & 2 \\ 0 & 2 & 1 \end{bmatrix} .$$

You notice that $B = A^t$ and $B^t = A$, which means that $(A^t)^t = A$. That is, taking the transpose twice gets us back to the matrix we started from. This always happens: Applying (5.12) twice we have $((A^t)^t)_{i,j} = (A^t)_{j,i} = A_{i,j}$, which means we always have

$$(A^t)^t = A . \tag{5.13}$$

Here is the main theorem:

Theorem 3 (The Transpose and the Dot Product) *Let A be an $m \times n$ matrix. Then (1) For any $\mathbf{x}$ in R^n and any $\mathbf{y}$ in R^m,*

$$\mathbf{y} \cdot A\mathbf{x} = (A^t\mathbf{y}) \cdot \mathbf{x} , \tag{5.14}$$

and A^t is the only $n \times m$ matrix with this property. That is, if B is any $n \times m$ matrix so that

$$(B\mathbf{y}) \cdot \mathbf{x} = \mathbf{y} \cdot A\mathbf{x} \tag{5.15}$$

holds for any $\mathbf{x}$ *in* R^n *and any* $\mathbf{y}$ *in* R^m, *then* $B = A^t$.

(2) If B *is any* $n \times p$ *matrix, then*

$$(AB)^t = B^t A^t .$$
(5.16)

Before we give the proof, note the resemblance of (5.16) to the formula (3.11) for the product of the inverse of a pair of invertible $n \times n$ matrices A and B, namely $(AB)^{-1} = B^{-1}A^{-1}$.

Proof: Let's try to derive conditions on an $n \times m$ matrix B that guarantee (5.15). To do this, write $\mathbf{x}$ and $\mathbf{y}$ out in terms of the standard basis vectors:

$$\mathbf{x} = \sum_{j=1}^{n} x_j \mathbf{e}_j \qquad \text{and} \qquad \mathbf{y} = \sum_{i=1}^{n} y_i \mathbf{e}_i .$$

(Notice that in the first expression $\mathbf{e}_j$ denotes a standard basis vector in R^n, and in the second one, $\mathbf{e}_i$ denotes a standard basis vector in R^m. Keep this "double duty" notation in mind, and it won't throw you off.) Then:

$$\mathbf{y} \cdot A\mathbf{x} = \left(\sum_{i=1}^{n} y_i \mathbf{e}_i \right) \cdot \left(\sum_{j=1}^{n} x_j A\mathbf{e}_j \right) = \sum_{i=1}^{n} \sum_{j=1}^{n} y_i x_j (\mathbf{e}_i \cdot A\mathbf{e}_j) .$$
(5.17)

$$(B\mathbf{y}) \cdot \mathbf{x} = \left(\sum_{i=1}^{n} y_i B\mathbf{e}_i \right) \cdot \left(\sum_{j=1}^{n} x_j \mathbf{e}_j \right) = \sum_{i=1}^{n} \sum_{j=1}^{n} y_i x_j (B\mathbf{e}_i \cdot \mathbf{e}_j) .$$
(5.18)

Comparing (5.17) and (5.18), we see that (5.15) holds if and only if

$$(\mathbf{e}_i \cdot A\mathbf{e}_j) = (B\mathbf{e}_i \cdot \mathbf{e}_j)$$
(5.19)

for each i and j. On the other hand, since $\mathbf{x}$ and $\mathbf{y}$ are arbitrary, if we fix any i and j, we can set $y_i = x_j = 1$ and set all other entries of $\mathbf{x}$ and $\mathbf{y}$ to be zero. Then there is only one term in each of the sums on the right side of (5.17) and (5.18), namely $(\mathbf{e}_i \cdot A\mathbf{e}_j)$ in the first and $(B\mathbf{e}_i \cdot \mathbf{e}_j)$ in the second. *Therefore,* $\mathbf{y} \cdot A\mathbf{x} = B\mathbf{y} \cdot \mathbf{x}$ *holds for all* $\mathbf{x}$ *and* $\mathbf{y}$ *if and only if (5.19) holds for each* i *and* j. Next, (5.19) is just the same as $A_{i,j} = B_{j,i}$, which in turn, is the same as $B = A^t$. This proves *(1)*.

For the second part, apply (5.14) in two steps,

$$\mathbf{y} \cdot AB\mathbf{x} = (A^t\mathbf{y}) \cdot B\mathbf{x} = (B^t(A^t(\mathbf{y}))) \cdot b x = ((B^t A^t)\mathbf{y}) \cdot \mathbf{x} .$$

Applying (5.14) all at once,

$$\mathbf{y} \cdot AB\mathbf{x} = ((AB)^t\mathbf{y}) \cdot \mathbf{x} .$$

1-53

Since $\mathbf{x}$ and $\mathbf{y}$ can be any vectors in R^p and R^m respectively, the uniqueness statement in part tells us we must have (5.16). ∎

We will make many applications of Theorem 3. Here is one we can make right now, though it too requires a small bit of introduction.

Transposes and isometries

Definition (Isometry) An $m \times n$ matrix A is an *isometry* in case for all $\mathbf{x}$ in R^n,

$$|A\mathbf{x}| = |\mathbf{x}| . \tag{5.20}$$

In other words, A is an isometry if the length of the output is the same as the length of the input, though keep in mind that the input vector is in $I\!R^n$, and the output vector is in $I\!R^m$.

The term isometry comes from the classical greek for "same measure". Isometries preserve the lengths of vectors by definition. However, they also preserve angles between vectors. To see why this is the case, let A be an $n \times n$ matrix that is an isometry, and let $\mathbf{x}$ and $\mathbf{y}$ be any two non–zero vectors in R^n. Then, $|A(\mathbf{x} - \mathbf{y})|^2 = |\mathbf{x} - \mathbf{y}|^2$ since A is an isometry. Since we also have $A(\mathbf{x} + \mathbf{y}) = A\mathbf{x} + A\mathbf{y}$, it follows that

$$|A\mathbf{x} - A\mathbf{y}|^2 = |\mathbf{x} - \mathbf{y}|^2 . \tag{5.21}$$

Now writing the left side out in terms of the dot product,

$$\begin{aligned}
|A\mathbf{x} - A\mathbf{y}|^2 &= (A\mathbf{x} - A\mathbf{y}) \cdot (A\mathbf{x} - A\mathbf{y}) \\
&= A\mathbf{x} \cdot A\mathbf{x} + A\mathbf{y} \cdot A\mathbf{y} - 2A\mathbf{x} \cdot A\mathbf{y} \\
&= |A\mathbf{x}|^2 + |A\mathbf{y}|^2 - 2A\mathbf{x} \cdot A\mathbf{y} \\
&= |\mathbf{x}|^2 + |\mathbf{y}|^2 - 2A\mathbf{x} \cdot A\mathbf{y} .
\end{aligned}$$

Much more directly, we have

$$|\mathbf{x} - \mathbf{y}|^2 = |\mathbf{x}|^2 + |\mathbf{y}|^2 - 2\mathbf{x} \cdot \mathbf{y} .$$

Now combining these computations with (5.21), we see that isometries also preserve dot products; that is

$$A\mathbf{x} \cdot A\mathbf{y} = \mathbf{x} \cdot \mathbf{y} . \tag{5.22}$$

From here it is a short step to the preservation of angles: Let θ be the angle between $\mathbf{x}$ and $\mathbf{y}$, and let ϕ be the angle between $A\mathbf{x}$ and $A\mathbf{y}$. Then

$$\cos(\theta) = \frac{\mathbf{x} \cdot \mathbf{y}}{|\mathbf{x}||\mathbf{y}|} \qquad \text{and} \qquad \cos(\phi) = \frac{(A\mathbf{x}) \cdot (A\mathbf{y})}{|A\mathbf{x}||A\mathbf{y}|} .$$

But $(A\mathbf{x}) \cdot (A\mathbf{y}) = \mathbf{x} \cdot \mathbf{y}$ and $|A\mathbf{x}||A\mathbf{y}| = |\mathbf{x}||\mathbf{y}|$. Therefore,

$$\frac{\mathbf{x} \cdot \mathbf{y}}{|\mathbf{x}||\mathbf{y}|} = \frac{(A\mathbf{x}) \cdot (A\mathbf{y})}{|A\mathbf{x}||A\mathbf{y}|} \ ,$$

and so $\theta = \phi$. This is what it means for A to be angle preserving.

In particular, if A is an isometry, then for any $\mathbf{x}$ and $\mathbf{y}$, $A\mathbf{x}$ and $A\mathbf{y}$ are orthogonal if and only if $\mathbf{x}$ and $\mathbf{y}$ are orthogonal.

• *A linear transformation that preserves lenghts, and hence distances, automatically preserves angles too.*

There is more that we can say about isometries: Suppose that $A = [\mathbf{v}_1, \mathbf{v}_2, \ldots, \mathbf{v}_n]$ is an $m \times n$ isometry. For each $j = 1, 2, \ldots, n$, $\mathbf{v}_j = A\mathbf{e}_j$. Then by (5.22), we have for each $i, j = 1, 2, \ldots, n$ that

$$\mathbf{v}_i \cdot \mathbf{v}_j = A\mathbf{e}_i \cdot A\mathbf{e}_j = \mathbf{e}_i \cdot \mathbf{e}_j = \begin{cases} 1 & \text{if } i = j \\ 0 & \text{if } i \neq j \end{cases} . \tag{5.23}$$

That is each $\mathbf{v}_j$ is a unit vector, and $\mathbf{v}_i$ and $\mathbf{v}_j$ are orthogonal for $i \neq j$. In other words, $\{\mathbf{v}_1, \mathbf{v}_2, \ldots, \mathbf{v}_n\}$ is an *orthonormal* set of vectors in $I\!\!R^m$.

Now by definition $\mathbf{v}_i$ is the ith row of A^t. Therefore, from (5.8) of Theorem 2 we have that

$$(A^t A)_{i,j} = \mathbf{v}_i \cdot \mathbf{v}_j = \begin{cases} 1 & \text{if } i = j \\ 0 & \text{if } i \neq j \end{cases} . \tag{5.24}$$

Notice that (5.24) holds if and only if the columns of A are an orthonormal set of vectors in $I\!\!R^m$.

On the right hand side we recognize the entries of the $n \times n$ identity matrix, and so if A is an isometry, we must have that $A^t A = I$. That is, if A is an isometry, then the transpose of A is a left inverse of A.

This is a very special property. Indeed, if A is any $m \times n$ matrix such that $A^t A = I$, then for any $\mathbf{x}$ in $I\!\!R^n$,

$$|A\mathbf{x}|^2 = A\mathbf{x} \cdot A\mathbf{x} = \mathbf{x} \cdot A^t A\mathbf{x} = \mathbf{x} \cdot \mathbf{x} = |\mathbf{x}|^2 \ ,$$

which means that A is an isometry. This proves the following theorem

Theorem 4 (Isometries and the Transpose) *Let A be an $m \times n$ matrix. Then A is an isometry if and only if A^t is a left inverse of A, that is*

$$A^t A = I \ , \tag{5.25}$$

which is the case if and only if the columns of A are an orthonormal set of vectors in $I\!\!R^m$.

Example 5 Let $A = \frac{1}{\sqrt{2}} \begin{bmatrix} 1 & 1 \\ -1 & 1 \end{bmatrix}$, so that $A^t = \frac{1}{\sqrt{2}} \begin{bmatrix} 1 & -1 \\ 1 & 1 \end{bmatrix}$. Doing the matrix multiplications, one finds

$$A^t A = \begin{bmatrix} 1 & 0 \\ 0 & 1 \end{bmatrix} \quad \text{and} \quad A A^t = \begin{bmatrix} 1 & 0 \\ 0 & 1 \end{bmatrix}.$$

The calculation on the left shows that A^t is a left inverse of A, and then by the theorem, A preserves lengths. The calculuation on the right shows us two more things: First that A^t is also a right inverse of A, and hence that A is invertible. Second, that A is a left inverse of A^t. Then since $(A^t)^t = A$, this means that A^t is also length preserving.

Section 6: Visualizing Linear Transformations in $\mathbb{R}^2$

A picture really can be worth a thousand words – and a good graph is worth, and often represents, *many* thousand computations. If f is a function of a single variable x, a lot of insight into the properties of f, and at least an approximate knowledge about the solutions of equations involving f, can be readily obtained from a graph of $y = f(x)$. For example, here is a graph of $y = f(x)$ for $f(x) = x^5 - x^2 + 1$ and $-1 \leq x \leq 1$, together with the horizontal line $y = 0.9$.

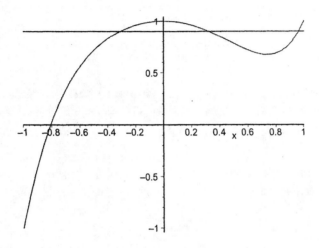

As you can see from the graph, there are three solutions to the equation $f(x) = 0.9$ with $-1 \leq x \leq 1$. In fact, you see that these solutions are approximately given by

$$x \approx -0.3 , \qquad x \approx 0.3 \qquad \text{and} \qquad x \approx 0.95 .$$

One could use this information as the starting point for an iterative solution of the equation through either Newton's method or the bisection algorithm, and a graph is often quite helpful whenever such methods are used. But the graph can do other things for us as well: You plainly see that the function has a local maximum near $x = 0$, and a local minimum near $x = 0.75$. (You can solve for these exactly; try it: The local maximum is indeed at $x = 0$, and the local minimum at $x = \sqrt[3]{2/5}$.)

In any case, the point is this: The graph gives us a concise visual description of *what it is that this function does*; that is, what effect it has on input.

Now consider a vector function, with vector input and vector output. More specifically, consider f_A, the linear transformation from R^2 to R^2 corresponding to the matrix

$$A = \begin{bmatrix} 1 & 1 \\ 0 & 1 \end{bmatrix} . \tag{6.1}$$

What sort of graph can we draw to represent the effect of this transformation – and others like it? Also, is there a graphical way to "see" the solutions, if any, of an equation like $A\mathbf{x} = \mathbf{b}$? These are is the central questions of this section.

Here is one approach to this. Let S be a set of points in the x, y plane. For example, S might be the unit circle, or the x–axis or the y axis. In principle it could be any set, but for our purposes it should be a familiar recognizable set. *What we are going to do is to draw a graph of the set that S gets transformed into under the linear transformation f_A.*

To do this, let's let $\mathbf{u} = \begin{bmatrix} u \\ v \end{bmatrix}$ denote the vector corresponding to the point (u, v) in the u, v plane. We think of f_A, which is a transformation of R^2 to R^2, as a transformation from the x, y plane to the u, v plane through

$$\begin{bmatrix} u \\ v \end{bmatrix} = f_A \left(\begin{bmatrix} x \\ y \end{bmatrix} \right) = \begin{bmatrix} x + y \\ y \end{bmatrix} . \tag{6.2}$$

Definition (Image of a Set) Let A be an $m \times n$ matrix, and let S be a subset of R^n. Then the *image of S under the linear transformation corresponding to A is the set of all vectors in R^m of the form $A\mathbf{x}$ for some $\mathbf{x}$ in S.*

Example 1 Let's take S to be the unit circle in the x, y plane. What is the image of S this time? Since S is given by an equation in the x, y plane, namely $x^2 + y^2 = 1$, it will help to invert the transformation (6.2). This is easy to do: We see from (6.2) that

$$u = x + y \quad \text{and} \quad v = y .$$

The second equation tells us that $y = v$, and substituting this into the first, we have $u = x + v$, or $x = u - v$. So if $\mathbf{u} = A\mathbf{x}$, then

$$\begin{bmatrix} x \\ y \end{bmatrix} = \begin{bmatrix} u - v \\ v \end{bmatrix} . \tag{6.3}$$

(Alternatively, we could use the formula (3.19) for 2×2 matrix inverses.)

Since $\mathbf{x}$ corresponds to a point in S if and only if $x^2 + y^2 = 1$. Using (6.3), this is the case if and only if $(u - v)^2 + v^2 = 1$. That is, $\begin{bmatrix} u \\ v \end{bmatrix}$ belongs to the image of S if and only if

$$u^2 - 2uv + 2v^2 = 1 . \tag{6.4}$$

This is the equation of an ellipse in the u, v plane:

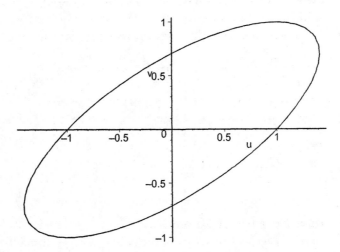

From this picture you see how the transformation f_A "stretches and pulls".

Example 2 We can learn more about how the transformation f_A "stretches and pulls" by transforming the whole collection of lines on a "graph-paper" grid: Consider the lines

$$x = i \quad \text{and} \cdot \quad y = j \quad \text{for } i, j = \ldots -3, -2, -1, 0, 1, 2, 3, \ldots \quad (6.5)$$

If we graph all of these lines at once in the x, y plane, we get a "graph-paper" grid with unit spacing:

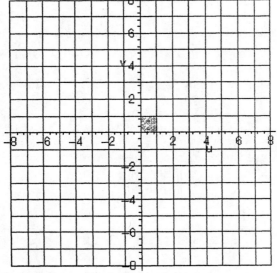

(We've shaded in the first square in the positive quadrant for reasons we'll soon explain.)

Let S be the set of all of these lines. What is the image of S? Here it is: (note the change in scale!)

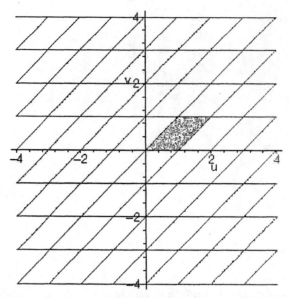

How do we know what lines we get in the image? By (6.3), the line image of the line $x = c$ is $u - v = c$ and the image of the line $y = c$ is $v = c$.

The whole u, v plane, of which we've drawn a piece here, is covered by "trapezoidal tiles", each one of which is the image of a square in our "graph-paper" grid in the x, y plane. The shaded "tile" in the second graph is the image of shaded "tile" in the first graph, as we'll soon check. Notice that all of the "tiles" are of the exact same size and shape, and they are all lined up so that if you were just shown where one of them was, you could put the rest in place without further thought.

• *All of the information contained in such a distorted graph paper picture is given in a graph of the image of the unit square in the positive quadrant of the x, y plane.*

This *unit square* is the square in our "graph–paper" grid that consists consists of all points (x, y) with

$$0 \le x \le 1 \quad \text{and} \quad 0 \le y \le 1 . \tag{6.6}$$

The corresponding vectors are the vectors $\mathbf{x}$ of the form

$$\mathbf{x} = x\mathbf{e}_1 + y\mathbf{e}_2 \tag{6.7}$$

where x and y satisfy (6.6). The unit square is shaded in the graph of the original "graph paper grid".

Now look at the "distorted graph paper grid" which is the image of the original one under our transformation. It would not be immediately clear, just looking at the graph, which of the four "tiles" around the origin is the image of the unit square if we hadn't shaded it in too. How can you tell we shaded in the right one?

You could check this by computing the outputs corresponding to the vertices of the "input square", namely

$$f_A\left(\begin{bmatrix}0\\0\end{bmatrix}\right) = \begin{bmatrix}0\\0\end{bmatrix} \quad , \quad f_A\left(\begin{bmatrix}1\\0\end{bmatrix}\right) = \begin{bmatrix}1\\0\end{bmatrix}$$

$$f_A\left(\begin{bmatrix}0\\1\end{bmatrix}\right) = \begin{bmatrix}1\\1\end{bmatrix} \quad , \quad f_A\left(\begin{bmatrix}1\\1\end{bmatrix}\right) = \begin{bmatrix}2\\1\end{bmatrix} .$$

These four output points are the vertices of the image, and to graph the image, you just put them in as dots, connect these dots with straight lines, and fill in the middle. (It's as simple as that, but ponder it a bit, and make sure you understand why it works.)

We've drawn a few graphs now. What can we learn from them about the linear transformation f_A? In particular, what can we learn about solutions of equations by looking at graphs of this type? Remember that at the beginning of this section we found three pretty good approximate solutions of a polynomial equation by looking at its graph. Can we use any of our graphs to do the same for vector equations, such as, say $f_A(\mathbf{x}) = \begin{bmatrix}1\\1\end{bmatrix}$?

Example 3 Let's try to look for a solution of

$$f_A\left(\begin{bmatrix}x\\y\end{bmatrix}\right) = \begin{bmatrix}1 & 1\\0 & 1\end{bmatrix}\begin{bmatrix}x\\y\end{bmatrix} = \begin{bmatrix}1\\1\end{bmatrix}$$

in some kind of graph.

To do this, write the vector equation as a system of eqautions

$$x + y = 1$$
$$y = 1 .$$

Each of these equations is the equation of a line in the x, y plane. Here is the graph of these two lines:

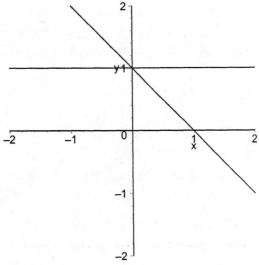

As you see, they cross at the $(1,0)$, the point corresponding to the solution we found above. The slanted line consists of all points (x, y) satisfying the equation $x + y = 1$. The horizontal line consists of all points satisfying the equation $y = 1$. The unique point where they meet is the unique point satisfying *both* equations, and hence corresponds to the solution of the vector equation $A\mathbf{x} = \begin{bmatrix} 1 \\ 1 \end{bmatrix}$. Hence the solution is $\begin{bmatrix} 0 \\ 1 \end{bmatrix}$. This graphical approach to solving vector equations can be very helpful, and we'll study it further in the next chapter.

All of the examples up to here have concerned a single linear transformation, f_A. To consolidate our gains, lets draw some of the corresponding graphs, and address some of the corresponding issues, for another linear transformation.

Example 4 Consider the linear transformation f_B where

$$B = \begin{bmatrix} 1 & 3 \\ -3 & -1 \end{bmatrix}. \tag{6.8}$$

In this example, we'll find the images of the lines bounding the unit square, and then the image of the "graph paper grid" given by (6.4).

First, lets find the image of the y-axis, which is the line $x = 0$, and the parallel line $x = 1$. Let's do it a little differently this time though. We know how to compute the inverse of a 2×2 matrix whenever it is invertible. All we have to do is to apply (3.19) to compute the inverse of B. The result is

$$B^{-1} = \frac{1}{8} \begin{bmatrix} -1 & -3 \\ 3 & 1 \end{bmatrix}. \tag{6.9}$$

This means that $\begin{bmatrix} u \\ v \end{bmatrix} = f_B\left(\begin{bmatrix} x \\ y \end{bmatrix}\right)$ exactly when

$$\begin{bmatrix} x \\ y \end{bmatrix} = \frac{1}{8} \begin{bmatrix} -1 & -3 \\ 3 & 1 \end{bmatrix} \begin{bmatrix} u \\ v \end{bmatrix} = \begin{bmatrix} -(u + 3v)/8 \\ (3u + v)/8 \end{bmatrix}.$$

This is equivalent to the pair of equations

$$x = -\frac{u + 3v}{8} \quad \text{and} \quad y = \frac{3u + v}{8}. \tag{6.10}$$

Now if S is given by the equation $x = 0$, we have from (6.10) that the image of S under f_B is given by

$$-\frac{u + 3v}{8} = 0 \quad \text{or what is the same,} \quad v = -\frac{1}{3}u. \tag{6.11}$$

Effectively, (6.10) is a "lexicon" that enables us to translate equations from x, y terms into u, v terms cooked up in terms of B such a way that if an equations describes a region S in the x, y plane, then the translated equation describes the images of S under f_B. Applying this translation procedure to the equation $x = 1$ yields

$$-\frac{u + 3v}{8} = 1 \qquad \text{or what is the same,} \qquad v = -\frac{1}{3}u - \frac{8}{3}.$$

Here is a graph showing these two lines:

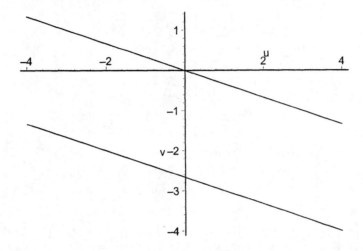

Now let's add to the same graph the images of the x–axis; that is, the line $y = 0$, and the parallel line $y = 1$. Using the "lexicon" (6.10), we translate these equations into $v = -u/3$ and $v = 8/3 - u/3$ respectively. All four lines are plotted below. (Since we did it in two stages we know which line is which.)

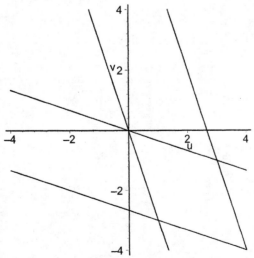

Next, let's draw the image of the lines in the "graph paper" grid, whose equations are given in (6.4). The result is

1-63

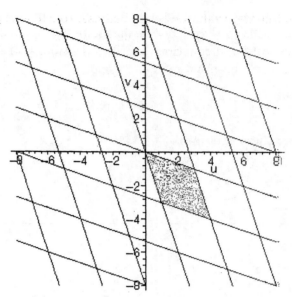

The "tile" that is the image of the unit square has been shaded in. We know which one it is by the graph drawn just above. Again, you see the "regularity and uniformity" of linear transformations. What the transformation does to the unit square in the x, y plane is just "stepped and repeated" throughout the whole plane.

Example 5 For the same matrix B, let's look at the image of the unit circle. This time, we'll not only draw it, but think harder about some of the "thousand words" it tells us.

Using the "lexicon" (6.10) to translate the equation for the unit circle, namely $x^2 + y^2 = 1$ into u, v terms, we get

$$\frac{1}{64}\left((u + 3v)^2 + (3u + v)^2\right) = 1 \qquad \text{or what is the same,} \qquad 5u^2 + 6uv + 5v^2 = 32 \ .$$

Again, this is the equation of an ellipse. Here is a graph of this ellipse, together with the centered circles of radius 2 and 4 in the u, v plane.

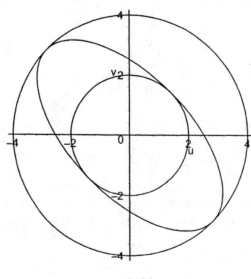

As you see, the larger circle circumscribes the ellipse, touching it along the line $v = -u$, while the smaller circle is inscribed within it, touching it along the line $v = u$. Later on, we will see how to compute the radii of such inscribed and circumscribed circles, but for now, suppose we just found then graphically, using a compass, or experimenting with various radii in a graphing program. Now that we've got them drawn in, what are they telling us?

• *The radius of the inner circle is 2, so for all unit vectors* **x**, *the length of* $f_B(\mathbf{x})$ *is at least 2.*

This is because when **x** is a unit vector, $f_B(\mathbf{x})$ is in the image of the unit circle, and hence somewhere on the ellipse. Every point on the ellipse has a distance of at least 2 from the origin, because, as you see, the circle of this radius is inscribed within it. Similarly, we conclude:

• *The radius of the outer circle is 4, so for all unit vectors* **x**, *the length of* $f_B(\mathbf{x})$ *is no more than 4.*

What about vectors with other lengths? Let **y** be an arbitrary non–zero vector in R^2, and define

$$\mathbf{x} = \frac{1}{|\mathbf{y}|}\mathbf{y} \qquad \text{so that} \qquad \mathbf{y} = |\mathbf{y}|\mathbf{x}$$

and **x** is a unit vector. Since f_B is linear and $|f_B(\mathbf{x})| \geq 2$,

$$|f_B(\mathbf{y})| = |f_B(|\mathbf{y}|\mathbf{x})| = ||\mathbf{y}|f_B(\mathbf{x})| = |\mathbf{y}||f_B(\mathbf{x})| \geq 2|\mathbf{y}| .$$

That is *the linear transformation* f_B *"stretches" every vector by a factor of at least 2.*

In the same way, we see that since the ellipse is inscribed within the circle of radius 4, $|f_B(\mathbf{x})| \leq 4$ for all unit vectors **x**, and so with the same notations,

$$|f_B(\mathbf{y})| = |f_B(|\mathbf{y}|\mathbf{x})| = ||\mathbf{y}|f_B(\mathbf{x})| = |\mathbf{y}||f_B(\mathbf{x})| \leq 4|\mathbf{y}| .$$

That is *the linear transformation* f_B *"stretches" every vector by a factor of at most 4.* Putting it all together, for all **y** in R^2,

$$2|\mathbf{y}| \leq |B\mathbf{y}| \leq 4|\mathbf{y}| . \tag{6.12}$$

These "minimum and maximum stretching factors", and especially their ratio, turn out to be an important characteristic of a linear transformation from a computational standpoint, as we shall see.

The previous graph suggests that 2 is the largest number so that the left inequality holds, and that 4 is the smallest number so that the right inequality holds. In fact, it *looks* like the ellipse and the circle of radius 2 touch at the point $(\sqrt{2}, \sqrt{2})$. Let's try to solve

$$f_B\left(\begin{bmatrix} x \\ y \end{bmatrix}\right) = B\begin{bmatrix} x \\ y \end{bmatrix} = \begin{bmatrix} x + 3y \\ -3x - y \end{bmatrix} = \begin{bmatrix} \sqrt{2} \\ \sqrt{2} \end{bmatrix} , \tag{6.13}$$

and see if this is the case.

As another illustration of the graphical solution method of Example 3, write the vector equation on the right in (6.13), namely $\begin{bmatrix} x + 3y \\ -3x - y \end{bmatrix} = \begin{bmatrix} \sqrt{2} \\ \sqrt{2} \end{bmatrix}$ as a pair of equations

$$x + 3y = \sqrt{2} \qquad \text{and} \qquad 3x + y = -\sqrt{2}. \tag{6.14}$$

Graphing the two lines in the x, y plane that the equations (6.14) describe, we get:

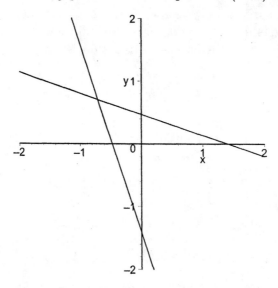

Remembering that $\sqrt{2}/2 \approx 0.707$, you see that the lines cross close to $(-\sqrt{2}/2, \sqrt{2}/2)$. And then you can easily check that indeed, we've found our solution:

$$f_B\left(\begin{bmatrix} -\sqrt{2}/2 \\ \sqrt{2}/2 \end{bmatrix}\right) = B\begin{bmatrix} -\sqrt{2}/2 \\ \sqrt{2}/2 \end{bmatrix} = \begin{bmatrix} \sqrt{2} \\ \sqrt{2} \end{bmatrix}.$$

Therefore, the input vector $\mathbf{y} = \begin{bmatrix} -\sqrt{2}/2 \\ \sqrt{2}/2 \end{bmatrix}$ satisfies $|B\mathbf{y}| = 2|\mathbf{y}|$, an so has its length stretched by a factor of 2. (Notice also that it gets rotated clockwise through an angle of $\pi/4$ radians.) Therefore, the factor of 2 in (6.12) cannot be replaced by any larger number. In the same way, you can see that the factor of 4 in in (6.12) cannot be replaced by any larger number – try it!

The effect of linear transformation on area

In the remaining examples we focus on the effect a linear transformation has on the area of a planar region.

Example 6 Let B be given once more by (6.8) and S to be the unit square. We've already seen that the image of S under f_B is the following parallelogram:

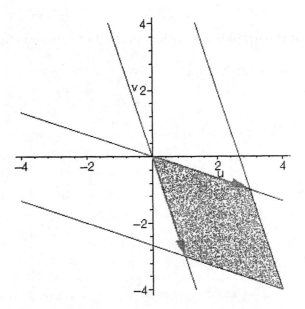

The two arrows drawn in the graph represent $f_B(\mathbf{e}_1)$ and $f_B(\mathbf{e}_2)$. Recall that the unit square is the set of all vectors of the form $x\mathbf{e}_1 + y\mathbf{e}_2$ with x and y satisfying (6.6), we see by the linearity of f_B that the image of the unit square is the set of all vectors of the form

$$x f_B(\mathbf{e}_1) + y f_B(\mathbf{e}_1)$$

with $0 \le x, y \le 1$. Let $\mathbf{v}_1$ and $\mathbf{v}_2$ denote $f_B(\mathbf{e}_1)$ and $f_B(\mathbf{e}_2)$ respectively, and recall that these are just the columns of B.

By elementary planar geometry, the area of the image parallelogram is

$$|\mathbf{v}_1||\mathbf{v}_2|| \sin(\theta)| \tag{6.15}$$

where θ is the interior angle in the vertex at the origin. This is because $|\mathbf{v}_1|$ and $|\mathbf{v}_2|$ are the side lengths of the parallelogram. (In this example, the angle is acute, and the absolute value on $\sin(\theta)$ is superfluous, but we have a better chance of learning something general if we use an expression that is valid for obtuse angles too.)

Now squaring (6.15) and using (4.16) we find

$$
\begin{aligned}
(\text{area})^2 &= |\mathbf{v}_1|^2|\mathbf{v}_2|^2 \sin^2(\theta) \\
&= |\mathbf{v}_1|^2|\mathbf{v}_2|^2(1 - \cos^2(\theta)) \\
&= |\mathbf{v}_1|^2|\mathbf{v}_2|^2 - |\mathbf{v}_1|^2|\mathbf{v}_2|^2 \cos^2(\theta) \\
&= |\mathbf{v}_1|^2|\mathbf{v}_2|^2 - (\mathbf{v}_1 \cdot \mathbf{v}_2)^2
\end{aligned}
\tag{6.16}
$$

We could plug in numbers from B at this point, but that would be premature. To keep our conclusions as general as possible, write

$$\mathbf{v}_1 = \begin{bmatrix} a \\ c \end{bmatrix} \quad \text{and} \quad \mathbf{v}_2 = \begin{bmatrix} b \\ d \end{bmatrix}. \tag{6.17}$$

Then

$$\begin{aligned}
|\mathbf{v}_1|^2|\mathbf{v}_2|^2 - (\mathbf{v}_1 \cdot \mathbf{v}_2)^2 &= (a^2 + c^2)(b^2 + d^2) - (ab + cd)^2 \\
&= (a^2b^2 + a^2d^2 + c^2b^2 + c^2d^2) - (a^2b^2 + c^2d^2 + 2abcd) \\
&= a^2d^2 + c^2b^2 - 2abcd \\
&= (ad - bc)^2 \ .
\end{aligned} \tag{6.18}$$

We've seen this before! Recall that the 2×2 matrix $[\mathbf{v}_1, \mathbf{v}_2] = \begin{bmatrix} a & b \\ c & d \end{bmatrix}$ is invertible if and only if $(ad - bc) \neq 0$, and that $(ad - bc)$ is the denominator in the inverse formula (3.19). Now we can understand this in geometric terms: Combining (6.16) and (6.18), we see that the image of the unit square under the linear transformation corresponding to $\begin{bmatrix} a & b \\ c & d \end{bmatrix}$ has area $|ad - bc|$. This quantity vanishes exactly when $\begin{bmatrix} a & b \\ c & d \end{bmatrix}$ "squashes" the unit square down to a line segment. But in this case $\theta = 0$, and the images of the bounding lines, $x = 0$ and $y = 0$, land on top of each other. Evidently in this case the transformation is not one–to–one, and can't be invertible. Otherwise, as long as the unit square doesn't get "squashed" to a line segment, the transformation will be invertible.

Finally, let's plug in some numbers For the matrix B given by (6.8), $|ad - bc| = |-1 - (-9)| = 8$. That is, this transformation magnifies the area of the unit square by a factor of 8.

Interestingly enough, we found in Example 6 that the image of the unit circle was an ellipse with major radius 4, and minor radius 2. The area included in such an ellipse is π times the product of the major and minor radii; i.e., 8π, while of course the area inside the unit circle is π. *This is the same factor of* 8. *Is this a coincidence?*

Actually not. *Let S be any subset of the plane for which the area is a well–defined number. Then the image of S under f_B has exactly 8 times as much area as S.* The same is true for any linear transformation of the plane, except of course that you have to replace 8 by the value of $|ad - bc|$ for the transformation at hand.

The details of the explanation for this are a bit subtle, and in fact although "area" seems to have a straightforward intuitive meaning, transforming that intuition into precise mathematics would take us far afield form linear algebra.

Consider the following region made up of 66 unit squares:

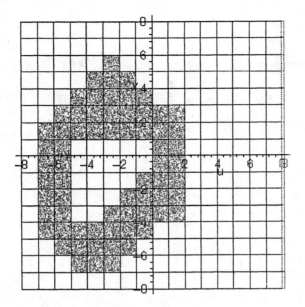

Evidently, its total area is 66 square units. What about its image under f_B? Then each one of these squares is transformed onto one of the "tiles" in the distorted graph paper of Example 4. *Each one of the 66 squares making up our region gets transformed into a "tile" whose area is 8 square units, and none of these "tiles" overlap.* Therefore, the total area of the image region is 8×66 square units. This happens to be 528 square units, but the important thing is that it is *exactly 8 times as much.*

- *The key point in the argument is that all of the transformed squares are congruent to one another, so we only have to work out the "area magnification factor" for one of them. This is a consequence of the "homogeneity and uniformity" of linear transformations, such as you see in the "distorted graph paper" graphs of Examples 2 and 4. It would not be true of a non–linear transformation.*

This argument applies to any region made up out of a finite number of little squares, and it doesn't matter what units we are using: Inches, millimeters, Angstroms ... anything we want. So the image of any such region will have it's area increased by a factor under f_B. Now not all regions are made up of a finite number of little squares –the circle is an example. But area is defined so that any region that has an area can be "approximated" arbitrarily closely by a region that is a finite collection of little squares. The remaining details belong to another subject (analysis), so we won't give them. But try to visualize a sequence of finer and finer approximation to the a circle using finer and finer squares, and you can probably develop a good picture of how the argument would go. In any case, as a check on this, we have already found by direct calculation that the area inside the unit circle is increased exactly by a factor of 8.

Let's summarize some of our conclusions in the a theorem:

Theorem 1 (Area and Linear Transformations in R^2) *Let* $A = \begin{bmatrix} a & b \\ c & d \end{bmatrix}$. *Then the area of the image of the unit square under the linear transformation corresponding to A is $|ad - bc|$, so that A is invertible if and only if this area is strictly positive.*

Let's also make one final definition:

Definition (Area Preserving Linear Transformation) Let f be a linear transformation from R^2 to R^2, and let $A_f = \begin{bmatrix} a & b \\ c & d \end{bmatrix}$ be the corresponding 2×2 matrix. Then f is an *area preserving* linear transformation in case $|ad - bc| = 1$.

As we've *proven* above, the condition $|ad - bc| = 1$ means that f preserves the area of the unit square, and hence any region made up of a finite number of squares. As we've *explained*, this condition actually means that f preserves the area of any region that has an area, so the name is appropriate.

Example 7 Let $A = \begin{bmatrix} 1 & 1 \\ 0 & 1 \end{bmatrix}$ be the matrix (6.1). For this matrix, $|ad - bc| = |1 - 0| = 1$, so f_A is an area preserving transformation.

Problems for Chapter One

Problems for Section 1

1.1.1 Let f be the following transformation from R^2 to R^3:

$$f\left(\begin{bmatrix} x \\ y \end{bmatrix}\right) = \begin{bmatrix} xy \\ x-y \\ x+y \end{bmatrix}$$

and let g be the following transformation from R^3 to R^2:

$$g\left(\begin{bmatrix} x \\ y \\ z \end{bmatrix}\right) = \begin{bmatrix} z \\ y \end{bmatrix}.$$

(a) Compute $f\left(\begin{bmatrix} 1 \\ 2 \end{bmatrix}\right)$.

(b) Compute $g\left(\begin{bmatrix} 2 \\ -1 \\ 3 \end{bmatrix}\right)$.

(c) Compute $g \circ f\left(\begin{bmatrix} 1 \\ 2 \end{bmatrix}\right)$.

(d) Compute a formula for the transformation $g \circ f$, like the formulas for f and g given above.

1.1.2 Define an appropriate transformation from R^n to R^m, and write the following systems of equations as single vector equations.

(a)
$$2x + 3y = 3$$
$$x/y = 1$$

(b)
$$xy + yz + zx = 3$$
$$x + y = 1$$
$$x - y + z = 2$$

(c)
$$x + y + x = 3$$
$$y + z = 2$$
$$x + z = 2$$
$$x + y = 1$$

1.1.3 Let f, g and h be the functions from $\{1, 2, 3, 4, 5\}$ to itself given by

$$f: \begin{array}{ccccc} 1 & 2 & 3 & 4 & 5 \\ \downarrow & \downarrow & \downarrow & \downarrow & \downarrow \\ 2 & 4 & 3 & 5 & 1 \end{array} \qquad g: \begin{array}{ccccc} 1 & 2 & 3 & 4 & 5 \\ \downarrow & \downarrow & \downarrow & \downarrow & \downarrow \\ 2 & 2 & 4 & 4 & 2 \end{array}$$

and

$$h: \begin{array}{ccccc} 1 & 2 & 3 & 4 & 5 \\ \downarrow & \downarrow & \downarrow & \downarrow & \downarrow \\ 5 & 4 & 3 & 2 & 1 \end{array}$$

(a) Give the tables specifying $f \circ g$, $g \circ f$, $h \circ g$ and $h \circ h$.

(b) Give the tables specifying $h \circ (g \circ f)$ and $(h \circ g) \circ f$. Do your results agree with the associativity of the composition product?

(c) Which of the functions f, g and h are one–to–one? Which are onto? Which are invertible? For those that are invertible, give the table specifying the inverses.

1.1.4 Let f, g and h be the functions from $\{1, 2, 3, 4, 5\}$ to itself given by

$$f: \begin{array}{ccccc} 1 & 2 & 3 & 4 & 5 \\ \downarrow & \downarrow & \downarrow & \downarrow & \downarrow \\ 2 & 5 & 4 & 2 & 1 \end{array} \qquad g: \begin{array}{ccccc} 1 & 2 & 3 & 4 & 5 \\ \downarrow & \downarrow & \downarrow & \downarrow & \downarrow \\ 1 & 5 & 3 & 4 & 2 \end{array}$$

and

$$h: \begin{array}{ccccc} 1 & 2 & 3 & 4 & 5 \\ \downarrow & \downarrow & \downarrow & \downarrow & \downarrow \\ 4 & 2 & 1 & 5 & 3 \end{array}$$

(a) Give the tables specifying $f \circ g$, $g \circ f$, $h \circ g$ and $h \circ h$.

(b) Give the tables specifying $h \circ (g \circ f)$ and $(h \circ g) \circ f$. Do your results agree with the associativity of the composition product?

(c) Which of the functions f, g and h are one–to–one? Which are onto? Which are invertible? For those that are invertible, give the table specifying the inverses.

Problems for Section 2

1.2.1 Let $\mathbf{a} = \begin{bmatrix} 1 \\ -1 \end{bmatrix}$ and let $\mathbf{b} = \begin{bmatrix} 2 \\ 1 \end{bmatrix}$.

(a) Compute $3\mathbf{a}$, $\mathbf{a} + \mathbf{b}$ and $\mathbf{a} - \mathbf{b}$.

(b) Draw a diagram in the plane showing $\mathbf{a}$ and $3\mathbf{a}$ as arrows, as in the examples of this section.

(c) Draw a diagram in the plane showing **a**, **b** and **a** + **b** as arrows, as in the examples of this section.

(d) Draw a diagram in the plane showing **a**, **b** and **a** − **b** as arrows, as in the examples of this section.

1.2.2 Let $\mathbf{a} = \begin{bmatrix} 1 \\ 1 \end{bmatrix}$ and let $\mathbf{b} = \begin{bmatrix} -2 \\ 3 \end{bmatrix}$.

(a) Compute 3**a**, **a** + **b** and **a** − **b**.

(b) Draw a diagram in the plane showing **a** and −2**a** as arrows, as in the examples of this section.

(c) Draw a diagram in the plane showing **a**, **a** and **a** + **b** as arrows, as in the examples of this section.

(d) Draw a diagram in the plane showing **a**, **a** and **a** − **b** as arrows, as in the examples of this section.

1.2.3 Consider the following transformations from R^2 to R^2. Which ones are linear? Explain your answers, and for those that are linear, write down the corresponding matrix.

$$f\left(\begin{bmatrix} x \\ y \end{bmatrix}\right) = \begin{bmatrix} y + x \\ y - x \end{bmatrix} \qquad g\left(\begin{bmatrix} x \\ y \end{bmatrix}\right) = \begin{bmatrix} |x| \\ y - x \end{bmatrix} \qquad h\left(\begin{bmatrix} x \\ y \end{bmatrix}\right) = \begin{bmatrix} x^2 - y^2 \\ x^2 + y^2 \end{bmatrix}$$

1.2.4 Consider the following transformations from R^3 to R^2. Which ones are linear? Explain your answers, and for those that are linear, write down the corresponding matrix.

$$f\left(\begin{bmatrix} x \\ y \\ z \end{bmatrix}\right) = \begin{bmatrix} xz \\ xy \end{bmatrix} \qquad g\left(\begin{bmatrix} x \\ y \\ z \end{bmatrix}\right) = \begin{bmatrix} z \\ y \end{bmatrix} \qquad h\left(\begin{bmatrix} x \\ y \\ z \end{bmatrix}\right) = \begin{bmatrix} 1 \\ x \end{bmatrix}$$

1.2.5 Consider the following transformations from R^2 to R^3. Which ones are linear? Explain your answers, and for those that are linear, write down the corresponding matrix.

$$f\left(\begin{bmatrix} x \\ y \end{bmatrix}\right) = \begin{bmatrix} y \\ 1 \\ x \end{bmatrix} \qquad g\left(\begin{bmatrix} x \\ y \end{bmatrix}\right) = \begin{bmatrix} y \\ 0 \\ x \end{bmatrix} \qquad h\left(\begin{bmatrix} x \\ y \end{bmatrix}\right) = \begin{bmatrix} 0 \\ 0 \\ 0 \end{bmatrix}$$

1.2.6 Compute the single vector specified by the following linear combinations:

(a) $2\begin{bmatrix} 1 \\ 2 \end{bmatrix} - 3\begin{bmatrix} 2 \\ 1 \end{bmatrix}$ (b) $2\begin{bmatrix} 1 \\ 2 \end{bmatrix} - 3\begin{bmatrix} 2 \\ 1 \end{bmatrix} + \frac{1}{2}\begin{bmatrix} -3 \\ 1 \end{bmatrix} + \begin{bmatrix} 0 \\ 1 \end{bmatrix}$

(c) $2 \begin{bmatrix} 1 \\ 2 \\ -1 \end{bmatrix} + 3 \begin{bmatrix} 2 \\ -1 \\ -1 \end{bmatrix}$ **(d)** $\begin{bmatrix} 5 \\ 0 \\ -1 \end{bmatrix} - \begin{bmatrix} 3 \\ 3 \\ -3 \end{bmatrix}$

1.2.7 Consider the following list of 4 matrices and 4 vectors. There are 16 different pairs of matrices and vectors. Say for which pairs the matrix–vector product is defined, for which it isn't, and compute it when it is.

$$A = \begin{bmatrix} 1 & 1 & 1 & 2 \\ 1 & 2 & 0 & -1 \end{bmatrix} \quad B = \begin{bmatrix} 1 & -1 & 1 \\ 2 & 1 & 0 \\ 4 & -1 & 2 \end{bmatrix} \quad C = \begin{bmatrix} 2 & 1 \\ 0 & 2 \\ 1 & -2 \\ 5 & 1 \end{bmatrix} \quad D = \begin{bmatrix} 0 & 1 \\ 1 & 0 \end{bmatrix}$$

$$\mathbf{v} = \begin{bmatrix} 1 \\ 1 \end{bmatrix} \quad \mathbf{x} = \begin{bmatrix} 1 \\ -1 \end{bmatrix} \quad \mathbf{y} = \begin{bmatrix} 2 \\ -1 \\ 2 \end{bmatrix} \quad \mathbf{z} = \begin{bmatrix} -1 \\ -1 \\ 3 \\ 2 \end{bmatrix}$$

1.2.8 Consider the following list of 4 matrices and 4 vectors. There are 16 different pairs of matrices and vectors. Say for which pairs the matrix–vector product is defined, for which it isn't, and compute it when it is.

$$A = \begin{bmatrix} 0 & 2 & 1 & 3 \\ 1 & 2 & 0 & -1 \\ 3 & -1 & 1 & 4 \end{bmatrix} \quad B = \begin{bmatrix} 0 & -2 & 1 \\ 2 & 1 & 0 \\ 3 & -1 & 2 \end{bmatrix} \quad C = \begin{bmatrix} 2 & 1 & 1 & 1 \\ 2 & 1 & 0 & -1 \\ 0 & -3 & 0 & 1 \\ 1 & 1 & 2 & -1 \end{bmatrix} \quad D = \begin{bmatrix} 1 & 1 \\ 2 & 0 \end{bmatrix}$$

$$\mathbf{v} = \begin{bmatrix} 1 \\ 2 \end{bmatrix} \quad \mathbf{x} = \begin{bmatrix} 2 \\ -1 \end{bmatrix} \quad \mathbf{y} = \begin{bmatrix} 1 \\ -3 \\ 1 \end{bmatrix} \quad \mathbf{z} = \begin{bmatrix} 2 \\ 1 \\ 3 \\ 4 \end{bmatrix}$$

1.2.9 Let $A = \begin{bmatrix} 1 & 1 & 1 & 2 \\ 1 & 2 & 0 & -1 \\ 2 & 0 & 0 & -1 \\ 3 & -2 & 0 & 2 \end{bmatrix}$ and $\mathbf{x} = \begin{bmatrix} -1 \\ -2 \\ 2 \\ 1 \end{bmatrix}$. Compute the third entry of $A\mathbf{x}$ without computing the whole vector $A\mathbf{x}$.

1.2.10 Let f be a linear transformation from $I\!R^2$ to $I\!R^2$. Suppose that

$$f(\mathbf{e}_1) = \begin{bmatrix} -1 \\ 2 \end{bmatrix} \quad \text{and} \quad f(\mathbf{e}_2) = \begin{bmatrix} -2 \\ -1 \end{bmatrix}.$$

Find the matrix A_f corresponding to f.

1.2.11 Let f be a linear transformation from $I\!R^2$ to $I\!R^2$. Suppose that

$$f(\mathbf{e}_1) = \begin{bmatrix} 1 \\ 3 \end{bmatrix} \quad \text{and} \quad f(\mathbf{e}_2) = \begin{bmatrix} -3 \\ 1 \end{bmatrix}.$$

Find the matrix A_f corresponding to f.

1.2.12 Let f be the linear transformation from $I\!R^2$ to $I\!R^2$ given by rotation in the counterclockwise direction through an angle of $\pi/3$ radians. Let g be the linear transformation from $I\!R^2$ to $I\!R^2$ given reflection about the line $y = x$. What is the matrix $A_{g \circ f}$ corresponding to the composition of g with f? What is the matrix $A_{f \circ g}$ corresponding to the composition of f with g?

1.2.13 Let f be the linear transformation from $I\!R^2$ to $I\!R^2$ given first reflecting about the line $y = x$, and then reflecting about the line $x = 0$. What is the matrix A corresponding to this linear transformation?

1.2.14 Let f be the linear transformation from $I\!R^2$ to $I\!R^2$ given first reflecting about the line $y = x$, and then reflecting about the line $y = -x$. What is the matrix A corresponding to this linear transformation?

1.2.15 Let f be the linear transformation from $I\!R^2$ to $I\!R^2$ given reflection about the line through the origin with slope s. Let g be the linear transformation from $I\!R^2$ to $I\!R^2$ given reflection about the x axis. What is the matirx $A_{g \circ f}$ corresponding to the composition of g with f? (Your answer should be a matrix with entries depending on s.) How does your answer change if instead g is reflection about the y–axis?

1.2.16 Let A and B be $m \times n$ matrices. If $A\mathbf{x} = B\mathbf{x}$ for all $\mathbf{x}$ in $I\!R^n$, does this mean that $A = B$? Explain why, or give a counterexample.

Problems for Section 3

1.3.1 Consider the four matrices A, B, C and D from problem 1.2.7. There are 16 ordered pairs of these matrices, allowing self pairing, namely

$$AA, \quad AB, \quad AC, \quad AD, \quad BA, \quad BB, \ldots$$

Which of these make sense as matrix products? Compute the product in each such case.

1.3.2 Consider the four matrices A, B, C and D from problem 1.2.8. There are 16 ordered pairs of these matrices, allowing self pairing, namely

$$AA, \quad AB, \quad AC, \quad AD, \quad BA, \quad BB, \ldots$$

Which of these make sense as matrix products? Compute the product in each such case.

1.3.3 Let $A = \begin{bmatrix} 1 & 1 & 1 & 2 \\ 1 & 2 & 0 & -1 \\ 2 & 0 & 0 & -1 \\ 3 & -2 & 0 & 2 \end{bmatrix}$ and let $B = \begin{bmatrix} 0 & -1 & 1 & 1 \\ 1 & 1 & 1 & -1 \\ 2 & 3 & 1 & -3 \\ 0 & -3 & 0 & 0 \end{bmatrix}$. Compute the third

column of AB by computing an appropriate matrix–vector product.

1.3.4 Let $A = \begin{bmatrix} 2 & 2 & 1 & 0 \\ 0 & 3 & 0 & -3 \\ 2 & 0 & 0 & -2 \\ 1 & -1 & 3 & 0 \end{bmatrix}$ and let $B = \begin{bmatrix} 0 & -6 & 6 & 2 \\ -1 & 1 & 1 & -2 \\ -2 & 3 & 1 & -1 \\ 0 & -3 & 0 & 1 \end{bmatrix}$. Compute the second column of AB by computing an appropriate matrix–vector product.

1.3.5 Let $A = \begin{bmatrix} 0 & 1 & 0 \\ 0 & 0 & 1 \\ 0 & 0 & 0 \end{bmatrix}$. Compute A^2 and A^3.

1.3.6 For any three numbers a, b and c, let $A = \begin{bmatrix} 0 & a & b \\ 0 & 0 & c \\ 0 & 0 & 0 \end{bmatrix}$. Compute A^2 and A^3.

1.3.7 For any three numbers a, b and c, let $A = \begin{bmatrix} a & b \\ 0 & c \end{bmatrix}$.

(a) Compute A^2.

(b) Find all possible values of a b and c so that $A^2 = B$ where $B = \begin{bmatrix} 1 & 3 \\ 0 & 4 \end{bmatrix}$. How many different sets of values for a b and c are there for which $A^2 = B$? (The matrices A that you compute here are *square roots* of the matrix B.)

1.3.8 For any five numbers a, b, c, d and e, let $A = \begin{bmatrix} a & b & c \\ 0 & c & d \\ 0 & 0 & e \end{bmatrix}$.

(a) Compute A^2.

(b) Find all possible values of a, b, c, d and e so that $A^2 = B$ where $B = \begin{bmatrix} 1 & 3 & 5 \\ 0 & 4 & 7 \\ 0 & 0 & 9 \end{bmatrix}$. How many different sets of values for a, b, c, d and e are there for which $A^2 = B$? (The matrices A that you compute here are *squaure roots* of the matrix B.)

1.3.9 Given two $n \times n$ matrices A and B, the *commutator of A and B* is defined to be the $n \times n$ matrix $AB - BA$, and is denoted by $[A, B]$. The reason for this is that $AB = BA$ if and only if $[A, B] = 0$. More generally, but also directly from the definition, $AB = BA + [A, B]$.

(a) Compute $[A, B]$ for $A = \begin{bmatrix} 0 & 1 \\ 1 & 0 \end{bmatrix}$ and $B = \begin{bmatrix} 1 & 0 \\ 0 & 2 \end{bmatrix}$.

(b) For the same matrices A and B, compute $[A, [A, B]]$.

(c) Using only the associative property of matrix multiplication, show that for any three $n \times n$ matircs A, B and C,

$$[A, [B, C]] + [B, [C, A]] + [C, [A, B]] = 0 .$$

This is known as *Jacobi's identity*. (We won't make use of it later. For our purposes, checking it is simply a good exercise in working with matrices.)

1.3.10 For any numbers a and b, let $A = \begin{bmatrix} 1 & a \\ 0 & 1 \end{bmatrix}$ and let $B = \begin{bmatrix} 1 & b \\ 0 & 1 \end{bmatrix}$.

(a) Compute the product AB.

(b) Show that A is always invertible, and find the inverse. (Your answer will be a matrix that, like A, depends on a.)

1.3.11 For any three numbers a, b and c, Let $A = \begin{bmatrix} 1 & a & b \\ 0 & 1 & c \\ 0 & 0 & 1 \end{bmatrix}$. For any three numbers u, v and w, let $B = \begin{bmatrix} 1 & u & v \\ 0 & 1 & w \\ 0 & 0 & 1 \end{bmatrix}$.

(a) Compute the product AB.

(b) Show that A is always invertible, and find the inverse. (Your answer will be a matrix that, like A, depends on a, b and c.)

1.3.12 Let $R(\theta)$ be the 2×2 rotation matrix $R(\theta) = \begin{bmatrix} \cos(\theta) & -\sin(\theta) \\ \sin(\theta) & \cos(\theta) \end{bmatrix}$. Check using the formula for inversion of 2×2 matrices that $(R(\theta))^{-1} = R(-\theta)$.

1.3.13 Suppose that f is a linear transformation from $I\!R^n$ to $I\!R^m$, and that g is a transformation from $I\!R^m$ to $I\!R^p$, but that g is not linear. Does it follow that $g \circ f$ is not linear, or can it be that $g \circ f$ is a linear linear transformation from $I\!R^n$ to $I\!R^p$? Does the answer depend on the values of n, m and p or not? Explain your answers.

1.3.14 Suppose that f is a transformation from $I\!R^n$ to $I\!R^n$, and that g is a transformation from $I\!R^n$ to $I\!R^n$, and that neither f nor g is linear, but both are invertible. Does it follow that $g \circ f$ is not linear, or can it be that $g \circ f$ is a linear linear transformation from $I\!R^n$ to $I\!R^n$? Explain your answer.

Problems for Section 4

1.4.1 Consider the following vectors:

$$\mathbf{v} = \begin{bmatrix} 1 \\ 1 \end{bmatrix} \qquad \mathbf{x} = \begin{bmatrix} 1 \\ -1 \end{bmatrix} \qquad \mathbf{y} = \begin{bmatrix} 2 \\ -1 \\ 2 \end{bmatrix} \qquad \mathbf{z} = \begin{bmatrix} -1 \\ -1 \\ 3 \\ 2 \end{bmatrix}$$

Compute the length of each of these vectors, and the dot product of each pair of vectors for which it is defined.

1.4.2 Why doesn't it make sense to even ask whether or not the dot product is associative?

1.4.3 Consider the vectors

$$\mathbf{v}_1 = \begin{bmatrix} 1 \\ 2 \end{bmatrix} \qquad \mathbf{v}_2 = \begin{bmatrix} 2 \\ 1 \end{bmatrix} \qquad \text{and} \qquad \mathbf{v}_3 = \begin{bmatrix} -2 \\ 1 \end{bmatrix}.$$

(a) Compute $\mathbf{v}_i \cdot \mathbf{v}_j$ for each $i, j = 1, 2, 3$.

(b) What are the lengths of each of the three vectors?

(c) What are the angles between each of the three pairs of vectors? Is any pair orthogonal?

(d) Draw a diagram showing each of the three vectors as arrows. Do the lengths and angles that you computed look right?

1.4.4 Consider the vectors

$$\mathbf{v}_1 = \begin{bmatrix} 3 \\ 1 \end{bmatrix} \qquad \mathbf{v}_2 = \begin{bmatrix} 2 \\ 2 \end{bmatrix} \qquad \text{and} \qquad \mathbf{v}_3 = \begin{bmatrix} -1 \\ 3 \end{bmatrix}.$$

(a) Compute $\mathbf{v}_i \cdot \mathbf{v}_j$ for each $i, j = 1, 2, 3$.

(b) What are the lengths of each of the three vectors?

(c) What are the angles between each of the three pairs of vectors? Is any pair orthogonal?

(d) Draw a diagram showing each of the three vectors as arrows. Do the lengths and angles that you computed look right?

1.4.5 Fix a number r with $-1 < r < 1$. For each n, let ℓ_n be the length of the vector
$$\begin{bmatrix} 1 \\ r \\ \mathbb{R}^2 \\ \vdots \\ \mathbb{R}^{n-1} \end{bmatrix} \quad \text{in } \mathbb{R}^n.$$

(a) Compute ℓ_n as a function of r.

(b) Now fix any other number s with $-1 < r < 1$. For each n, let α_n be the angle between the vectors $\begin{bmatrix} 1 \\ r \\ \mathbb{R}^2 \\ \vdots \\ \mathbb{R}^{n-1} \end{bmatrix}$ and $\begin{bmatrix} 1 \\ s \\ s^2 \\ \vdots \\ s^{n-1} \end{bmatrix}$ in $\mathbb{R}^n$. Compute α_n as a function of r and s.

(c) Compute $\lim_{n \to \infty} \ell_n$ and $\lim_{n \to \infty} \alpha_n$. Think about what the existence of these limits might say about the possibility of geometric considerations in infinitely many dimensions.

1.4.6 Consider the vectors $\mathbf{x} = \begin{bmatrix} 1 \\ 1 \\ 1 \end{bmatrix}$ and $\mathbf{y} = \begin{bmatrix} 1 \\ -2 \\ 1 \end{bmatrix}$.

(a) Compute $|\mathbf{x}|$, $|\mathbf{y}|$ and $\mathbf{x} \cdot \mathbf{y}$.

(b) Compute $|s\mathbf{x} + t\mathbf{y}|$ as a function of s and t.

(c) How does the angle between $\mathbf{x}$ and $\mathbf{x} + t\mathbf{y}$ depend on t?

1.4.7 Consider the vectors $\mathbf{a} = \begin{bmatrix} a \\ b \end{bmatrix}$. Find all vectors $\begin{bmatrix} c \\ d \end{bmatrix}$ that are orthogonal to $\mathbf{a}$. That is find conditions on c and d in therms of a and b that are necessary and sufficient for this orthogonality.

1.4.8 Let $\mathbf{u}$ be any unit vector in $I\!\!R^2$. Define a transformation f from $I\!\!R^2$ to $I\!\!R^2$ by

$$f(\mathbf{x}) = \mathbf{x} - 2(\mathbf{u} \cdot \mathbf{x})\mathbf{u} .$$

(a) Show that this transformation is linear and is *length preserving*; that is, for all $\mathbf{x}$, the length of the output $|f(\mathbf{x})|$ equals the length of the input $|\mathbf{x}|$.

(b) Show that $f \circ f$ is the indentity transformation.

(c) Specifically let $\mathbf{u} = \begin{bmatrix} \cos(\theta) \\ \sin(\theta) \end{bmatrix}$, and find the matrix A_f.

1.4.9 Let $\mathbf{x}$ and $\mathbf{y}$ be two non–zero vectors in $I\!\!R^n$. Suppose that

$$|\mathbf{x} + \mathbf{y}|^2 = |\mathbf{x}|^2 + |\mathbf{y}|^2 .$$

What is the angle between these two vectors?

1.4.10 Let $\mathbf{x}$ and $\mathbf{y}$ be any two vectors in $I\!\!R^n$. Show that the parallelogram identity, namely

$$|\mathbf{x} + \mathbf{y}|^2 + |\mathbf{x} - \mathbf{y}|^2 = 2|\mathbf{x}|^2 + 2|\mathbf{y}|^2$$

always holds.

1.4.11 The set of vectors $\begin{bmatrix} x \\ y \\ z \end{bmatrix}$ that are orthogonal to the fixed vector $\begin{bmatrix} 1 \\ 2 \\ -1 \end{bmatrix}$ is a plane in $I\!\!R^3$. Find an equation specifying this plane.

1.4.12 Consider the vectors $\mathbf{x} = \begin{bmatrix} t \\ 1 \\ 1 \end{bmatrix}$ and $\mathbf{y} = \begin{bmatrix} 1 \\ -2 \\ 1 \end{bmatrix}$. For which values of t, if any, are these vectors orthogonal? For which values of t, if any, are these vectors parallel?

Problems for Section 5

1.5.1 Let $A = \begin{bmatrix} 1 & 2 & 3 \\ 2 & 0 & 2 \\ 0 & 1 & 1 \\ 1 & 2 & 3 \end{bmatrix}$ and let $\mathbf{v} = \begin{bmatrix} 1 \\ 2 \\ 1 \end{bmatrix}$. Use a single dot product to compute the second entry of $A\mathbf{v}$.

1.5.2 Let $A = \begin{bmatrix} 1 & 0 & 1 \\ 4 & 0 & 2 \\ 2 & 3 & 1 \\ 1 & 0 & 1 \end{bmatrix}$ and let $\mathbf{v} = \begin{bmatrix} 2 \\ 3 \\ 1 \end{bmatrix}$. Use a single dot product to compute the third entry of $A\mathbf{v}$.

1.5.3 Consider the matrices

$$A = \begin{bmatrix} 1 & 2 & 0 & 1 \\ 1 & 3 & 1 & 2 \\ 1 & 2 & 2 & 2 \end{bmatrix} \quad \text{and} \quad B = \begin{bmatrix} 1 & 1 & 1 \\ 2 & 0 & 2 \\ 0 & 0 & 1 \\ 3 & 2 & 1 \end{bmatrix}.$$

(a) Compute $(AB)_{2,3}$ without computing the whole matrix product AB.

(b) Write the second column of AB as a linear combination of the columns of A. That is, find numbers a, b, c and d so that the third column of AB eqauls

$$a \begin{bmatrix} 1 \\ 1 \\ 1 \end{bmatrix} + b \begin{bmatrix} 2 \\ 3 \\ 2 \end{bmatrix} + c \begin{bmatrix} 0 \\ 1 \\ 2 \end{bmatrix} + d \begin{bmatrix} 1 \\ 2 \\ 2 \end{bmatrix}.$$

(c) Write the second row of AB as a linear combination of the rows of B.

1.5.4 Consider the matrices

$$A = \begin{bmatrix} 1 & 2 & 4 \\ 2 & 2 & 4 \\ 0 & 2 & -1 \end{bmatrix} \quad \text{and} \quad B = \begin{bmatrix} 1 & 2 & 3 \\ 2 & 1 & 3 \\ 1 & 1 & 2 \end{bmatrix}.$$

(a) Compute $(AB)_{2,2}$ and $(BA)_{2,2}$ without computing the whole matrix products AB and BA.

(b) Write the first column of AB as a linear combination of the columns of A. (See the previous problem, part (b).)

(c) Write the first row of AB as a linear combination of the rows of B.

1.5.5 Let A be a 3×3 matrix whose column representation is $A = [\mathbf{v}_1, \mathbf{v}_2, \mathbf{v}_3]$ Find an explicit numerical 3×3 matrix B so that

$$AB = [\mathbf{v}_2 + \mathbf{v}_3, \mathbf{v}_1 + \mathbf{v}_2, \mathbf{v}_1 + \mathbf{v}_2] .$$

(You are looking for a single B that works no matter what $\mathbf{v}_1$, $\mathbf{v}_2$ and $\mathbf{v}_3$ happen to be.

1.5.6 Let B be a 3×3 matrix whose row representation is $BA = \begin{bmatrix} \mathbf{v}_1 \\ \mathbf{v}_2 \\ \mathbf{v}_3 \end{bmatrix}$ Find an explicit numerical 3×3 matrix A so that

$$AB = \begin{bmatrix} \mathbf{v}_2 + \mathbf{v}_3 \\ \mathbf{v}_1 + \mathbf{v}_2 \\ \mathbf{v}_1 + \mathbf{v}_2 \end{bmatrix} .$$

(You are looking for a single B that works no matter what $\mathbf{v}_1$, $\mathbf{v}_2$ and $\mathbf{v}_3$ happen to be.

1.5.7 (a) Let A be the $n \times n$ diagonal matrix $A = \mathrm{diag}(a_1, a_2, \ldots, a_j)$ in which, for some i with $1 \leq i \leq n$, $A_i = 1$, but $a_j = 0$ for $j \neq i$. In other words, $A_{i,i} = 1$, and every other entry is zero. Let B be any $n \times n$ matrix. Describe the rows of AB and the columns of BA.

(b) Is there any *non-diagonal* $n \times n$ matrix B that commutes with A (that is $AB = BA$) for all $n \times n$ diagonal matrices A? (We already know that each diagonal $n \times n$ matrix B commutes with every diagonal matrix. The question is: are there any others?) Give na example, or explain why not.

1.5.8 Let A be an $m \times n$ matrix, and B be an $n \times p$ matrix. If A has no zero columns, does it then follow that AB has no zero columns? Explain why, or give a counter example.

1.5.9 Let A be an $m \times n$ matrix, and B be an $n \times p$ matrix. If B has at least one zero column, can it ever be the case that AB has no zero columns? Explain why not, or give an example.

1.5.10 Find 2×2 matrices A and B such that $(A + B)^2 \neq A^2 + 2AB + B^2$. Can this happen for any two diagonal matrices?

1.5.11 Let A be the matrix $A = \begin{bmatrix} 1 & 0 & 0 & 0 \\ 0 & 1 & 0 & 0 \\ 0 & 0 & 0 & 0 \\ 0 & 0 & 0 & 0 \end{bmatrix}$. Let B be any other 4×4 matrix.

(a) Which rows of B, if any, can be freely modified without affecting the product AB?

(b) Which columns of B, if any, can be freely modified without affecting the product AB?

1.5.12 Let A be the matrix $A = \begin{bmatrix} 1 & 0 & 0 & 0 \\ 0 & 1 & 0 & 0 \\ 0 & 0 & 0 & 0 \\ 0 & 0 & 0 & 0 \end{bmatrix}$. Let B be any other 4×4 matrix.

(a) Which rows of B, if any, can be freely modified without affecting the product BA? (Note that B was n the right in the previous problem.)

(b) Which columns of B, if any, can be freely modified without affecting the product BA?

1.5.13 Let C be a 2 by 2 matrix such that

$$C \begin{bmatrix} 1 \\ 2 \end{bmatrix} = \begin{bmatrix} 2 \\ 1 \end{bmatrix} \quad \text{and} \quad C \begin{bmatrix} 2 \\ 1 \end{bmatrix} = \begin{bmatrix} -1 \\ 1 \end{bmatrix}$$

Using the given information, find 2×2 matrices A and B so that $CA = B$, and then solve for C.

1.5.14 Let C be a 2 by 2 matrix such that

$$C \begin{bmatrix} 1 \\ 2 \end{bmatrix} = \begin{bmatrix} 2 \\ 1 \end{bmatrix} \quad \text{and} \quad C^2 \begin{bmatrix} 1 \\ 2 \end{bmatrix} = \begin{bmatrix} -1 \\ 1 \end{bmatrix}$$

Using the given information, fins 2×2 matrices A and B so that $CA = B$, and then solve for C.

1.5.15 Are there any $m \times n$ length preserving matrices with $n > m$? Give an example or explain why not.

1.5.16 Let A be an $m \times n$ matrix and m a positive integer. How is the mth power of A realted to the mth power of A^t? Is it always true that $(A^m)^t = (A^m)^t$? Explain why, or give a counterexample.

1.5.17 Consider the matrices

$$A = \frac{1}{\sqrt{2}} \begin{bmatrix} 1 & 0 \\ 0 & \sqrt{2} \\ 1 & 0 \end{bmatrix} \qquad B = \begin{bmatrix} 0 & -1 & 1 \\ \sqrt{2} & 0 & 0 \\ 0 & 1 & 1 \end{bmatrix} \qquad C = \frac{1}{\sqrt{6}} \begin{bmatrix} \sqrt{3} & -1 \\ 0 & 2 \\ \sqrt{3} & 1 \end{bmatrix}$$

Which, if any, are length preserving?

Problems for Section 6

1.6.1 Let A be the matrix $A = \begin{bmatrix} 1 & 2 \\ 0 & 3 \end{bmatrix}$. Find the equation describing the image of S under the linear transformation corresponding to A, and graph this image of S, where:

1-82

(a) S is the line y-axis.

(b) S is the line $x + y = 3$.

(c) S the unit circle.

1.6.2 Let A be the matrix $A = \begin{bmatrix} 2 & -3 \\ 1 & 3 \end{bmatrix}$. Find the equation describing the image of S under the linear transformation corresponding to A, and graph this image of S, where:

(a) S is the line x-axis.

(b) S is the line $x - y = 0$.

(c) S the unit circle.

1.6.3 Let A be the matrix $A = \begin{bmatrix} 1 & 2 \\ 0 & 3 \end{bmatrix}$, and let f_A be the corresponding linear transformation.

(a) Graph the image of the unit square under f_A, and compute the area of this image.

(b) Let S be the triangle with vertices $(1, 1)$, $(1, 4)$ and $(3, -5)$. What is the area of S? Graph the image of S under f_a, and compute the area of this image.

1.6.4 Let A be the matrix $A = \begin{bmatrix} 2 & -3 \\ 1 & 3 \end{bmatrix}$.

(a) Graph the image of the unit square under the corresponding linear transformation, and compute the area of this image.

(b) Let S be the triangle with vertices $(-1, 1)$, $(1, 1)$ and $(0, -1)$. What is the area of S? Graph the image of S under f_a, and compute the area of this image.

1.6.5 Which, if either, of the following matrices corresponds to an area preserving transformation?

$$A = \begin{bmatrix} 2 & -3 \\ 1 & 3 \end{bmatrix} \qquad B = \begin{bmatrix} 2 & -3 \\ 1 & 3 \end{bmatrix} .$$

1.6.6 Let $A = \begin{bmatrix} a & b \\ c & d \end{bmatrix}$ be an invertible 2×2 matrix. We have asserted that the image of the unit circle under such a transformation is always an ellipse. The object here is to derive the equation of this ellipse in the u, v plane in terms of the entries of A.

(a) Explain why $\begin{bmatrix} u \\ v \end{bmatrix}$ is in the image of the the unit circle if and only if

$$\left| A^{-1} \begin{bmatrix} u \\ v \end{bmatrix} \right| = 1 .$$

(b) Explain why $\begin{bmatrix} u \\ v \end{bmatrix}$ is in the image of the the unit circle if and only if

$$\begin{bmatrix} u \\ v \end{bmatrix} \cdot (A^{-1})^t A^{-1} \begin{bmatrix} u \\ v \end{bmatrix} = 1 \ .$$

(c) Show that the equation for the image of S is

$$(d^2 + c^2)u^2 + (a^2 + b^2)v^2 - 2(bd + ac)uv = (ad - bc)^2 \ .$$

1.6.7 Let $A = \begin{bmatrix} 1 & 2 \\ 3 & 4 \end{bmatrix}$. By the results of problem 1.6.6, the image of the unit circle under A is an ellipse. Find the equation of the ellipse and graph it.

1.6.8 Let $A = \begin{bmatrix} 2 & 1 \\ 2 & 4 \end{bmatrix}$. By the results of problem 1.6.6, the image of the unit circle under A is an ellipse. Find the equation of the ellipse and graph it.

Overview of Chapter Two

This chapter is concerned with solving systems of linear equations, or, what is the same thing, equations of the form $A\mathbf{x} = \mathbf{b}$ where A is a given $m \times n$ matrix and $\mathbf{b}$ is a given vector in $I\!R^n$. The set of all vectors in $I\!R^m$, if any, that satisfy this equation is called its *solution set*.

- *Solving an equation means finding a clear description of its solution set.*

The first major issue arises right here: What do we mean by a "clear description" of the solution set? Can't we just list the solutions? In general, no. It turns out, as we'll soon see, that equations of the form $A\mathbf{x} = \mathbf{b}$ have either no solutions, exactly one solution, or infinitely many solutions. If there are infinitely many solutions, we clearly cannot list them. What do we do instead?

There is a good substitute for a list in this context: We can *parameterize* the solution set. This means finding vectors $\mathbf{x}_0, \mathbf{v}_1, \mathbf{v}_2, \ldots, \mathbf{v}_r$ so that for each set of values for the parameters $t_1, t_2, \ldots, t_r$,

$$\mathbf{x}_0 + t_1 \mathbf{v}_1 + t_1 \mathbf{v}_2 + \cdots + t_r \mathbf{v}_r$$

we get a solution of the equation, and each solution can be written in this way for *exactly one choice* of values for the parameters $t_1, t_2, \ldots, t_r$. A parameterization with this uniqueness property is a *one–to–one parameterization*. In this context, they are the only useful kind.

This chapter is all about finding parameterizations of the solution sets of systems of linear equations. There is an *algorithm* for doing this: The *row reduction algorithm*. Analysis of this algorithm leads to a systematic way of determining the solution set of an arbitrary linear systems of equations, no matter how many variables there are, and no matter how many equations there are. So in this chapter, we will solve one of the main problems of linear algebra!

Section 1: Solving One Linear Equation

This chapter is mainly about solving general systems of linear equations in n variables, but we begin by considering the special case of a single linear equation in n variables.

A *linear form* in n variables $x_1, x_2, \ldots, x_n$ is a polynomial in these variables in which every term is exactly first degree. Hence it has the form

$$a_1 x_1 + a_2 x_2 + \cdots + a_n x_n \,. \tag{1.1}$$

Any linear form defines a transformation f from $I\!R^n$ to $I\!R$ through

$$f(\mathbf{x}) = a_1 x_1 + a_2 x_2 + \cdots + a_n x_n \,. \tag{1.2}$$

As you can check directly from (1.2), f is additive; i.e., $f(\mathbf{x}+\mathbf{y}) = f(\mathbf{x}) + f(\mathbf{y})$ for all $\mathbf{x}$ and $\mathbf{y}$ in $I\!R^n$. Also, f is homogeneous; i.e., $f(t\mathbf{x}) = tf(\mathbf{x})$ for all $\mathbf{x}$ in $I\!R^n$ and all numbers t. Together, these facts imply that the function f defined by (1.2) is a linear transformation from $I\!R^n$ to $I\!R$.

Conversely, suppose that f is any linear transformation from $I\!R^n$ to $I\!R$. Then we know

$$f(\mathbf{x}) = A_f \mathbf{x}$$

where A_f is the $1 \times n$ matrix $A_f = [f(\mathbf{e}_1), f(\mathbf{e}_2), \ldots, f(\mathbf{e}_n)]$. If we define $a_j = f(\mathbf{e}_j)$ for each $j = 1, 2 \ldots, n$, we have

$$A_f \mathbf{x} = [a_1, a_2, \ldots, a_n] \begin{bmatrix} x_1 \\ x_2 \\ \vdots \\ x_n \end{bmatrix} = a_1 x_1 + a_2 x_2 + \cdots + a_n x_n \,. \tag{1.3}$$

Therefore, we see that every linear form determines a unique linear transformation from $I\!R^n$ to $I\!R$, and *vice–versa*: A linear form is just one way of writing down a linear transformation from $I\!R^n$ to $I\!R$.

There is still one more way to write a linear form. Introduce the vector $\mathbf{a} = \begin{bmatrix} a_1 \\ a_2 \\ \vdots \\ a_n \end{bmatrix}$ so that

$$\mathbf{a} \cdot \mathbf{x} = a_1 x_1 + a_2 x_2 + \cdots + a_n x_n \,. \tag{1.4}$$

Combining (1.3) and (1.4), we see that every linear transformation f from $I\!R^n$ to $I\!R$ can be written in terms of the dot product as $f(\mathbf{x}) = \mathbf{a} \cdot \mathbf{x}$ for a vector $\mathbf{a}$ which is specified in (1.3). This simple fact is useful, and worth recording as a theorem:

Theorem 1 (Representation Theorem for Linear Forms) *If f is any linear transformation from $I\!R^n$ to $I\!R$, there is a vector $\mathbf{a}$ in $I\!R^n$ so that $f(\mathbf{x}) = \mathbf{a} \cdot \mathbf{x}$ for all $\mathbf{x}$ in $I\!R^n$. This vector is unique, and is given by $a_j = f(\mathbf{e}_j)$ for each $j = 2, 1 \ldots, n$.*

Example 1 Consider the linear form f on $I\!\!R^3$ given by $f(\mathbf{x}) = x_1 - 2x_2 + 3x_3$. For which a is $f(\mathbf{x}) = \mathbf{a} \cdot \mathbf{x}$? By definition, $f(\mathbf{e}_1) = 1$, $f(\mathbf{e}_2) = -2$ and $f(\mathbf{e}_3) = 3$, so

$$\mathbf{a} = \begin{bmatrix} f(\mathbf{e}_1) \\ f(\mathbf{e}_2) \\ f(\mathbf{e}_3) \end{bmatrix} = \begin{bmatrix} 1 \\ -2 \\ 3 \end{bmatrix} .$$

Even more directly, just from the definition of the dot product, you can see that

$$x_1 - 2x_2 + \mathbf{x}_3 = \mathbf{a} \cdot \mathbf{x}$$

for this choice of $\mathbf{a}$.

A *linear equation* is the statement obtained by setting a linear form equal to some constant b:

$$a_1 x_1 + a_2 x_2 + \cdots + a_n x_n = b . \tag{1.5}$$

Using (1.4), we can write this same equation as

$$\mathbf{a} \cdot \mathbf{x} = b . \tag{1.6}$$

The advantage of the expression (1.6) over (1.5) lies in the geometric meaning attached to the dot product; the expression (1.6) opens the way to a geometric understanding of linear equations and their solutions.

Definition The *solution set* of the linear equation $\mathbf{a} \cdot \mathbf{x} = b$ in n variables is the set of all vectors $\mathbf{x}$ in $I\!\!R^n$ that satisfy this equation.

- *When we talk about "solving an equation", we mean finding an explicit description of its solution set.*

Example 2 Consider the linear equation

$$3x_1 + 2x_2 = 4 \tag{1.7}$$

in the two variables x_1 and x_2. You recognize this as the equation of a line in the x_1, x_2 plane. Solving for x_2, we can rewrite (1.7) as

$$x_2 = -\frac{3}{2}x_1 + 2 . \tag{1.8}$$

This tells us that the solution set is the line whose slope is $-3/2$, and whose x_2 intercept is 2. This gives a *geometric* description of the solution set. Here is a picture:

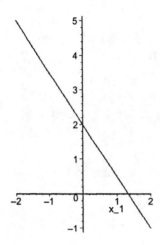

Since there are infinitely many points on the line, we can't specify the solution set by simply listing all of the vectors in it. But there is something that is almost as good as a list in this case, namely a *parametric description*.

The parametric description can be obtained from (1.8). The essential difference between (1.8) and (1.7) is that while both variables enter (1.7) on an equal footing, in (1.8), we regard x_2 as a *dependent variable*, and x_1 as an *independent variable*, with (1.8) giving the rule for how x_2 depends on x_1. To obtain the parametric description, simply use (1.8) to eliminate x_2 from $\begin{bmatrix} x_1 \\ x_2 \end{bmatrix}$: Since $x_2 = 2 - (3/2)x_1$,

$$\begin{bmatrix} x_1 \\ x_2 \end{bmatrix} = \begin{bmatrix} x_1 \\ 2 - (3/2)x_1 \end{bmatrix} = \begin{bmatrix} 0 \\ 2 \end{bmatrix} + x_1 \begin{bmatrix} 1 \\ -3/2 \end{bmatrix} . \qquad (1.9)$$

It is usual to do one more thing: Change the name of x_1 to t (or something like that). We'll come back to the reasons for this. But for now, let's just do it. Define

$$t = x_1 , \qquad (1.10)$$

and then write (1.9) as

$$\begin{bmatrix} x_1 \\ x_2 \end{bmatrix} = \begin{bmatrix} 0 \\ 2 \end{bmatrix} + t \begin{bmatrix} 1 \\ -3/2 \end{bmatrix} . \qquad (1.11)$$

The description of the solution set provided by (1.11) is called a *parametric description*: As the parameter t varies over the real line, the vectors $\mathbf{x} = \begin{bmatrix} x_1 \\ x_2 \end{bmatrix}$ in (1.11) cover the whole solution set, and each different t gives a different solution. Let's picture this:

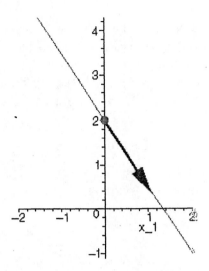

We've drawn in the "base point" $\begin{bmatrix} 0 \\ 2 \end{bmatrix}$ with a dot, and we've drawn in the "direction

vector" $\begin{bmatrix} 1 \\ -3/2 \end{bmatrix}$ as an arrow, transported so its tail lies at the base point. As t varies in

(1.11), $\mathbf{x} = \begin{bmatrix} x_1 \\ x_2 \end{bmatrix}$ ranges over the line through $\begin{bmatrix} 0 \\ 2 \end{bmatrix}$ in the direction $\begin{bmatrix} 1 \\ -3/2 \end{bmatrix}$. (Negative

values of t give points in the "backwards" direction).

It wasn't really necessary to change the name of x_1 to t. But when we passed from (1.7) to (1.8), we broke the symmetry between the two varibles. They both enter (1.7) on an equal footing, but it is natural to regard (1.8) as expressing a dependence of x_2 on x_1. In this equation, x_2 is something to be computed, while x_1 is something to which we can assign values at will. This is how we often think of *parameters*. Therefore, we change the name of the independent variable to t to emphasize its special status as a parameter.

The parametric description of the solution set of (1.7) that we gave in Example 2 is not unique; there are always many ways to go about the parameterization. One thing we could have done differently would have been to solve (1.7) for x_1 instead of x_2. This gives

$$x_1 = \frac{4}{3} - \frac{2}{3}x_2 , \qquad (1.12)$$

which in turn gives us the parameterization

$$\begin{aligned}
\begin{bmatrix} x_1 \\ x_2 \end{bmatrix} &= \begin{bmatrix} (4/3) - (2/3)x_2 \\ x_2 \end{bmatrix} = \begin{bmatrix} 4/3 \\ 0 \end{bmatrix} + x_2 \begin{bmatrix} -2/3 \\ 1 \end{bmatrix} \\
&= \begin{bmatrix} (4/3) - (2/3)x_2 \\ x_2 \end{bmatrix} = \begin{bmatrix} 4/3 \\ 0 \end{bmatrix} + t \begin{bmatrix} -2/3 \\ 1 \end{bmatrix} ,
\end{aligned} \qquad (1.13)$$

where in the second equation we have renamed x_2 by putting $x_2 = t$. Draw a picture showing the "base point" and "direction vector" for this parameterization.

We now have two different parameterizations of the solution set for the equation of Example 2. But there are as many ways to parameterize a line as there are ways to pick a pair of distinct points from it – infinitely many. Since it is clear that the problem of parameterizing lines in $I\!\!R^2$ is going to come up all the time when we want to parameterize the solutions set of a linear equation in $I\!\!R^2$, let's pause to take a general look at this problem.

Parameterizing lines

Recall that given two distinct points $\mathbf{x}_0$ and $\mathbf{x}_1$, there is a unique line running through them. The points on this line are given by

$$\mathbf{x}_0 + t(\mathbf{x}_1 - \mathbf{x}_0) = (1 - t)\mathbf{x}_0 + t\mathbf{x}_1 \tag{1.14}$$

as t varies over the real line. We can think of $\mathbf{x}_0$ as a "base point" and $\mathbf{v} = \mathbf{x}_1 - \mathbf{x}_0$ as specifying the "direction" of the line. We could then write out the parameterization in the form $\mathbf{x}_0 + t\mathbf{v}$.

Now suppose $\tilde{x}_0$ and $\tilde{\mathbf{v}}$ are another given pair of vectors. As t varies, $\mathbf{x}_0 + t\mathbf{v}$ traces out a line, and as s varies, $\tilde{\mathbf{x}}_0 + s\tilde{\mathbf{v}}$ traces out a line. *When are these two lines the same?* Here is the answer:

- *The lines traced out by $\mathbf{x}_0 + t\mathbf{v}$ as t varies and by $\tilde{\mathbf{x}}_0 + s\tilde{\mathbf{v}}$ as s varies are the same if and only if $\tilde{\mathbf{x}}_0 - \mathbf{x}_0$, $\tilde{\mathbf{v}}$ and $\mathbf{v}$ are all proportional.*

Let's see why this is true. Assume first that the two lines are the same. Then both $\tilde{\mathbf{x}}_0$ and $\tilde{\mathbf{x}}_0 + \tilde{\mathbf{v}}$ lie on the line given by $\mathbf{x}_0 + t\mathbf{v}$, and so there are numbers t_1 and t_2 so that

$$\tilde{\mathbf{x}}_0 = \mathbf{x}_0 + t_1\mathbf{v} \qquad \text{and} \qquad \tilde{\mathbf{x}}_0 + \tilde{\mathbf{v}} = \mathbf{x}_0 + t_2\mathbf{v}$$

The first equation above can be written as

$$\tilde{\mathbf{x}}_0 - \mathbf{x}_0 = t_1\mathbf{v}$$

which says that $\tilde{\mathbf{x}}_0 - \mathbf{x}_0$ is proportional to $\mathbf{v}$. The second equation can be rewritten as

$$\tilde{\mathbf{v}}_2 = (\mathbf{x}_0 - \tilde{\mathbf{x}}_0) + t_2\mathbf{v} = (t_1 - t_2)\mathbf{v}$$

which says that $\tilde{\mathbf{v}}$ and $\mathbf{v}$ are proportional. Thus, when the two lines are the same, $\tilde{\mathbf{x}}_0 - \mathbf{x}_0$, $\tilde{\mathbf{v}}$ and $\mathbf{v}$ are all proportional.

Conversely, suppose that $\tilde{\mathbf{x}}_0 - \mathbf{x}_0$, $\tilde{\mathbf{v}}$ and $\mathbf{v}$ are all proportional, say if $\tilde{\mathbf{x}}_0 - \mathbf{x}_0 = a\mathbf{v}$ and $\tilde{\mathbf{v}} = b\mathbf{v}$ for some numbers a and b. Then $\tilde{\mathbf{x}}_0 = \mathbf{x}_0 + a\mathbf{v}$ and $\tilde{\mathbf{x}}_0 + \tilde{\mathbf{v}} = \mathbf{x}_0 + (a + b)\mathbf{v}$, so both vectors are on the line given by $\mathbf{x}_0 + t\mathbf{v}$, and hence the two lines are the same. This validates our assertion above.

Example 3 Consider the lines given by $\mathbf{x}_0 + t\mathbf{v}$ and $\tilde{\mathbf{x}}_0 + s\tilde{\mathbf{v}}$ where

$$\mathbf{x}_0 = \begin{bmatrix} 0 \\ 2 \end{bmatrix} \qquad \text{and} \qquad \mathbf{v} = \begin{bmatrix} 1 \\ -3/2 \end{bmatrix}$$

and where

$$\tilde{\mathbf{x}}_0 = \begin{bmatrix} 4/3 \\ 0 \end{bmatrix} \quad \text{and} \quad \tilde{\mathbf{v}} = \begin{bmatrix} -2/3 \\ 1 \end{bmatrix} .$$

Then, to see if these are two parameterizations of the same line, we check to see if $\tilde{\mathbf{x}}_0 - \mathbf{x}_0$, $\tilde{\mathbf{v}}$ and $\mathbf{v}$ are all proportional or not:

$$\tilde{\mathbf{x}}_0 - \mathbf{x}_0 = \begin{bmatrix} 4/3 \\ -2 \end{bmatrix} = \frac{4}{3}\mathbf{v} \quad \text{and} \quad \tilde{\mathbf{v}} = \begin{bmatrix} -2/3 \\ 1 \end{bmatrix} = -\frac{2}{3}\mathbf{v} .$$

Hence these two lines are the same.

Everything we've said in the last few paragraphs applies in $I\!\!R^n$ for any n – nowhere in the discussion did we refer to $I\!\!R^2$. The formula (1.14) gives the parametric description of the line in $I\!\!R^n$ passing through two points $\mathbf{x}_0$ and $\mathbf{x}_1$. It will be very useful in higher dimensions, when we get there.

Back to the equation

Now let's return to our two parameterizations of the solution set of (1.7). Write the right hand side of (1.11) as

$$\mathbf{x}_0 + t\mathbf{v} \quad \text{where} \quad \mathbf{x}_0 = \begin{bmatrix} 0 \\ 2 \end{bmatrix} \quad \text{and} \quad \mathbf{v} = \begin{bmatrix} 1 \\ -3/2 \end{bmatrix} , \qquad (1.15)$$

and write the right hand side of (1.13) as

$$\tilde{\mathbf{x}}_0 + t\tilde{\mathbf{v}} \quad \text{where} \quad \tilde{\mathbf{x}}_0 = \begin{bmatrix} 4/3 \\ 0 \end{bmatrix} \quad \text{and} \quad \tilde{\mathbf{v}} = \begin{bmatrix} -2/3 \\ 1 \end{bmatrix} . \qquad (1.16)$$

We know these two lines must be two parameterizations of the same line because we got them from the same equation. But notice that Example 3 provides an independent check on this.

Now let's consider the following question:

Let $\mathbf{a} \cdot \mathbf{x} = b$ be a given linear equation written in the geometric form (1.6). Let $\mathbf{x}_0 + t\mathbf{v}$ be the parameterization of a line. *When is $\mathbf{x}_0 + t\mathbf{v}$ a line of solutions for $\mathbf{a} \cdot \mathbf{x} = b$?* That is, when does $\mathbf{x}_0 + t\mathbf{v}$ solve $\mathbf{a} \cdot \mathbf{x} = b$ for every value of t?

Taking $t = 0$, we see that $\mathbf{x}_0$ is a particular solution of this equation. Then

$$b = \mathbf{a} \cdot (\mathbf{x}_0 + t\mathbf{v}) = \mathbf{a} \cdot \mathbf{x}_0 + t(\mathbf{a} \cdot \mathbf{v}) = b + t(\mathbf{a} \cdot \mathbf{v}) . \qquad (1.17)$$

Equation (1.17) will hold for all t if and only if $\mathbf{a} \cdot \mathbf{v} = 0$.

We answered that question pretty easily! Still it turns out that there are some very useful ideas involved here, and it is worthwhile to introduce some terminology. The equation $\mathbf{a} \cdot \mathbf{x} = 0$, is called the *homogeneous equation* corresponding to $\mathbf{a} \cdot \mathbf{x} = b$. The term "homogeneous" is used because if $\mathbf{x}$ is a solution of $\mathbf{a} \cdot \mathbf{x} = 0$, then so is $s\mathbf{x}$ for any multiple s. As far as satisfying this equation is concerned, all multiples of any given vector are the same.

The following theorem summarizes some useful points concerning the relation of solutions of $\mathbf{a} \cdot \mathbf{x} = b$ to the corresponding homogeneous equation $\mathbf{a} \cdot \mathbf{x} = 0$.

Theorem 2: (Solutions of the Homogeneous Equation and the Inhomogeneous Equation) *For any vector* $\mathbf{a}$ *in* $I\!\!R^n$ *and any number* b, *the following are true:*

(1) The vector $\mathbf{x}_0$ *satisfies* $\mathbf{a} \cdot \mathbf{x} = b$, *and* $\mathbf{v}$ *satisfies the corresponding homogeneous equation* $\mathbf{a} \cdot \mathbf{x} = 0$, *if and only if* $\mathbf{x}_0 + t\mathbf{v}$ *is the parameterization of a line of solutions of the original equation* $\mathbf{a} \cdot \mathbf{x} = b$.

(2) If $\mathbf{v}_1, \mathbf{v}_2, \ldots, \mathbf{v}_r$ *is any collection of solutions of the homogeneous equation* $\mathbf{a} \cdot \mathbf{x} = 0$, *any linear combination*

$$\mathbf{w} = t_1\mathbf{v}_1 + t_2\mathbf{v}_2 + \cdots + t_r\mathbf{v}_r$$

of these vectors is also a solution of $\mathbf{a} \cdot \mathbf{x} = 0$.

(3) If $\mathbf{x}_1$ *and* $\mathbf{x}_0$ *are any two solutions of* $\mathbf{a} \cdot \mathbf{x} = b$, *their difference* $\mathbf{x}_1 - \mathbf{x}_0$ *is a solution of the corresponding homogeneous equation* $\mathbf{a} \cdot \mathbf{x} = 0$.

Proof: For part (1), $\mathbf{a} \cdot (\mathbf{x}_0 + \mathbf{w}) = \mathbf{a} \cdot \mathbf{x}_0 + \mathbf{a} \cdot \mathbf{w} = b + 0 = b$. For part (2),

$$\mathbf{a} \cdot (t_1\mathbf{v}_1 + t_2\mathbf{v}_2 + \cdots + t_{n-1}\mathbf{v}_{n-1}) = t_1(\mathbf{a} \cdot \mathbf{v}_1) + t_2(\mathbf{a} \cdot \mathbf{v}_2) + \cdots + t_{n-1}(\mathbf{a} \cdot \mathbf{v}_{n-1})$$

$$= t_1 0 + t_2 0 + \cdots t_{n-1} 0$$

$$= 0 \ .$$

For part (3), $\mathbf{a} \cdot (\mathbf{x}_1 - \mathbf{x}_0) = \mathbf{a} \cdot \mathbf{x}_1 - \mathbf{a} \cdot \mathbf{x}_0 = b - b = 0$. ∎

Now consider an example in $I\!\!R^3$:

Example 4 You may recognize the equation

$$3x_1 + 2x_2 - 2x_3 = 4 \tag{1.18}$$

in the three variables x_1, x_2 and x_3 as the equation of a plane in $I\!\!R^3$. To parameterize the solution set S of this equation, choose one of the variables to be the dependent variable. We will choose x_1, and solving for this variable in terms of the other two we find

$$x_1 = -\frac{2}{3}x_2 + \frac{2}{3}x_2 + \frac{4}{3} \ . \tag{1.19}$$

The solution set S therefore consists of the vectors

$$\begin{bmatrix} x_1 \\ x_2 \\ x_3 \end{bmatrix} = \begin{bmatrix} -(2/3)x_2 + (2/3)x_3 + 4/3 \\ x_2 \\ x_3 \end{bmatrix} = \begin{bmatrix} 4/3 \\ 0 \\ 0 \end{bmatrix} + x_2 \begin{bmatrix} -2/3 \\ 1 \\ 0 \end{bmatrix} + x_3 \begin{bmatrix} 2/3 \\ 0 \\ 1 \end{bmatrix} \ . \tag{1.20}$$

We now rename the independent variables x_2 and x_3 as $x_2 = t_1$ and $x_3 = t_2$ as a badge of their status as parameters. We then have the following parameterization of the solution set S:

$$x_0 + t_1 \mathbf{v}_1 + t_2 \mathbf{v}_2 \tag{1.21}$$

where

$$\mathbf{x}_0 = \begin{bmatrix} 4/3 \\ 0 \\ 0 \end{bmatrix} \quad , \quad \mathbf{v}_1 = \begin{bmatrix} -2/3 \\ 1 \\ 0 \end{bmatrix} \quad \text{and} \quad \mathbf{v}_2 = \begin{bmatrix} 2/3 \\ 0 \\ 1 \end{bmatrix} . \tag{1.22}$$

Notice that this time two parameters are required to describe S, which accords with our intuitive understanding of planes as "two dimensional", while lines are "one dimensional".

Introduce $\mathbf{a} = \begin{bmatrix} 3 \\ 2 \\ -2 \end{bmatrix}$ so that (1.18) can be written as $\mathbf{a} \cdot \mathbf{x} = 4$. Then you can easily check that

$$\mathbf{a} \cdot \mathbf{x}_0 = 4 \quad , \quad \mathbf{a} \cdot \mathbf{v}_1 = 0 \quad \text{and} \quad \mathbf{a} \cdot \mathbf{v}_2 = 0 ,$$

so that once again $\mathbf{x}_0$ is a particular solution of (1.18), while $\mathbf{v}_1$ and $\mathbf{v}_2$ are solutions of the corresponding homogeneous equation, $\mathbf{a} \cdot \mathbf{x} = 0$. Again, this had to be: For example, the line parameterized by $\mathbf{x}_0 + t_1 \mathbf{v}_1$ as t_1 varies is a line of solutions, and by Theorem 2, this happens if and only if $\mathbf{a} \cdot \mathbf{v}_1 = 0$.

We see from (1.19) that there is exactly one vector in S for each pair of values for the independent variables x_2 and x_3. We see from (1.20) that this solution is included in our parameterization. Moreover, since x_2 and x_3 correspond directly to the parameters t_1 and t_2, (1.21) gives us a one to one correspondence between pairs of values for the parameters and vectors in the solution sets S.

It is important to note that we could have solved for any of the three variables as a function of the other two. Each choice would give us a different, but equivalent, parameterization.

A general procedure in $I\!R^n$

Now let's try to draw some general conclusions from these two examples that will apply to the general linear equation $\mathbf{a} \cdot \mathbf{x} = b$ in $I\!R^n$. First, let's exclude the case $\mathbf{a} = 0$. We will say that the equation $\mathbf{a} \cdot \mathbf{x} = b$ is *trivial* in case $\mathbf{a} = 0$. The point is that if $\mathbf{a} = 0$ but $b \neq 0$, then the equation can never be satisfied for any $\mathbf{x}$, and the solution set is the empty set $\emptyset$. On the other hand, if $b = 0$ as well as $\mathbf{a}$, the equation $\mathbf{a} \cdot \mathbf{x} = b$ reduces to the tautology $0 = 0$, and the solution set is all of $I\!R^n$. These cases are not very interesting, and from now on, we focus on non–trivial equations.

Here is the general form of the procedure used in Examples 2 and 4 to parameterize the solution set of a non–trivial linear equation:

Procedure for Parameterizing the Solution Set of a Linear Equation: Suppose that the equation $\mathbf{a} \cdot \mathbf{x} = b$ is non–trivial, or else there is nothing to do. Define the integer j_p by

$$j_p = \min\{i = 1, 2, \dots, n \mid a_i \neq 0\} \tag{1.23}$$

This is the index of the first non–zero entry in $\mathbf{a}$. We call x_{j_p} the *pivotal variable* for the equation $\mathbf{a} \cdot \mathbf{x} = b$, and the $n - 1$ variables x_i with $i \neq j_p$ are called *non–pivotal variables*. ("Pivotal" is just the traditional adjective used to describe dependent variables in this context). In Example 4, x_1 is the pivotal variable.

Now solve the equation for the pivotal variable:

$$x_{j_p} = \frac{1}{a_{j_p}} \left(b - \sum_{j=1, j \neq j_p}^{n} a_j x_j \right) , \qquad (1.24)$$

and use (1.24) to eliminate the pivotal variable from $\begin{bmatrix} x_1 \\ x_2 \\ \vdots \\ x_n \end{bmatrix}$, just as we used (1.19) to

eliminate x_3 from $\begin{bmatrix} x_1 \\ x_2 \\ x_3 \end{bmatrix}$ in (1.20).

The result is a vector whose entries depend linearly on the non–pivotal variables. Change their names to $t_1, t_2, \ldots, t_{n-1}$ to get a vector whose entries depend linearly on $t_1, t_2, \ldots, t_{n-1}$. Expand this result as

$$\mathbf{x}_0 + t_1 \mathbf{v}_1 + t_2 \mathbf{v}_2 + \cdots + t_{n-1} \mathbf{v}_{n-1} , \qquad (1.25)$$

where $\mathbf{x}_0, \mathbf{v}_1, \mathbf{v}_2, \ldots, \mathbf{v}_{n-1}$ are *constant* vectors, and where for each $j = 1, 2, \ldots, n - 1$, $\mathbf{v}_j$ is whatever is multiplied by t_j, just as in (1.20). This is the parameterization.

Example 5 Let's do an example with some more variables – this will reveal something interesting. Consider the equation

$$x_1 + 2x_2 + 3x_3 + 4x + 4 + 5x_5 = 6 .$$

Evidently, $j_p = 1$, so x_1 is the pivotal variable. Then (1.24) becomes

$$x_1 = 6 - 2x_2 - 3x_3 - 4x_4 - 5x_5 .$$

Hence the solutions are

$$\begin{bmatrix} x_1 \\ x_2 \\ x_3 \\ x_4 \\ x_5 \end{bmatrix} = \begin{bmatrix} 6 - 2x_2 - 3x_3 - 4x_4 - 5x_5 \\ x_2 \\ x_3 \\ x_4 \\ x_5 \end{bmatrix} .$$

Defining $t_1 = x_2$, $t_2 = x_3$, $t_3 = x_4$, and $t_4 = x_5$, we get our parameterization:

$$\begin{bmatrix} 6 - 2t_1 - 3t_2 - 4t_3 - 5t_4 \\ t_1 \\ t_2 \\ t_3 \\ t_4 \end{bmatrix} = \begin{bmatrix} 6 \\ 0 \\ 0 \\ 0 \\ 0 \end{bmatrix} + t_1 \begin{bmatrix} -2 \\ 1 \\ 0 \\ 0 \\ 0 \end{bmatrix} + t_2 \begin{bmatrix} -3 \\ 0 \\ 1 \\ 0 \\ 0 \end{bmatrix} + t_3 \begin{bmatrix} -4 \\ 0 \\ 0 \\ 1 \\ 0 \end{bmatrix} + t_4 \begin{bmatrix} -5 \\ 0 \\ 0 \\ 0 \\ 1 \end{bmatrix} .$$

So the parameterized solution has the form

$$\mathbf{x}_0 + t_1\mathbf{v}_1 + t_2\mathbf{v}_2 + t_3\mathbf{v}_3 + t_4\mathbf{v}_4 \ .$$

A few remarks are in order: For one thing, the choice of the pivotal variable as the *first* one with a non–zero coefficient is arbitrary. However, taking the first one is a good systematic rule, and that is what we'll do.

Next notice that each of the vectors $\mathbf{v}_k$ has at most two non–zero entries. The entry corresponding to the kth non–pivotal variable itself is 1, and the only other possible place for a non zero entry is the place of the pivotal variable, as in (1.22).

This is because in the beginning, when we start with $\begin{bmatrix} x_1 \\ x_2 \\ \vdots \\ x_n \end{bmatrix}$, each variable appears in just one place – its own. Then after eliminating the pivotal variable using (1.24), each of the non–pivotal variables can now show up in one more place, namely the place of the pivotal variable. Because the kth non–pivotal variable shows up in only in its own place and possibly the j_pth place too, the vector $\mathbf{v}_k$ multiplying t_k in (1.25) has a 1 in the place of the kth non–pivotal variable, and every other entry, except possibly the j_pth, is zero.

Also note that $\mathbf{x}_0$ is an even simpler vector: It is obtained by setting all of the non–pivotal variables equal to zero, and so its *only* non–zero entry is the j_pth, as in (1.22).

Since only $\mathbf{v}_k$ has a non–zero entry in the place of the kth non–pivotal variable, when we choose t_k we are choosing the value of the kth non–pivotal variable. Since according to (1.24) there is exactly one solution for each choice of values for the non–pivotal variables, we get a one–to–one correspondence between choices of values of the parameters and vectors in S.

- *The parameterization procedure that we have just described always provides a one–to–one correspondence between parameter values and vectors in the solution set S of a non–trivial linear equation $\mathbf{a} \cdot \mathbf{x} = b$ in $I\!R^n$.*

Let's conclude this section by looking at (1.25) in matrix terms. Indeed, if we introduce the $n \times (n-1)$ matrix B by

$$B = [\mathbf{v}_1, \mathbf{v}_2, \ldots, \mathbf{v}_{n-1}] \tag{1.26}$$

and the vector $\mathbf{t} = \begin{bmatrix} t_1 \\ t_2 \\ \vdots \\ t_{n-1} \end{bmatrix}$ in $I\!R^{n-1}$, we have

$$B\mathbf{t} = [\mathbf{v}_1, \mathbf{v}_2, \ldots, \mathbf{v}_{n-1}] \begin{bmatrix} t_1 \\ t_2 \\ \vdots \\ t_{n-1} \end{bmatrix} = t_1\mathbf{v}_1 + t_2\mathbf{v}_2 + \cdots + t_{n-1}\mathbf{v}_{n-1}$$

and hence we can write (1.25) in the form

$$\mathbf{x}_0 + B\mathbf{t} \ . \tag{1.27}$$

As $\mathbf{t}$ varies over $\mathbb{R}^{n-1}$, $\mathbf{x}_0 + B\mathbf{t}$ varies over the solution set S of $\mathbf{a} \cdot \mathbf{x} = b$ in a one–to–one manner. The matrix B is called a *parameterization matrix* for this equation.

Example 6 For the system in Example 4, the parameterization matrix is

$$B = \frac{1}{3} \begin{bmatrix} -2 & 2 \\ 3 & 0 \\ 3 & 3 \end{bmatrix} \ .$$

Section 2: Lines, Planes and Solution Sets

In the previous section we saw how to parameterize the solution set S of a single linear equation $\mathbf{a} \cdot \mathbf{x} = b$. In this section, we'll go a bit more deeply into the geometry of these solution sets.

First note that if $\mathbf{x}_0$ is any particular solution of $\mathbf{a} \cdot \mathbf{x} = b$, and if $\mathbf{x}$ is any vector in $I\!\!R^n$, then

$$\mathbf{a} \cdot (\mathbf{x} - \mathbf{x}_0) = \mathbf{a} \cdot \mathbf{x} - \mathbf{a} \cdot \mathbf{x}_0 = \mathbf{a} \cdot \mathbf{x} - b$$

and therefore, $\mathbf{a} \cdot \mathbf{x} = b$ if and only if

$$\mathbf{a} \cdot (\mathbf{x} - \mathbf{x}_0) = 0 \qquad (2.1)$$

Because of the way the dot product expresses orthogonality, this gives us a geometric description of the solution set:

- *For the linear equation $\mathbf{a} \cdot \mathbf{x} = b$ in $I\!\!R^n$, if $\mathbf{a} \neq 0$ and $\mathbf{x}_0$ is some solution, then the solution set S of this equation is exactly the set of vectors $\mathbf{x}$ such that $\mathbf{x} - \mathbf{x}_0$ is orthogonal to $\mathbf{a}$. In $I\!\!R^2$, S is the line passing through $\mathbf{x}_0$ that is orthogonal to $\mathbf{a}$, while in $I\!\!R^3$, S is the plane passing through $\mathbf{x}_0$ that is orthogonal to $\mathbf{a}$.*

Moreover, finding a "base point" $\mathbf{x}_0$ is easy. Besides the method discussed in the last section, we can always take

$$\mathbf{x}_0 = \frac{b}{|\mathbf{a}|^2} \mathbf{a} \,, \qquad (2.2)$$

since with this definition, $\mathbf{a} \cdot \mathbf{x}_0 = (b/|\mathbf{a}|^2)(\mathbf{a} \cdot \mathbf{a}) = b$.

Example 1: Consider the equation from Example 1 of the last section. This can be written in the form $\mathbf{a} \cdot \mathbf{x} = b$ where $\mathbf{a} = \begin{bmatrix} 3 \\ 2 \end{bmatrix}$ and $b = 4$. The normal line is the line parameterized by $t\mathbf{a}$. Since $\mathbf{a}$ itself is on the line, the slope is $2/3$, so the equation of the normal line is

$$x_2 = \frac{2}{3} x_1 \,.$$

If we use (2.2) to get our base point, we first work out $|\mathbf{a}|^2 = 9 + 4 = 13$ so we get

$$\frac{4}{13} \begin{bmatrix} 3 \\ 2 \end{bmatrix} \,.$$

Alternatively, we can get a base point by choosing any value for one of the variables, and solving for the other. If we set $x_1 = 0$, the equation says $x_2 = 2$. So $\begin{bmatrix} 0 \\ 2 \end{bmatrix}$ is a solution.

Here is the graph, with the base point $\begin{bmatrix} 0 \\ 2 \end{bmatrix}$ is another solution, and we can use it as a base point as well. Let's use it, since it is simpler. Here is a picture showing the normal line and the base point $\begin{bmatrix} 0 \\ 2 \end{bmatrix}$

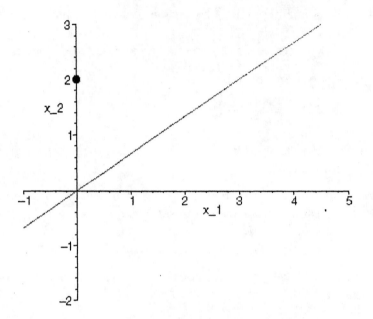

The solution set of our equation $\mathbf{a} \cdot \mathbf{x} = b$ is the line through the base point $\begin{bmatrix} 0 \\ 2 \end{bmatrix}$ that is perpendicular to the normal line. We could draw in the perpendicular line using a dratfman's square, but can't we also find the coordinates of a vector that points in the orthogonal direction? Yes, since we've already found the matrix for the linear transformation given by rotation through an angle of $\pi/2$ in the counterclockwise direction. In Example 7 of Section 2.2 we found that this matrix is $\begin{bmatrix} 0 & -1 \\ 1 & 0 \end{bmatrix}$. Hence the vector

$$\begin{bmatrix} -2 \\ 3 \end{bmatrix} = \begin{bmatrix} 0 & -1 \\ 1 & 0 \end{bmatrix} \begin{bmatrix} 3 \\ 2 \end{bmatrix}$$

is orthogonal to $\mathbf{a}$, as is its opposite, $\begin{bmatrix} 2 \\ -3 \end{bmatrix}$

Now we are ready to draw the solution set! Add on the vector $\begin{bmatrix} 2 \\ -3 \end{bmatrix}$ with its tail at the base point, and then draw in the line along the arrow. Here it is:

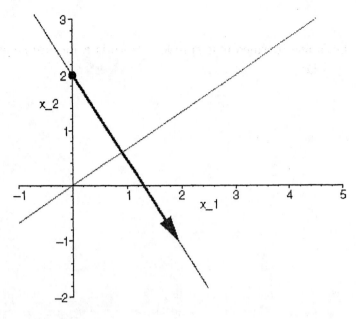

In the process, we've found a parameterization of the solution set. It is the set of vectors given by

$$\begin{bmatrix} 0 \\ 2 \end{bmatrix} + t \begin{bmatrix} 2 \\ -3 \end{bmatrix} ,$$

where t ranges over the real line.

The level set point of view

There is another way to visualize solution sets. It is often useful to think of the solution set S of $\mathbf{a} \cdot \mathbf{x} = b$ as a *level set* of the real valued function $f(\mathbf{x}) = \mathbf{a} \cdot \mathbf{x}$. Given any real valued function h on $I\!\!R^2$, think of $h(x_1, x_2)$ as representing the altitude at some point whose coordinates on a map are given by x_2 and x_2. For any number c, the set of all points (x_1, x_2), if any, such that $h(x_1, x_2) = c$ is called the *level set of h at height c.* These are points that have the same altitude, and on a topographical map, they would be represented by drawing contour lines. If you are familiar with topographical maps, you are familiar with level sets.

Example 2 Consider the equation $\mathbf{a} \cdot \mathbf{x} = b$ with $\mathbf{a} = \begin{bmatrix} 3 \\ 2 \end{bmatrix}$ and $b = 4$, as in Example 1. Then if we put $h(x_1, x_2) = \mathbf{a} \cdot \mathbf{x}$, we get

$$h(x_1, x_2) = 3x_1 + 2x_2 .$$

Here is a graph showing the level sets on which $h = -3$, $h = -2$, and so on, up to $h = 11$:

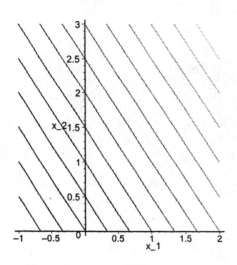

In these terms, the solution set S of $\mathbf{a} \cdot \mathbf{x} = b$ is just the level set of $f(\mathbf{x}) = \mathbf{a} \cdot \mathbf{x}$ at height b. In this case, it is the line passing through the x_2 axis at $x_2 = 2$, as we found in Example 1.

The normal line N also has an informative interpretation in this context:

- *The direction of the normal line gives the direction of steepest ascent for the function $f(\mathbf{x}) = \mathbf{a} \cdot \mathbf{x}$.*

In the picture below, you see the circle of points that are at unit distance from the point $\begin{bmatrix} 1 \\ 1 \end{bmatrix}$. Which of these points is at the highest altitude, where altitude is given by $h(x_1, x_2)$ as above? The one at the tip of the arrow pointing in the direction of the normal line:

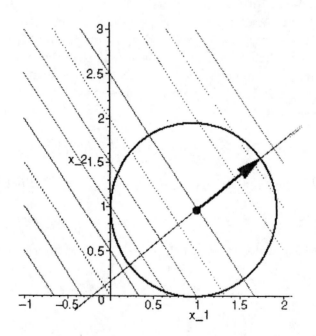

To get to the highest ground you can reach moving one unit, move "uphill" in the direction of the normal line.

To see this in general, using the equation instead of a picture, suppose you are at a point $\mathbf{x}_0$ with altitude b; that is, at a point in the solution set of $\mathbf{a} \cdot \mathbf{x} = b$. Suppose you wish to move one unit of distance, and want to move to higher ground, as high as possible. In which direction should you move?

The points you can move to if you move a unit horizontal distance are $\mathbf{x}_0 + \mathbf{u}$ where $\mathbf{u}$ is any unit vector. (These are the ones on the circle above). The altitude at $\mathbf{x}_0 + \mathbf{u}$ is

$$f(\mathbf{x}_0 + \mathbf{u}) = \mathbf{a} \cdot (\mathbf{x}_0 + \mathbf{u}) = \mathbf{a} \cdot \mathbf{x}_0 + \mathbf{a} \cdot \mathbf{u} = b + \mathbf{a} \cdot \mathbf{u} \, .$$

Therefore to increase our altitude as much as possible, we should choose $\mathbf{u}$ to maximize $\mathbf{a} \cdot \mathbf{u}$. But since $\mathbf{u}$ is a unit vector,

$$\mathbf{a} \cdot \mathbf{u} = |\mathbf{a}||\mathbf{u}| \cos(\theta) = |\mathbf{a}| \cos(\theta) \, ,$$

where θ is the angle between $\mathbf{a}$ and $\mathbf{u}$. Clearly we get the best result with $\theta = 0$, which means that $\mathbf{u}$ is parallel to $\mathbf{u}$, and so

$$\mathbf{u} = \frac{1}{|\mathbf{a}|} \mathbf{a} \, .$$

Since $\mathbf{x}_0 + t\mathbf{u}$ traces out a line parallel to the normal line, we see that indeed, the direction of the normal line is the direction of steepest ascent, as we asserted.

The distance to the nearest solution

The level set picture helps us answer the following question: "Given a linear equation in $I\!R^2$ and a point $\mathbf{y}$ in $I\!R^2$, what is the solution $\mathbf{w}$ of the equation that is closest to $\mathbf{y}$?" Since every line in $I\!R^2$ is the solution of some linear equation, this is the essentially the same question as: "Given a line ℓ and a vector $\mathbf{y}$, what is the distance between $\mathbf{y}$ and ℓ?". So how do we find the distance between a point $\mathbf{y}$ and a line ℓ?

Here is a picture showing the solution set of our equation $3x_1 + 2x_2 = 4$, as in the previous examples. The solution set S is the line drawn in bold, and there are some more level sets as well. You also see that the point $\mathbf{y} = \begin{bmatrix} 1 \\ 0 \end{bmatrix}$ is not on the solution set. But if we go out from $\mathbf{y}$ along the line in the direction $\mathbf{a}$, we run into $\mathbf{w}$ which is the closest solution.

Here is the picture:

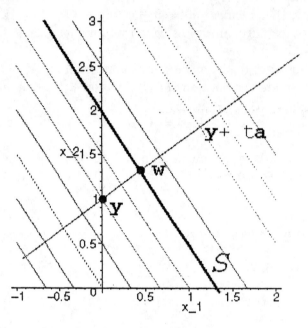

Let's try to compute a formula for $\mathbf{w}$. Here how this goes:

(1) We know $\mathbf{w}$ lies on the line given by $\mathbf{y} + t\mathbf{a}$ for some values of t.

(2) Since $\mathbf{w}$ also satisfies the equation, we can solve $\mathbf{a} \cdot (\mathbf{y} + t\mathbf{a}) = b$ for t.

(3) Once we know the values of t, we just evaluate $\mathbf{w} = \mathbf{y} + t\mathbf{a}$ to get $\mathbf{w}$.

This works in general since if $\mathbf{y}$ is not already on S, then $\mathbf{y}$ is at the "wrong altitude". Either $\mathbf{a} \cdot \mathbf{y} > b$ or $\mathbf{a} \cdot \mathbf{y} < b$. But moving in the direction of the normal line, we adjust our altitude in the most direct way!

Now let's do it. Plugging $\mathbf{y} + t\mathbf{a}$ into the equation,

$$\mathbf{a} \cdot (\mathbf{y} + t\mathbf{a}) = b \ .$$

This is the same as $\mathbf{a} \cdot \mathbf{y} + t|\mathbf{a}|^2 = b$, and so you hit the altitude b at

$$t = \frac{\mathbf{a} \cdot \mathbf{y} - b}{|\mathbf{a}|^2} \ . \tag{2.3}$$

It follows that the solution of $\mathbf{a} \cdot \mathbf{x} = b$ that is closest to $\mathbf{y}$ is $\mathbf{y} + t\mathbf{a}$ for this value of t. Written out in full, this is

$$\mathbf{w} = \mathbf{y} + \frac{\mathbf{a} \cdot \mathbf{y} - b}{|\mathbf{a}|^2}\mathbf{a} \ . \tag{2.4}$$

Since $\mathbf{w}$ is the vector in S that is closest to $\mathbf{y}$, it is natural therefore to call $|\mathbf{y} - \mathbf{w}|$ the *distance from* $\mathbf{y}$ *to* S, or what is the same thing, the distance from $\mathbf{y}$ to S.

Evidently from (2.4), $|\mathbf{y} - \mathbf{w}|$, the distance from $\mathbf{y}$ to S is given by

$$\frac{|\mathbf{a} \cdot \mathbf{y} - b|}{|\mathbf{a}|}\mathbf{a} \ . \tag{2.5}$$

The formula we've just derived for the distance from a point $\mathbf{y}$ in $\mathbb{R}^2$ to the solution set S of $\mathbf{a} \cdot \mathbf{x} = b$ in $\mathbb{R}^2$ is also valid, without any change, in $\mathbb{R}^n$ for all n.

Theorem 1 (Distance to Solution Sets) *Let S be the solution set of a non–trivial linear equation $\mathbf{a} \cdot \mathbf{x} = b$ in $\mathbb{R}^n$. Then for any vector $\mathbf{y}$ in $\mathbb{R}^n$,*

$$\mathbf{y} - \frac{\mathbf{a} \cdot \mathbf{y} - b}{|\mathbf{a}|^2}\mathbf{a} \tag{2.6}$$

belongs to S, and is closer to $\mathbf{y}$ than any other vector in S. The distance between this closest vector and $\mathbf{y}$, which is by definition the distance between S and $\mathbf{y}$, is

$$\frac{|\mathbf{a} \cdot \mathbf{y} - b|}{|\mathbf{a}|} \ .$$

Proof: For $n = 2$, we know this and deduced it geometrically using the level set point of view. Now that we've found it, it is easy to check that it works in $\mathbb{R}^n$ for any n.

First of all,

$$\mathbf{a} \cdot \left(\mathbf{y} - \frac{\mathbf{a} \cdot \mathbf{y} - b}{|\mathbf{a}|^2}\mathbf{a}\right) = \mathbf{a} \cdot \mathbf{y} - (\mathbf{a} \cdot \mathbf{y} - b) = b \ ,$$

and so the vector in (2.6) is a solution of the equation. Let's give it a name since we'll be working with it. Let's call it $\mathbf{w}_0$.

Next, the vector $((\mathbf{a} \cdot \mathbf{y} - b)/|\mathbf{a}|^2)\mathbf{a}$ is orthogonal to any solution of $\mathbf{a} \cdot \mathbf{x} = 0$. Indeed, if $\mathbf{x} \cdot \mathbf{a} = 0$, then $\mathbf{x}$ is orthogonal to $\mathbf{a}$, and hence to $((\mathbf{a} \cdot \mathbf{y} - b)/|\mathbf{a}|^2)\mathbf{a}$, which is just a multiple of $\mathbf{a}$. Notice that this vector $((\mathbf{a} \cdot \mathbf{y} - b)/|\mathbf{a}|^2)\mathbf{a}$ is just $\mathbf{y} - \mathbf{w}_0$. So what we've observed is that $\mathbf{y} - \mathbf{w}_0$ is orthogonal to every solution of the homogoneous equation $\mathbf{a} \cdot \mathbf{x} = 0$.

Now let $\mathbf{w}$ be any solution of $\mathbf{a} \cdot \mathbf{x} = b$. Then

$$\mathbf{y} - \mathbf{w} = (\mathbf{y} - \mathbf{w}_0) + (\mathbf{w}_0 - \mathbf{w}) \ .$$

By part *(3)* of Theorem 2 of the previous section, $\mathbf{w}_0 - \mathbf{w}$ satisifes the homogeneous equation $\mathbf{a} \cdot \mathbf{x} = 0$ and is therefore orthogonal to $\mathbf{y} - \mathbf{w}_0$. Hence by the Pythagorean Theorem,

$$|\mathbf{y} - \mathbf{w}|^2 = |\mathbf{y} - \mathbf{w}_0|^2 + |\mathbf{w} - \mathbf{w}_0|^2$$

and so the distance from $\mathbf{y}$ to $\mathbf{w}$ is smallest when $\mathbf{w} = \mathbf{w}_0$. ∎

Example 3 Let's find the distance from the point $(1, 2, 3)$ to the plane given by

$$2x_1 - x_2 + 2x_3 = 3 \ . \tag{2.7}$$

To do this, introduce $\mathbf{a} = \begin{bmatrix} 2 \\ -1 \\ 2 \end{bmatrix}$ and $b = 3$ so that we may write the equation for the

plane in the form $\mathbf{a} \cdot \mathbf{x} = b$. Then with $\mathbf{y} = \begin{bmatrix} 1 \\ 2 \\ 3 \end{bmatrix}$, $\mathbf{a} \cdot \mathbf{y} = 6$ and $|\mathbf{a}|^2 = 9$, so the distance is

$$\frac{|\mathbf{a} \cdot \mathbf{y} - b|}{|\mathbf{a}|} = \frac{6 - 3}{\sqrt{9}} = 1 \ .$$

The point on the plane closest to $(1, 2, 3)$ corresponds to the vector

$$\mathbf{w} = \mathbf{y} - \frac{\mathbf{a} \cdot \mathbf{y} - b}{|\mathbf{a}|^2} \mathbf{a} = \begin{bmatrix} 1 \\ 2 \\ 3 \end{bmatrix} - \frac{6 - 3}{9} \begin{bmatrix} 2 \\ -1 \\ 2 \end{bmatrix} = \frac{1}{3} \begin{bmatrix} 2 \\ 7 \\ 7 \end{bmatrix} \ .$$

Example 4 Let's find the solution $\mathbf{w}$ of (2.7) that has least length; that is, with $|\mathbf{w}| < |\mathbf{v}|$ for any other solution $\mathbf{v}$. As soon as we recognize that the length $|\mathbf{v}|$ equals the distance from the origin, $|\mathbf{v} - 0|$, we are ready to go. We just apply Theorem 1 with $\mathbf{y} = 0$, and the $\mathbf{a}$ and b from Example 3. Since $\mathbf{y} = 0$, this is quite easy and we find

$$\mathbf{w} = \frac{b}{|\mathbf{a}|^2} \mathbf{a} = \frac{1}{3} \begin{bmatrix} 2 \\ -1 \\ 2 \end{bmatrix} \ .$$

There is a general conclusion to be drawn here: Notice that before we plugged in any numbers, we found that the least length solution was given by $(b/|\mathbf{a}|^2)\mathbf{a}$, which is the particular solution given in (2.2).

• *The particular solution (2.2) is the one that is closest to the origin, or what is the same, has the least length.*

A linear transformation point of view

It is useful to look at (2.6) from a linear transformation point of view. Let S be the solution set of any non–trivial linear equation $\mathbf{a} \cdot \mathbf{x} = b$ in $I\!\!R^n$. For any vector $\mathbf{y}$ in $I\!\!R^n$, we know that

$$\mathbf{y} - \frac{\mathbf{a} \cdot \mathbf{y}}{|\mathbf{a}|^2} \mathbf{a}$$

is the vector in S that is closer to $\mathbf{y}$ than any other vector in S. Since this vector is uniquely determined – we have a formula for it, after all – we can define a transformation g_S from $I\!\!R^n$ to $I\!\!R^n$ by putting

$$g_S(\mathbf{y}) = \mathbf{y} - \frac{\mathbf{a} \cdot \mathbf{y} - b}{|\mathbf{a}|^2} \mathbf{a} \tag{2.8}$$

Transformations of this type are very important in graphics, among other applications. Here is a graphics problem: Suppose you were viewing a collection of N points in $\mathbb{R}^3$ given by vectors $\mathbf{y}_1, \mathbf{y}_2, \ldots, \mathbf{y}_N$. You want to draw a picture of them, and your picture is going to be on a screen or a piece of paper, but in any case, something we can identify with a subset of $\mathbb{R}^2$. How do you represent the points in $\mathbb{R}^3$ down in $\mathbb{R}^2$ so that you capture what you see?

That depends on your viewing angle, and how far away you are. Suppose that you were viewing them from a good distanace* in the direction along the vector $\mathbf{a}$. If you were to draw on a screen (or a sheet of paper) what you saw, where would you put the N dots representing the N points in space? You'd identify your screen (or piece of paper) with a piece of the plane S, and you'd orthogonally project the points $\mathbf{y}_1, \mathbf{y}_2, \ldots, \mathbf{y}_N$ onto S. This would place the dots on your screen. The projection transformation is given by (2.8).

If $b = 0$, (2.8) becomes

$$g_S(\mathbf{y}) = \mathbf{y} - \frac{\mathbf{a} \cdot \mathbf{y}}{|\mathbf{a}|^2} \mathbf{a} . \tag{2.9}$$

(In this case, our screen lies in a plane through the origin). It is easy to check, using the properties of the dot product, that g_S is additive and homogenous, and hence it is linear. (Check this now)!

Therefore, there is an $n \times n$ matrix P_S so that for all $\mathbf{y}$ in $\mathbb{R}^n$,

$$g_S(\mathbf{y}) = P_S \mathbf{y} . \tag{2.10}$$

Notice that P_S is always a symmetric matrix. Indeed,

$$(P_S)_{i,j} = \mathbf{e}_i \cdot P_S \mathbf{e}_j = (\mathbf{e}_i \cdot \mathbf{e}_j) - \frac{(\mathbf{e}_i \cdot \mathbf{a})(\mathbf{e}_j \cdot \mathbf{a})}{|\mathbf{a}|^2} = (P_S)_{j,i} \tag{2.11}$$

That is,

$$(P_S)_{i,j} = \begin{cases} 1 - (a_i a_j)/|\mathbf{a}|^2 & \text{if } i = j \\[2mm] -(a_i a_j)/|\mathbf{a}|^2 & \text{if } i \neq j \end{cases} . \tag{2.12}$$

As you see from (2.12), P_S is symmetric:

$$(P_S)^t = P_S . \tag{2.13}$$

Furthermore, if $\mathbf{y}$ is in S, then $\mathbf{a} \cdot \mathbf{y} = 0$, and so from (2.9), $P_S \mathbf{y} = \mathbf{y}$. This can be understood as follows: $P_s \mathbf{y}$ is the closest member of S to $\mathbf{y}$. If $\mathbf{y}$ happens to be a member of S already, then it is closest to itself.

In any case, since $P_S \mathbf{y}$ is in S for any vector $\mathbf{y}$, we have

$$P_S^2 \mathbf{y} = P_S(P_S \mathbf{y}) = P_S \mathbf{y} ,$$

---

* This assumption that they are a reasonable distance away simplifies some problems about perspective that we don't want to go into here.

which means that

$$P_S^2 = P_S \ . \tag{2.14}$$

Any transformation from $I\!\!R^n$ to $I\!\!R^n$ that is equal to its own square, as in (2.14), is called a *projection*. We will be working with projection matrices frequently, so let's make a formal definition.

Definition (Projections) An $n \times n$ matrix P is called a *projection* if $P^2 = P$. It is an orthogonal projection in case it is also symmetric; that is, $P = P^t$.

Notice that if P is any projection, then so is $I - P$ since

$$(I - P)^2 = I - 2P + P^2 = I - P \ .$$

Things are especially interesting for projection matrices that are symmetric, like P_S. Here is why this is interesting: Suppose that P is an $n \times n$ projection matrix and P is symmetric. Let $\mathbf{x}$ and $\mathbf{y}$ be any two vectors in $I\!\!R^n$.

$$P\mathbf{x} \cdot ((I - P)\mathbf{y}) = \mathbf{x} \cdot (P(I - P)\mathbf{y})$$
$$= \mathbf{x} \cdot ((P - P^2)\mathbf{y}) = 0 \ .$$

That is, $P\mathbf{x}$ and $(I - P)\mathbf{y}$ are orthogonal for every $\mathbf{x}$ and $\mathbf{y}$. For this reason, it is usual to denote $I - P$ by $P^\perp$ when P is an orthogonal projection. Next note that for any $\mathbf{y}$, $P\mathbf{y}$ and $P^\perp\mathbf{y}$ add up to $\mathbf{y}$:

$$\mathbf{y} = \mathbf{y} - P\mathbf{y} + P\mathbf{y}$$
$$= (I - P)\mathbf{y} + P\mathbf{y}$$
$$= P^\perp\mathbf{y} + P\mathbf{y} \ .$$

- *If P is a symmetric $n \times n$ projection matrix, then for any $\mathbf{y}$ in $I\!\!R^n$, the formula $\mathbf{y} = P\mathbf{y} + P^\perp\mathbf{y}$, with $P^\perp = I - P$, expresses $\mathbf{y}$ as a sum of two orthogonal vectors.*

Here is an important special case: If $n = 3$ and P_S is the projection P_S onto the solution set of the equation $\mathbf{a} \cdot \mathbf{x} = 0$, then $P_S\mathbf{y}$ is the orthognal projection of $\mathbf{y}$ onto S, and $P_S^\perp\mathbf{y}$ is the component of $\mathbf{y}$ along the normal line.

Note that since $P_S^\perp = I - P_S$, it follows from (2.12) that

$$P_S^\perp\mathbf{y} = \frac{\mathbf{a} \cdot \mathbf{y}}{|\mathbf{a}|^2}\mathbf{a} \ . \tag{2.15}$$

By Theorem 1, $|P_S^\perp\mathbf{y}|$ is the distance from $\mathbf{y}$ to the solution set of $\mathbf{a} \cdot \mathbf{x} = 0$.

Example 5 Let S be the solution set of $2x_1 - x_2 + 2x_3 = 0$. This can be written in the form $\mathbf{a} \cdot \mathbf{x} = 0$ where $\mathbf{a} = \begin{bmatrix} 2 \\ -1 \\ 2 \end{bmatrix}$, as in Example 3. Let's find the projection onto the solution sets S of this equation.

Since $|a|^2 = 9$, and since by (2.15) or by (2.12) and the relaltion $P_S^\perp = I - P_S$,

$$(P_S^\perp)_{i,j} = \frac{a_i a_j}{|\mathbf{a}|^2} \ .$$

Evaluating this we find

$$P_S^\perp = \frac{1}{9} \begin{bmatrix} 4 & -2 & 4 \\ -2 & 1 & -2 \\ 4 & -2 & 4 \end{bmatrix} \ .$$

It is a good idea to check at this point that $(P_S^\perp)^2 = P_S^\perp$, as we claimed it would. Check this out!

Now since $P_S^\perp = I - P_S$, we also have $P_S = I - P_S^\perp$. So

$$\begin{aligned} P_S &= \begin{bmatrix} 1 & 0 & 0 \\ 0 & 1 & 0 \\ 0 & 0 & 1 \end{bmatrix} - \frac{1}{9} \begin{bmatrix} 4 & -2 & 4 \\ -2 & 1 & -2 \\ 4 & -2 & 4 \end{bmatrix} \\ &= \frac{1}{9} \begin{bmatrix} 9 & 0 & 0 \\ 0 & 9 & 0 \\ 0 & 0 & 9 \end{bmatrix} - \frac{1}{9} \begin{bmatrix} 4 & -2 & 4 \\ -2 & 1 & -2 \\ 4 & -2 & 4 \end{bmatrix} \\ &= \frac{1}{9} \begin{bmatrix} 5 & 2 & -4 \\ 2 & 8 & 2 \\ -4 & 2 & 5 \end{bmatrix} \end{aligned}$$

It is a good idea to check directly that indeed $P_S^2 = P_S$ for the matrix we just computed.

Now you can compute the orthogonal projection of any vector $\mathbf{y}$ onto S just by doing a matrix–vector product; namely, by computing $P_S \mathbf{y}$. The vector $P_S^\perp \mathbf{y}$ is the orthogonal component of $\mathbf{y}$.

Take for example $\mathbf{y} = \begin{bmatrix} 1 \\ 1 \\ 1 \end{bmatrix}$. Then

$$P_S \mathbf{y} = \frac{1}{3} \begin{bmatrix} 1 \\ 4 \\ 1 \end{bmatrix} \quad \text{and} \quad P_S^\perp \mathbf{y} = \frac{1}{3} \begin{bmatrix} 2 \\ -1 \\ 2 \end{bmatrix} \ .$$

As you can see, these are two orthogonal vectors whose sum is $\mathbf{y}$.

Section 3: Systems of Linear Equations

A system of m linear equations in $I\!R^n$ is simply a list of m such equations:

$$\mathbf{a}_1 \cdot \mathbf{x} = b_1$$

$$\mathbf{a}_2 \cdot \mathbf{x} = b_2$$

$$\vdots \quad \vdots$$

$$\mathbf{a}_m \cdot \mathbf{x} = b_m$$

(3.1)

There are a number of ways of writing such a system, each with its own merits. One way is to introduce the vectors $\mathbf{x} = \begin{bmatrix} x_1 \\ x_2 \\ \vdots \\ x_n \end{bmatrix}$ and $\mathbf{b} = \begin{bmatrix} b_1 \\ b_2 \\ \vdots \\ b_m \end{bmatrix}$ and the matrix

$$A = \begin{bmatrix} \mathbf{a}_1 \\ \mathbf{a}_2 \\ \vdots \\ \mathbf{a}_m \end{bmatrix} = \begin{bmatrix} A_{1,1} & A_{1,2} & \cdots & A_{1,n} \\ A_{2,1} & A_{2,2} & \cdots & A_{2,n} \\ & & \vdots & \\ A_{m,1} & A_{m,2} & \cdots & A_{m,n} \end{bmatrix}.$$

Notice that $\mathbf{x}$ is in $I\!R^n$, and $\mathbf{b}$ is in $I\!R^m$, and that $A_{i,j}$ is the jth entry of $\mathbf{a}_i$. Then (3.1) can be rewritten in the compact form

$$A\mathbf{x} = \mathbf{b} .$$

(3.2)

Going in the other direction, we can write this out in full detail as

$$A_{1,1}x_1 + A_{1,2}x_2 + \cdots + A_{1,n}x_n = b_1$$

$$A_{2,1}x_1 + A_{2,2}x_2 + \cdots + A_{2,n}x_n = b_2$$

$$\vdots \qquad\qquad\qquad \vdots$$

$$A_{m,1}x_1 + A_{m,2}x_2 + \cdots + A_{m,n}x_n = b_m$$

(3.3)

It should be clear that (3.1), (3.2) and (3.3) are just three ways of expressing the same thing, and that you can pass back and forth between them easily.

Definition (Solution Set of a Linear System) The *solution set S* of a linear system such as (3.1) is simply the set of all vectors that satisfy each of the equations. In other words, it is the intersection of the solution sets of the individual linear equations.

To solve a system of linear equations is to find an explicit description of its solution set. As with a single linear equation, this will usually mean giving a parametric description of S in the form

$$\mathbf{x}_0 + t_1\mathbf{v}_1 + t_2\mathbf{v}_2 + \cdots + t_r\mathbf{v}_r \tag{3.4}$$

in terms of some number r of parameters.

The case $m = n$, where the number of equations equals the number of variables is especially important.

Two equations in two variables

Let's assume none of the equations is trivial. Then, if $m = n = 2$, each of the equations describes a line, and the solution set is the intersection of these two lines. If these lines are not parallel, they will intersect in exactly one point $\mathbf{x}_0$, and in this case the solution set consists of a single vector $\mathbf{x}_0$, as in the picture below. *In this case there are no parameters required.*

If however, the lines are parallel, then either they don't intersect, and so the solution set is empty, or else the two lines are the same, and *every point* on this line is a solution.

Altogether then, there are either no solutions, exactly one solution, or infinitely many solutions. (There are never exactly two solutions, for example – something quite common with quadratic equations in one variable.)

Example 1: Consider the system of equations

$$\begin{aligned} 2x_1 + x_2 &= 3 \\ x_1 + 2x_2 &= 3 \ . \end{aligned} \tag{3.5}$$

Each of these equations is the equation of a line, the first one with slope -2, and the second one with slope $-1/2$. Since the slopes are different, the two lines will intersect.

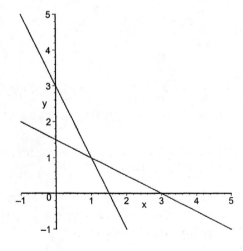

As you see in the graph, and can easily check, the point of intersection is $\mathbf{x}_0 = \begin{bmatrix} 1 \\ 1 \end{bmatrix}$,

which is the unique solution of this system.

On the other hand, consider the system

$$2x_1 + x_2 = 3$$
$$4x_1 + 2x_2 = 3 \ .$$

(3.6)

Each of these equations is the equation of a line, but this time both lines have the same slope, namely -2. Since they have different x_2 intercepts, they are distinct parallel lines, and there is no point of intersection, and hence no solution.

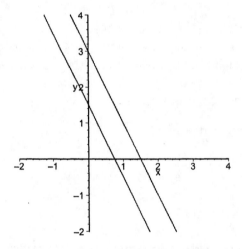

This is an example of a system for which the solution set is the empty set $\emptyset$.

The third possibility is exemplified by

$$2x_1 + x_2 = 3$$
$$4x_1 + 2x_2 = 6 \ .$$

(3.7)

There aren't really two equations here: The second one is just a multiple of the first. Both equations describe the same line, and so every point on this line is a solution of both equations. Here is the picture:

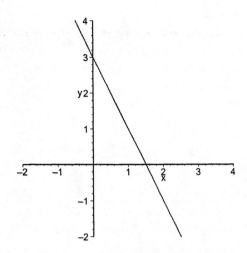

Two or three equations in three variables

Since a non-trivial linear equation in three variables describes a plane, the solution set of a system of two such equations is the intersection of two planes. If the planes are parallel, they either coincide, in which case every vector in the plane is a solution, or else they do not intersect at all, in which case there is no solution.

Usually though the planes will not be parallel, and they will intersect in a line. The vectors on this line are the ones satisfying both equations, so this line is the solution set.

Example 2 Consider the system of equations

$$2x_1 + x_2 + x_3 = 3$$
$$x_1 + 2x_2 - x_3 = 1 \, .$$

(3.8)

Each of these equations is the equation of a plane. Here is a graph showing their intersection:

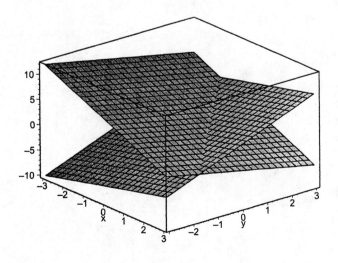

We can *see* the line, but how can we write it down? Here is how: Multiply the second equation through by 2, and subtract it from the first. This eliminates x_1, and we get $-3x_2 + 3x_3 = 1$. This says

$$x_2 = x_3 - \frac{1}{3} \ . \tag{3.9}$$

The second equation says $x_1 = 1 - 2x_2 - x_3$, and we can use (3.9) to eliminate x_2, giving us

$$x_1 = 1 - 2 \left(x_3 - \frac{1}{3} \right) + x_3 = \frac{5}{3} - x_3 \ . \tag{3.10}$$

Together, (3.9) and (3.10) express x_1 and x_2 as linear functions of the variable x_3. This gives us a parameterization of the line of solutions: Setting $x_3 = t$, just to emphasize its status as a parameter, we have from (3.9) and (3.10) that

$$\begin{bmatrix} x_1 \\ x_2 \\ x_3 \end{bmatrix} = \begin{bmatrix} 5/3 - t \\ t - 1/3 \\ t \end{bmatrix} = \frac{1}{3} \begin{bmatrix} 5 \\ -1 \\ 0 \end{bmatrix} + t \begin{bmatrix} -2 \\ 1 \\ 1 \end{bmatrix} \ . \tag{3.11}$$

We've just found our first parameterizatin of the solution set of a system of equations. In the next few sections we're going to develop a systematic way of doing this that works no matter how many variables or equations there are. The key, as above, will be to separate the variables into a set of independent variables, or free variables which become parameters, and dependent variables. Here we expressed x_1 and x_2 as depending on x_3, which is free; (3.9) and (3.10) express the dependence.

When there are three equations in three variables, the solution set is the intersection of three planes. If no two of these are parallel, the intersection is a single point:

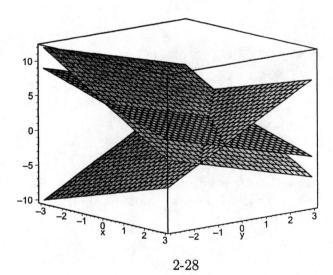

The three planes shown in this graph come from the system

$$2x_1 + x_2 + x_3 = 3$$
$$x_1 + 2x_2 - x_3 = 1 \qquad\qquad (3.12)$$
$$x_1 + x_2 + x_3 = 3 \ .$$

Since the first two equations are the same as in the ones we considered just above, we know that any vector $\begin{bmatrix} x_1 \\ x_2 \\ x_3 \end{bmatrix}$ satisfying these two equations has the form (3.11). Plugging this into the third equation, we get

$$(5/3 - t) + (t - 1/3) + t = 3$$

which reduces to $t = 4/9$. So the third equation is also satisfied if and only if t has this value. Setting $t = 5/3$ in (3.11), we find the unique solution:

$$\begin{bmatrix} x_1 \\ x_2 \\ x_3 \end{bmatrix} = \frac{1}{3} \begin{bmatrix} 0 \\ 4 \\ 5 \end{bmatrix} \ .$$

As you can check, this does solve all three equations. There is just one solution; no parameters are needed.

As you see having more equations can lead to a simpler solution set – a single vector, or possibly even the empty set. However, having many equations generally makes it more work to figure out what the solution set is. We are going to approach this problem in a systematic way. The first step is to identify some nice systems of equations. The second step is going to be to transform any system into a nice system!

A nice sort of system

There is one case in which a system of equations can be dealt with very simply, *one equation at a time*. Recall that the key to parameterizing the solution set of a single equation is separating the variables into one pivotal variable, and $n - 1$ non–pivotal variables. The values of the non pivotal variables can be freely assigned, and then the equation determines the value of the pivotal variable.

Everything is very nice when each equation in a system is non–trivial, and determines a *different* pivotal variable. Let

$$j_p(i) = \min\{k = 1, 2, \ldots, n \mid (\mathbf{a}_i)_k \neq 0\} \qquad\qquad (3.13)$$

In other words, the first non–zero coefficient of the ith equation is in the $j_p(i)$th place. Then $x_{j_p(i)}$ is the pivotal variable for the ith equation. We can assign arbitrary values to the other variables now, and then our ith equation becomes

$$a_{j_p(i)} x_{j_p(i)} = \text{something}$$

which, no matter what the right hand side is, can be solved for $x_{j_p(i)}$ by dividing the right side by $a_{j_p(i)}$, and that can be done since $a_{j_p(i)} \neq 0$. Setting $x_{j_p(i)}$ to this value solves the ith equation.

• *No matter how the values of the other variables are assigned, we can always find a value for $x_{j_p(i)}$ so that the ith equation will be satisfied.*

Next, as long as

$$i \neq k \Rightarrow j_p(i) \neq j_p(k) \tag{3.14}$$

no pair of our equations will "insist" on two different values of the same variable. Single linear equations only "insist" on the value of one pivotal variable each, so if all of the equations have different pivotal variables, there is no conflict, and they can all be satisfied.

Example 3 Consider the system

$$\begin{aligned} 2x_1 + 2x_2 + x_3 &= 4 \\ x_2 - x_3 &= -1 \ . \end{aligned} \tag{3.15}$$

For the first equation, the coefficient of x_1 is non–zero, so $j_p(1) = 1$. For the second equation, x_2 is the first variable with a non–zero coefficient, so that $j_p(2) = 2$. The pivotal variable for the first equation is x_1, and the pivotal variable for the second equation is x_2. The variable x_3 is non–pivotal for both.

We therefore set

$$x_3 = t \ , \tag{3.16}$$

where t is a parameter. The second equation then determines its pivotal variable, and gives us

$$x_2 = t - 1 \ . \tag{3.17}$$

Now that x_2 and x_3 are determined in terms of t, as in (3.16) and (3.17), the first equation determines the value of x_1 in terms of t:

$$\begin{aligned} x_1 &= (4 - 2x_2 - x_3)/2 \\ &= 2 - (t - 1) - t/2 \\ &= 3 - (3/2)t \ . \end{aligned} \tag{3.18}$$

Putting (3.16), (3.17) and (3.18) together, the solutions of this system are given by

$$\begin{bmatrix} x_1 \\ x_2 \\ x_3 \end{bmatrix} = \begin{bmatrix} 3 - (3/2)t \\ t - 1 \\ t \end{bmatrix} = \begin{bmatrix} 3 \\ -1 \\ 0 \end{bmatrix} + t \begin{bmatrix} -3/2 \\ 1 \\ 1 \end{bmatrix} \ . \tag{3.19}$$

If we introduce $\mathbf{x}_0 = \begin{bmatrix} 3 \\ -1 \\ 0 \end{bmatrix}$ and $\mathbf{v} = \begin{bmatrix} -3/2 \\ 1 \\ 1 \end{bmatrix}$, we can write the solution in the form (3.4) with a single parameter t:

$$\mathbf{x}_0 + t\mathbf{v} \ . \tag{3.20}$$

This is the parametric representation of a line, namely the line through $\mathbf{x}_0$ in the direction $\mathbf{v}$. Evidently, there are infinitely many solutions.

Each of the two equations in (3.15) is the equation of a plane. These planes intersect in a line, and (3.20) is the parametric representation of this line of intersection.

The procedure we have just used is very important, so let's look at another example with more variables.

Example 4 Consider the system

$$
\begin{aligned}
x_1 + 2x_2 + x_3 + x_5 &= 4 \\
x_3 - x_5 &= -1 \\
x_4 + x_5 &= 0 \; .
\end{aligned}
\tag{3.21}
$$

This time, $j_p(1) = 1$, $j_p(2) = 3$, and $j_p(3) = 4$. Therefore, x_1 is pivotal for the first equation, x_3 is pivotal for the second equation, and x_4 is pivotal for the third equation. Each equation has a different pivotal variable, and we can deal with them one at a time. The variables x_2 and x_5 are non-pivotal for each of the equations, and so we can set their values freely. Hence we introduce parameters t_1 and t_2 with

$$
x_2 = t_1 \quad \text{and} \quad x_5 = t_2 \; .
\tag{3.22}
$$

The third equation determines the value of its pivotal variable, x_4:

$$
\begin{aligned}
x_4 &= -x_5 \\
&= -t_2 \; .
\end{aligned}
\tag{3.23}
$$

The second equation determines the value of its pivotal variable, x_3:

$$
\begin{aligned}
x_3 &= -1 + x_5 \\
&= t_2 - 1 \; .
\end{aligned}
\tag{3.24}
$$

Finally, the first equation determines the value of its pivotal variable, x_1:

$$
\begin{aligned}
x_1 &= 4 - 2x_2 - x_3 - x_5 \\
&= 4 - 2t_1 - (t_2 - 1) - t_2 \\
&= 5 - 2t_1 - 2t_2 \; .
\end{aligned}
\tag{3.25}
$$

Combining (3.22) through (3.25), we find for the solutions

$$
\begin{bmatrix} x_1 \\ x_2 \\ x_3 \\ x_4 \\ x_5 \end{bmatrix}
=
\begin{bmatrix} 5 - 2t_1 - 2t_2 \\ t_1 \\ t_2 - 1 \\ -t_2 \\ t_2 \end{bmatrix}
=
\begin{bmatrix} 5 \\ 0 \\ -1 \\ 0 \\ 0 \end{bmatrix}
+ t_1
\begin{bmatrix} -2 \\ 1 \\ 0 \\ 0 \\ 0 \end{bmatrix}
+ t_2
\begin{bmatrix} -2 \\ 0 \\ 1 \\ -1 \\ 1 \end{bmatrix} \; .
$$

This can be written in the form $\mathbf{x}_0 + t_1\mathbf{v}_1 + t_2\mathbf{v}_2$.

These examples lead to the following definition:

Definition (Disentangled System) A system of linear equations as in (3.1) is *disentangled* in case

$$i < k \Rightarrow j_p(i) < j_p(k) \tag{3.26}$$

where $j_p(i)$ is the place of the pivotal variable in the ith equation, and $j_p(k)$ is the place of the pivotal variable in the kth equation, as defined in (3.13).

The systems of equations in Examples 3 and 4 were both disentangled, and for this reason, we were easily able to parameterize their solution sets. (Actually, all we needed was (3.14) instead of (3.26) which is stronger). As long as no two equations have the same pivotal variable, which is what (3.14) implies, we can successively solve the equations one at a time, as in Examples 3 and 4. This makes disentangled systems easy to deal with. However:

• *We are not claiming that only disentangled systems are nice, just that disentangled systems are always nice.*

And as we will soon see, we can transform any system into a disentangled sytem without changing the solution set, so we don't need to worry about what else is nice – disentanged systems are enough. Here though is an example of a nice system that isn't disentangled

Example 5 Consider the system

$$\begin{aligned}2x_1 + x_2 &= 3 \\ x_1 &= 1\;.\end{aligned} \tag{3.27}$$

This is not disentagled; x_1 is pivotal for both equations. All the same the second equation plainly says that $x_1 = 1$, and substituting this into the first equation, we get $2 + x_2 = 3$, so $x_2 = 1$ also. Hence the solution is $\begin{bmatrix} x_1 \\ x_2 \end{bmatrix} = \begin{bmatrix} 1 \\ 1 \end{bmatrix}$. So this system was pretty nice.

Note that while (3.27) is not disentagled, it would be if we just changed the order of the variables: If we define $y_1 = x_2$ and $y_2 = x_1$, it is disentangled in the new variables.

The back substitution method for disentangled systems

If a system is disentangled, we can solve it by the *back substitution method*, which is just what we did in Examples 3 and 4. Review those examples after reading the next paragraph.

Back Substitution Method for Solving Disentangled Systems: If a system of m linear equations in n variables as in (3.1) is disentangled, then there are a total of m pivotal variables, one for each equation, and $n - m$ non–pivotal variables. The non–pivotal variables can be set equal to parameters $t_1, t_2, \ldots, t_{n-m}$, and then the m equations can be solved for the values of the $n - m$ pivotal variables, starting from the last equation, and working backwards to the first. The result is a vector depending in the parameters $t_1, t_2, \ldots, t_{n-m}$, which can be written in the form

$$\mathbf{x}_0 + t_1 \mathbf{v}_1 + t_2 \mathbf{v}_2 + \cdots + t_{n-m} \mathbf{v}_{n-m} , \qquad (3.28)$$

where $\mathbf{x}_0, \mathbf{v}_1, \ldots, \mathbf{v}_{n-m}$ are constant vectors, independent of the parameters.

Theorem 1 (Back Substitution Yields One–to–One Parameterizations) *Given a disentangled system of linear equations, the back substitution method always produces a one–to–one parameterization of the solution set. That is, each different set of values for the parameters yields a different solution.*

Proof: The argument is much the same as with a single linear equation. The kth vector $\mathbf{v}_k$ in the parameterization (3.28) has a 1 in the place of the kth non–pivotal variable, and 0 in the place of every other non–pivotal variable. Moreover, $\mathbf{x}_0$ has 0 in every entry corresponding to a non–pivotal variable. Thus to see what the value of t_k is for any solution, just look in the place of the kth pivotal variable. What you see there is t_k since this entry of $\mathbf{v}_k$ is 1, and no other vector in (3.28) has a non–zero entry in this place. ∎

Now suppose we are given a system to solve, and it isn't disentangled. What should we do? Disentangle it! The fact is that one can change a system of equations without changing the solution set, and this makes the back substitution method a *general method*, and not just a method for dealing with a special class of systems. We will develop this idea in the next section, and we close this one with some general observations on linear systems.

Uniqueness and the homogeneous equation

Recall that when we discussed a single linear equation $\mathbf{a} \cdot \mathbf{x} = b$, the homogeneous equation played an important role. This is even more the case with systems. The equation $A\mathbf{x} = 0$ is called the *homogeneous equation* associated to $A\mathbf{x} = \mathbf{b}$.

Theorem 2 (Uniqueness and the Homogeneous Equation) *The equation $A\mathbf{x} = \mathbf{b}$ has at most one solution if and only if the homogeneous equation $A\mathbf{x} = 0$ has only the zero solution. In fact:*

(1) Given any two distinct solutions $\mathbf{x}_0$ and $\mathbf{x}_1$ of $A\mathbf{x} = \mathbf{b}$, their difference, $\mathbf{v} = \mathbf{x}_1 - \mathbf{x}_0$ is a non–zero solution of the homogeneous equation $A\mathbf{x} = 0$.

(2) Conversely, given any non–zero solution $\mathbf{v}$ of the homogeneous equation $A\mathbf{x} = 0$, and any solution $\mathbf{x}_0$ of $A\mathbf{x} = \mathbf{b}$, their sum $\mathbf{x}_0 + \mathbf{v}$ is a second solution of $A\mathbf{x} = \mathbf{b}$.

Proof: First $A(\mathbf{x}_1 - \mathbf{x}_0) = A\mathbf{x}_1 - A\mathbf{x}_0 = \mathbf{b} - \mathbf{b} = 0$, so that $\mathbf{v} = \mathbf{x}_1 - \mathbf{x}_0$ solves $A\mathbf{x} = 0$. Next, if $A\mathbf{x}_0 = \mathbf{b}$ and $A\mathbf{v} = 0$, $A(\mathbf{x}_0 + \mathbf{v}) = A\mathbf{x}_0 + A\mathbf{v} = \mathbf{b}$ ∎

Section 4: Row Reduction

In the last section we mentioned that because one can change a system of equations without changing the solution set, the back substitution method is a *general method*, and not just a method for dealing with a special class of systems. To proceed with this idea, let's make the following definition:

Definition (Equivalent Systems) Two systems of linear equations in the same n variables are *equivalent* in case they have the same solution set.

It is *not* required that both systems have the same number of equations; just that they have the same solution sets.

If we are trying to solve a system, which means trying to find its solution set, we are free to change it as long as the result is an equivalent system. If we can change it into an equivalent system that is also a disentangled system, we can use the back substitution method to find the solution set of the disentangled system, which, by the equivalence, is also the solution set of the original system.

We've seen examples of this strategy in practice, though the ideas behind it have not been spelled out. Let's look at an example, and examine it against this background.

Example 1 Consider the system of equations

$$
\begin{aligned}
x_1 + x_2 &= 2 \\
2x_1 + 3x_2 &= 3 \ .
\end{aligned}
\tag{4.1}
$$

This system is not disentangled; x_1 is the pivotal variable for both equations.

But we can fix this: First, multiply the top equation in (4.1) through by 2. We get the new system

$$
\begin{aligned}
2x_1 + 2x_2 &= 4 \\
2x_1 + 3x_2 &= 3 \ .
\end{aligned}
\tag{4.2}
$$

and we haven't changed its solution set since clearly $x + y = 2$ and $2x + 2y = 4$ describe the exact same line. So these two systems, (4.1) and (4.2) have the same solution set. Now we're set to eliminate x_1 from the bottom equation. Subtract the top equation in (4.2) from the bottom one to obtain

$$
\begin{aligned}
2x_1 + 2x_2 &= \ \ 4 \\
x_2 &= -1 \ .
\end{aligned}
\tag{4.3}
$$

Subtracting equals from equals yields equals. So anything that was in the solution set of (4.2) is in the solution set of (4.3). But the passage from (4.2) to (4.3) is *reversible*, and by the same means too: If you add the first equation in (4.3) to the second, you recover (4.2). Since *adding equals to equals yields equals*, anything in the solution set of (4.3) is in the solution set of (4.2). So these two solution sets contain each other, which means they are the same.

We already knew that (4.1) had the same solution sets as (4.2), and we've just seen that (4.2) has the same solution set as (4.3). Therefore (4.1) has the same solution set as (4.3). *All three systems are equivalent in the sense of the definition.*

Moreover, (4.3) is disentangled since $j_p(1) = 1$ and $j_p(2) = 2$. That is x_1 is pivotal for the first equation, x_2 is pivotal for the second equation, and there are no non–pivotal variables. Hence there will be no parameters. Proceeding by back substitution, we start with the second equation in (4.3), which tells us right away that $x_2 = -1$. Then substituting $x_2 = -1$ into the first equation in (4.3), we have $2x_1 - 2 = 4$ which tells us right away that $x_1 = -3$. Thus the system (4.3) has only one solution, namely

$$\begin{bmatrix} x_1 \\ x_2 \end{bmatrix} = \begin{bmatrix} -3 \\ -1 \end{bmatrix}. \tag{4.4}$$

But equivalent systems have the same solution set, *so we've found the solution set of our original system as well.*

There is a really great strategy here, whose utility extends far beyond the present problem. To bring it into focus, let's go back through the steps in a more compact notation. First let's simplify things by putting the variables in **x** out of sight.

Definition (Augmented Matrix) Given any system of m equations in n variables, written in the vector form $A\mathbf{x} = \mathbf{b}$, the $m \times (n+1)$ matrix $[A|\mathbf{b}]$ whose first n columns are the corresponding columns of A, and whose $(n+1)$st column is $\mathbf{b}$ is called the *augmented matrix of the system* $A\mathbf{x} = \mathbf{b}$. We will refer to the solution set of the system corresponding to $[A|\mathbf{b}]$ simply as *the solution set of* $[A|\mathbf{b}]$ We say *two augmented matrices are equivalent* if they have the same solution set in this sense.

Example 2 Let's write down the augmented matrix descriptions of each of the equivalent systems (4.1) and (4.3): These are, respectively,

$$\begin{bmatrix} 1 & 1 & 2 \\ 2 & 3 & 3 \end{bmatrix} \tag{4.5}$$

and

$$\begin{bmatrix} 2 & 2 & 4 \\ 0 & 1 & -1 \end{bmatrix}. \tag{4.6}$$

By the definition and the fact that (4.1) and (4.3) are equivalent systems, these are equivalent augmented matrices. Since the jth column of A corresponds to the jth variable x_j, we aren't losing any information in passing to an augmented matrix description of a system. You can easily recover (4.1) and (4.3) from (4.5) and (4.6). We are just using "placeholding notation" for a clean expression of the system.

Now that we have introduced augmented matrices into the game, an $m \times n$ matrix with $n > 1$ has two possible interpretations. For example, the 2×3 matrix (4.5) could represent the system of equations (4.1), if we regard it as an augmented matrix, which is what we were doing in Example 2. But (4.5) could also represent a linear transformation from $I\!R^3$ to $I\!R^2$, as in the first chapter. *However, whether a matrix represents a system of equations or a linear transformation will usually be clear from context, so usually no special notation is needed to distinguish the two cases.* But sometimes, when we want to emphasize that, say (4.5), represents a system of equations instead of a linear transformation, we will write

$$\begin{bmatrix} 1 & 1 & | & 2 \\ 2 & 3 & | & 3 \end{bmatrix} \tag{4.7}$$

The bar before the last column emphasizes that this is short for

$$\begin{bmatrix} 1 & 1 \\ 2 & 3 \end{bmatrix} \begin{bmatrix} x_1 \\ x_2 \end{bmatrix} = \begin{bmatrix} 2 \\ 3 \end{bmatrix}, \tag{4.8}$$

which in turn is short for (4.1). The shortest notation of all is just (4.5) without the bar, and that is what we will use when the context is clear.

The following theorem tells us when two augmented matrices correspond to equivalent systems:

Theorem 1 (Augmented Matrices and Equivalence) *Two augmented matrices $[A|\mathbf{b}]$ and $[\tilde{A}|\tilde{\mathbf{b}}]$ are equivalent in case each row of $[\tilde{A}|\tilde{\mathbf{b}}]$ is a linear combination of the rows of $[A|\mathbf{b}]$, and vice versa.*

Notice that for *any* row of $[\tilde{A}|\tilde{\mathbf{b}}]$ to be a linear combination of the rows of $[A|\mathbf{b}]$, both $[\tilde{A}|\tilde{\mathbf{b}}]$ and $[A|\mathbf{b}]$ must have the same number of columns, or, in other words, refer to the same number of variables. However, the conditions of the theorem can easily hold when the two matrices have a different number of rows, or, in other words, represent a different number of equations.

Proof: Given $\mathbf{x}$ in $I\!R^n$, define the vector $\hat{\mathbf{x}}$ in $I\!R^{n+1}$ by

$$\hat{\mathbf{x}} = \begin{bmatrix} x_1 \\ x_2 \\ \vdots \\ x_n \\ -1 \end{bmatrix} \qquad \text{where} \qquad \mathbf{x} = \begin{bmatrix} x_1 \\ x_2 \\ \vdots \\ x_n \end{bmatrix}.$$

Then from the row representation of matrix–vector multiplication, you can see that

$$[A|\mathbf{b}]\hat{\mathbf{x}} = 0 \iff A\mathbf{x} = \mathbf{b}.$$

Therefore, $\mathbf{x}$ is in the solution set of $A\mathbf{x} = \mathbf{b}$ if and only if $[A|\mathbf{b}]\hat{\mathbf{x}} = 0$. But again from the row representation of matrix–vector multiplication, you can see that $[A|\mathbf{b}]\hat{\mathbf{x}} = 0$ if and only if $\mathbf{r}_i \cdot \hat{\mathbf{x}} = 0$ for each row $\mathbf{r}_i$ of $[A|\mathbf{b}]$.

This does it. For if each row of $[\tilde{A}|\tilde{\mathbf{b}}]$ is a linear combination of the rows of $[A|\mathbf{b}]$, it has the form

$$c_1\mathbf{r}_1 + c_2\mathbf{r}_2 + \cdots + c_m\mathbf{r}_m$$

for some numbers $c_1, c_2, \ldots, c_m$. But then by the distributive property of the dot product,

$$(c_1\mathbf{r}_1 + c_2\mathbf{r}_2 + \cdots + c_m\mathbf{r}_m) \cdot \hat{\mathbf{x}} = c_1\mathbf{r}_1 \cdot \hat{\mathbf{x}} + c_2\mathbf{r}_2 \cdot \hat{\mathbf{x}} + \cdots + c_m\mathbf{r}_m \cdot \hat{\mathbf{x}} = 0$$

so that $\hat{\mathbf{x}}$ is orthogonal to this row, and hence every row, of $[\tilde{A}|\tilde{\mathbf{b}}]$. Thus $[\tilde{A}|\tilde{\mathbf{b}}]\hat{\mathbf{x}} = 0$, and $\mathbf{x}$ is in the solution set of $[\tilde{A}|\tilde{\mathbf{b}}]$

We've just shown that if each row of $[\tilde{A}|\tilde{\mathbf{b}}]$ is a linear combination of the rows of $[A|\mathbf{b}]$, then each solution of $A\mathbf{x} = \mathbf{b}$ is also a solution of $\tilde{A}\mathbf{x} = \tilde{\mathbf{b}}$. If we also have that each row of $[A|\mathbf{b}]$ is a linear combination of the rows of $[\tilde{A}|\tilde{\mathbf{b}}]$, then each solution of $\tilde{A}\mathbf{x} = \tilde{\mathbf{b}}$ is also a solution of $A\mathbf{x} = \mathbf{b}$, and hence the systems are equivalent. ■

This theorem tells us pretty much exactly what we can do to the augmented matrix of a system without changing its solution set. It leads straight to the following list:

Theorem 2 (Elementary Row Operations) $[A|\mathbf{b}]$ *be the augmented matrix of some system, and let* $[\tilde{A}|\tilde{\mathbf{b}}]$ *be the augmented matrix otained from it by any of the following operations:*

(1) Exchanging the places of two rows.

(2) Multiplying one row through by a non–zero constant.

(3) Adding a multiple of one row to another.

Then $[A|\mathbf{b}]$ *and* $[\tilde{A}|\tilde{\mathbf{b}}]$ *are equivalent.*

Proof: It is quite clear that in case (1), $[A|\mathbf{b}]$ and $[\tilde{A}|\tilde{\mathbf{b}}]$ actually have the *same* set of rows.

Consider case (3): If we produce $[\tilde{A}|\tilde{\mathbf{b}}]$ by adding a multiple of ith row of $[A|\mathbf{b}]$ to the jth, $j \neq i$, then only the jth row has changed at all, and it is clearly a linear combination of the rows of $[A|\mathbf{b}]$. Therefore, each row of $[\tilde{A}|\tilde{\mathbf{b}}]$ is linear combination of the rows of $[A|\mathbf{b}]$.

Next, we we can recover the jth row of $[A|\mathbf{b}]$ by subtracting the same multiple of the ith row of $[\tilde{A}|\tilde{\mathbf{b}}]$, which equals the ith row of $[A|\mathbf{b}]$, back off the jth row of $[\tilde{A}|\tilde{\mathbf{b}}]$. We are subtracting off exactly what we added on, so we get back the original jth row. Hence each row of $[A|\mathbf{b}]$ is a linear combination of rows of $[\tilde{A}|\tilde{\mathbf{b}}]$, and $[A|\mathbf{b}]$ and $[\tilde{A}|\tilde{\mathbf{b}}]$ are equivalent.

Case (2) is very much the same, and left as an exercise. ■

Definition (Row Operations) We use the term *row operation* to refer to any one of the three operations on the rows of a matrix listed in Theorem 2. These are sometimes called *elementary row operations*, but since they are the only kind we will consider, we use the short form of the name.

Example 3 Consider the augmented matrix

$$[A|\mathbf{b}] = \begin{bmatrix} 3 & 2 & | & 1 \\ 1 & 2 & | & 3 \end{bmatrix} . \tag{4.9}$$

By swapping the two rows, we obtain $\begin{bmatrix} 1 & 2 & | & 3 \\ 3 & 2 & | & 1 \end{bmatrix}$. Next, multiplying the first row through by -3, and adding this to the second row, we obtain

$$\begin{bmatrix} 1 & 2 & | & 3 \\ 0 & -4 & | & -8 \end{bmatrix} . \tag{4.10}$$

Since we only used operations from the list in Theorem 2, we know that at each step along the way, we had an augmented matrix equivalent to the original one. In particular, (4.10) is equivalent to (4.9).

This is good since the augmented matrix (4.10) represents a disentangled system, namely

$$\begin{aligned} x_1 + 2x_2 &= 3 \\ -4x_2 &= -8 . \end{aligned} \tag{4.11}$$

There are no pivotal variables, and back substitution tells us that the unique solution is $\mathbf{x}_0 = \begin{bmatrix} -1 \\ 2 \end{bmatrix}$. Therefore, this is also the unique solution to the system represented by (4.9).

It is easy to tell when an augmented matrix $[A|\mathbf{b}]$ represents a disentangled system just by looking at the pattern of zeros in it.

First, let's introduce some notation. If B is an $m \times n$ matrix whose ith row is not all zeros, let $j(i)$ be the number of the column that contains the first non–zero entry in the ith row of B. That is,

$$j(i) = \min\{k = 1, 2, \ldots, n \mid B_{i,k} \neq 0\} . \tag{4.12}$$

If the ith row is all zeros, $j(i)$ is undefined.

Compare this definition with the definition (3.13) of $j_p(i)$ for a system. You see that if all of the equations in the system $A\mathbf{x} = \mathbf{b}$ are non–trivial, then for each i, $j(i)$ for the matrix $B = [A|\mathbf{b}]$ equals $j_p(i)$ for the system $A\mathbf{x} = \mathbf{b}$. The only difference is in case the ith equation is the trivial equation $0 \cdot \mathbf{x} = b$, $b \neq 0$, in which case $j(i) = n$, and $j_p(i)$ is undefined – there is no pivotal variable for such an equation.

Definition (Row Reduced Form, Pivots and Pivotal Columns) An $m \times p$ matrix B is in *row reduced form* in case the first non–zero entry in each row comes strictly to the left of the first non–zero entry in the row below it. In more detail, in a row reduced matrix B, any rows that are all zeros are at the bottom of the matrix, and if there are r rows that are not all zeros,

$$i < k \leq r \Rightarrow j(i) < j(k) . \tag{4.13}$$

The r entries $B_{i,j(i)}$ are called the *pivots*. A column is called a *pivotal column* in case it contains a pivot, so that columns $j(1), j(2), \ldots, j(r)$ are the pivotal columns.

- *Notice that there is exactly one pivot for each row that is not all zeros.*

The pattern of non–zero entries in a matrix that is in row reduced form has a sort of "staircase pattern" to it, such as

$$\begin{bmatrix} 0 & \bullet & * & * & * & * & * \\ 0 & 0 & \bullet & * & * & * & * \\ 0 & 0 & 0 & 0 & \bullet & * & * \\ 0 & 0 & 0 & 0 & 0 & \bullet & * \\ 0 & 0 & 0 & 0 & 0 & 0 & 0 \end{bmatrix} . \tag{4.14}$$

The bullets mark the pivots, entries that are definitely not zero, while the asterisks mark entries that might or might not be zero. The reason for the term "row reduced form" will become clear shortly.

Theorem 3 (Recognizing Disentangled Systems) *Let $[A|\mathbf{b}]$ be an $m \times (n+1)$ augmented matrix representing a system of m linear equations in n variables. Suppose that no row of $[A|\mathbf{b}]$ consists of all zeros. Then $[A|\mathbf{b}]$ represents a disentangled system if and only if:*

(1) $[A|\mathbf{b}]$ is in row reduced form

   and

(2) There is no pivot in the final column of $[A|\mathbf{b}]$.

If condition (1) is satisfied, but (2) is not, then $[A|\mathbf{b}]$ represents a system that has no solution.

Before going into the proof, notice that if any row of $[A|\mathbf{b}]$ is all zeros, it represents a trivial equation for which every vector in $I\!\!R^n$ is the solution. Such equations are irrelevant, and such rows can be ignored. Therefore, the first step in analyzing an augmented matrix with rows that are all zeros is to cross these rows out.

Proof: Let $j(i)$ be the position of the first non–zero entry in the ith row of $[A|\mathbf{b}]$, as above.

If $j(i) = n+1$ for any i, then all the entries in the ith row are zero except the $(n+1)$st, then the ith row represents the impossible equation

$$0 \cdot \mathbf{x} = b$$

with $b \neq 0$, where b is the $(n+1)$st entry in this row. In this case, the system has no solution.

But if $j(i) \leq n$, then by (3.13), $j_p(i) = j(i)$, and the pivotal variable for the ith equation is $x_{j(i)}$. When $j_p(i) = j(i)$ for each i, the condition (4.13) in the definition of row reduced for is the same as the condition (4.15) in the definition of disentangled. So if there are no pivots in the final column, then $[A|\mathbf{b}]$ being row reduced corresponds exactly to $A\mathbf{x} = \mathbf{b}$ being disentangled. ∎

Example 4 Consider the following matrices:

$$A = \begin{bmatrix} 1 & 6 & 5 & 1 \\ 2 & 1 & 0 & 3 \\ 0 & 0 & 1 & 0 \\ 0 & 0 & 0 & 1 \end{bmatrix} \quad B = \begin{bmatrix} 1 & 3 & 1 & 2 \\ 0 & 1 & 4 & 1 \\ 0 & 0 & 1 & 1 \\ 0 & 0 & 0 & 0 \end{bmatrix} \quad C = \begin{bmatrix} 1 & 0 & 2 & 0 \\ 0 & 1 & 1 & 1 \\ 0 & 0 & 0 & 1 \end{bmatrix}$$

For the matrix A, $j(1) = 1$, $j(2) = 1$, $j(3) = 3$ and $j(4) = 4$. Since it is not the case that $j(2) > j(1)$, A is not in row reduced form, and considered as an augmented matrix, A does not represent a disentangled system.

For the matrix B, $j(1) = 1$, $j(2) = 2$, $j(3) = 3$, and $j(4)$ is undefined since this row is all zeros. Ignoring the third row, conditions (1) and (2) of Theorem 3 are both satisfied, and so B is in row reduced form, and considered as an augmented matrix, B does represent a disentangled system.

For the matrix C, $j(1) = 1$, $j(2) = 2$, and $j(3) = 4$. Since $j(1) < j(2) < j(3)$, C is in row reduced form, but there is a pivot in the final column so considered as an augmented matrix, C does not represent a disentangled system. In fact, C represents a system with no solution at all.

From these examples you see that it is straightforward to decide whether an augmented matrix represents a disentangled system or not.

The row reduction algorithm

Here is the key point: We can start from *any* system of m linear equations, write down its augmented matrix, and then using fewer than $m(m + 1)/2$ of the row operations described in Theorem 2, we can transform it into an equivalent augmented matrix that is in row reduced form. Then if there is a pivot in the last row, the solution set is empty. Otherwise the system is now disentangled, and back substitution will provide us with a one–to–one parameterization of the solution set.

The proof is given in some detail. We are careful to count the number of steps it might use. We do not just say that the procedure always terminates in success, we *calculate* an upper bound on the number of steps it could take to do this in the worst case.

This is a crucial issue in the analysis of algorithms. If you are going to write a program implementing one, you certainly want to be sure that you don't write an infinite loop. A proof that the procedure always terminates in a finite number of steps provides that. But if you have a formula for the number of steps it can take the algorithm to terminate in the worst case, then you do more; you can say something useful about the "run time" of the program. So the following *proof* is an example of an type of *calculation* that is very important in applications.

Theorem 4 (Row Reduction) *Let $[A|\mathbf{b}]$ be an $m \times (n + 1)$ augmented matrix. Then there is an equivalent augmented matrix $[\tilde{A}|\tilde{\mathbf{b}}]$ in row reduced form that can be obtained from $[A|\mathbf{b}]$ through a sequence of at most $m(m + 1)/2$ elementary row operations of the type listed in Theorem 2.*

Proof: We suppose the matrix isn't all zeros, or there is nothing to do.

Now go to the first column from the left that has a non-zero entry in it; there is such a column by the initial hypothesis. Call this the *first pivotal column*. If the entry in this column in the first row is zero, swap two rows to bring a non-zero entry in this column into the first row. *That's one row operation.* Now subtract appropriate multiple of the first row from all the rows beneath it so that you "clean out" the rest of the first pivotal column; that is, make every entry below the first row in the first pivotal column equal to zero. *There are at most $m - 1$ rows to deal with, so that's at most $m - 1$ more row operations at this stage.* The total number of row operations used so far is at most m. At this stage the matrix looks something like

$$\begin{bmatrix} 0 & \bullet & * & * & * & * \\ 0 & 0 & * & * & * & * \\ 0 & 0 & * & * & * & * \\ 0 & 0 & * & * & * & * \\ 0 & 0 & * & * & * & * \end{bmatrix}.$$

In the case represented above, the second column would be the first pivotal column. The bullet represents a non–zero entry, and the asterisks can be either zero or not.

If the matrix is all zeros below the first row we are done. Otherwise, go to the first column that contains a non–zero entry *below the first row*. Call this column the *second pivotal column*. Observe that the second pivotal column is necessarily to the right of the first. If the entry in this column in the second row is zero, swap two rows to bring a non-zero entry in this column into the second row. *That's one more row operation.* Now subtract appropriate multiple of the second row from all the rows beneath (but not above) it so that you "clean out" the rest of the second pivotal column as before. *There are $m - 2$ rows to deal with, so that's at most $m - 2$ more row operations at this stage.* Altogether we used at most $m - 1$ row operations dealing with the j_2nd column, and the running total of row operations used so far is

$$m + (m - 1) \, .$$

At this stage the matrix looks something like

$$\begin{bmatrix} 0 & \bullet & * & * & * & * \\ 0 & 0 & \bullet & * & * & * \\ 0 & 0 & 0 & * & * & * \\ 0 & 0 & 0 & * & * & * \\ 0 & 0 & 0 & * & * & * \end{bmatrix} \quad \text{or perhaps} \quad \begin{bmatrix} 0 & \bullet & * & * & * & * \\ 0 & 0 & 0 & \bullet & * & * \\ 0 & 0 & 0 & 0 & * & * \\ 0 & 0 & 0 & 0 & * & * \\ 0 & 0 & 0 & 0 & * & * \end{bmatrix} ,$$

among other possibilities, depending on how far to the right we have to go to find a non–zero entry below the first row.

If the matrix is all zeros below the second row we are done. Otherwise, go to the first column that contains a non–zero entry below the second row. Call this the third pivotal column, and observe that it is necessarily to the right of the second pivotal column. If the entry in this column in the third row is zero, swap two rows to bring a non-zero entry in this column into the third row. *That's one more row operation.* Now subtract appropriate

2-41

multiple of the third row from all the rows beneath (but not above) it so that you "clean out" the rest of the j_3rd column as before. *There are at most $m-3$ rows to deal with, so that's at most $m-2$ more row operations at this stage.* Altogether we used at most $m-2$ row operations dealing with the j_3rd column, and the running total of row operations used so far is

$$m + (m-1) + (m-2) \; .$$

Now just repeat this procedure until you get to the bottom of the matrix, or run out of non-zero rows. This takes at most $m-1$ stages so the total row operation count is

$$m + (m-1) + (m-2) + \cdots + 2 = \frac{m(m+1)}{2} - 1 \; .$$

After we're done we have a produced a matrix that is in row reduced form. And since we just used elementary row operations to effect the transformation, the resulting augmented matrix is equivalent to the one from which we started. ∎

The procedure described in the proof of Theorem 4 is often called *row reduction* or *the row reduction algorithm.*

Row reduction gives us an explicit procedure – an algorithm – for finding the solution set of any linear system of equations:

(1) First, write down the corresponding augmented matrix.

(2) Apply the procedure described in the proof of the Row Reduction Theorem to reduce the augmented matrix to row reduced form.

(3) Now convert back to a system, and solve it by back substitution.

We close this section with one more example.

Example 5 Let's put everything together, and find the solution set to the system

$$\begin{aligned}
x_2 - x_3 + 2x_4 &= 1 \\
x_1 + 2x_2 + x_3 - x_4 &= 3 \\
x_1 - x_2 + x_4 &= -2 \; .
\end{aligned} \tag{4.16}$$

The corresponding augmented matrix is

$$\left[\begin{array}{cccc|c}
0 & 1 & -1 & 2 & 1 \\
1 & 2 & 1 & -1 & 3 \\
1 & -1 & 0 & 1 & -2
\end{array} \right] .$$

The first column contains non-zero entries, but not in the first row. So we swap rows one and two to get

$$\left[\begin{array}{cccc|c}
1 & 2 & 1 & -1 & 3 \\
0 & 1 & -1 & 2 & 1 \\
1 & -1 & 0 & 1 & -2
\end{array} \right] .$$

(We could just as well have swapped rows one and three, since that too would have brought a non–zero entry into the first row). Now subtract the first row from the third (or add minus one times the first row to the third, if you like) to obtain

$$\left[\begin{array}{cccc|c} 1 & 2 & 1 & -1 & 3 \\ 0 & 1 & -1 & 2 & 1 \\ 0 & -3 & -1 & 2 & -5 \end{array}\right] .$$

We've just dealt with the first pivotal column; now for the second. The second column has nonzero entries below the first row, so it is the second pivotal column. It even already has a non–zero entry in the second row, so no swapping of rows is required this time. We can add 3 times the second row to the third to obtain

$$\left[\begin{array}{cccc|c} 1 & 2 & 1 & -1 & 3 \\ 0 & 1 & -1 & 2 & 1 \\ 0 & 0 & 2 & -1 & 4 \end{array}\right] .$$

This represents the equivalent disentangled system

$$x_1 + 2x_2 + x_3 - x_4 = 3$$
$$x_2 - x_3 + 2x_4 = 1 \tag{4.17}$$
$$2x_3 - x_4 = 4 .$$

We set the non pivotal variable x_4 equal to a parameter t:

$$x_4 = t . \tag{4.18}$$

We substitute (4.18) into the last equation, and solve for x_3:

$$x_3 = (t+4)/2 . \tag{4.19}$$

We substitute (4.18) and (4.19) into the second equation, and solve for x_2:

$$x_2 = 1 + (t+4)/2 - 2t . \tag{4.20}$$

We substitute (4.18), (4.19) and (4.20) into the first equation, and solve for x_1:

$$x_1 = 3 - 2(1 + (t+4)/2 - 2t) - (t+4)/2 + t . \tag{4.21}$$

Simplifying, we have

$$\begin{bmatrix} x_1 \\ x_2 \\ x_3 \\ x_4 \end{bmatrix} = \begin{bmatrix} -5 + (7/2)t \\ 3 - (3/2)t \\ 2 + t/2 \\ t \end{bmatrix} = \begin{bmatrix} -5 \\ 3 \\ 2 \\ 0 \end{bmatrix} + t \begin{bmatrix} 7/2 \\ -3/2 \\ 1/2 \\ 1 \end{bmatrix} .$$

Section 5: The Importance of Being Pivotal

We now have a general method for finding the solution set S of any linear system $A\mathbf{x} = \mathbf{b}$: Row reduction of $[A|\mathbf{b}]$ to obtain a disentangled ststem, followed by back substitution.

As we have seen, once $[A|\mathbf{b}]$ is row reduced, an easy visual inspection makes it clear whether or not the solution set is empty: The solution set is empty exactly when there is a pivot in the final column. Recall the generic "look" of a row reduced matrix, as explained in the previous section:

$$\begin{bmatrix} 0 & \bullet & * & * & * & * & * \\ 0 & 0 & \bullet & * & * & * & * \\ 0 & 0 & 0 & 0 & \bullet & * & * \\ 0 & 0 & 0 & 0 & 0 & \bullet & * \\ 0 & 0 & 0 & 0 & 0 & 0 & 0 \end{bmatrix}. \tag{5.1}$$

The bullets represent pivots, and there is one for each row that is not all zeros. Here there is no pivot in the final column, so the solution set is not empty.

At this point, we know how to tell whether or not $A\mathbf{x} = \mathbf{b}$ has solutions. Let's move on to the next question: Supposing $A\mathbf{x} = \mathbf{b}$ does have solutions, is there only one, or are there more than one?

How many solution does $A\mathbf{x} = \mathbf{b}$ have?

Look at (5.1), thought of as the row reduced form of $[A|b]$ for some system $A\mathbf{x} = \mathbf{b}$. There is no pivot in the final column, so the solution set is not empty. *How many solutions are there?*

Here is the key observation: There is exactly one pivotal column for each row that is not all zeros. There are four such rows, corresponding to four equations in six variables.

Of the six varibles, four will be pivotal, since there are four pivots. In the usual notation, the pivotal variables would be x_2, x_3, x_5 and x_6. The variables x_1 and x_4 are non–pivotal. This means that in the back substitution procedure, x_1 and x_4 become parameters t_1 and t_2. The solution will have the form

$$\mathbf{x}_0 + t_1 \mathbf{v}_1 + t_2 \mathbf{v}_2$$

and so there will be infinitely many solutions. Notice that to figure this out, we didn't need specific numbers; we just needed to know the positions of the pivots. This leads directly to the following theorem:

Theorem 1 (Pivots and the Solution Set) *Let A be and $m \times n$ matrix, and let $[A|\mathbf{b}]$ be the augmented matrix corresponding to $A\mathbf{x} = \mathbf{b}$. Suppose it row reduces to $[\tilde{A}|\tilde{\mathbf{b}}]$, and there are r pivots in $[\tilde{A}|\tilde{\mathbf{b}}]$. Then $r \leq m$ and $r \leq n + 1$, and when there is no pivot in the final column, $r \leq n$. Furthermore*

(1) If the final column of $[\tilde{A}|\tilde{\mathbf{b}}]$ is pivotal there is no solution to $A\mathbf{x} = \mathbf{b}$.

(2) If the final column of $[\tilde{A}|\tilde{\mathbf{b}}]$ is not pivotal, and $r = n$, then there are no non–pivotal variables, and $A\mathbf{x} = \mathbf{b}$ has a unique solution.

(3) If the final column of $[\tilde{A}|\tilde{\mathbf{b}}]$ is not pivotal, and $r < n$, then there are $n - r$ non–pivotal variables, and $A\mathbf{x} = \mathbf{b}$ has infinitely many solutions.

Proof: First, there is exactly one pivot in each non–zero row, as in (5.1), so $r \leq m$. Also, since $[\tilde{A}|\tilde{\mathbf{b}}]$ is in row reduced form, each pivot is in a different column, as in (5.1), so there can't be more than $n + 1$ pivots. If no pivot is in the last column, there can be no more than n of them.

Next, we already know (1). For (2) and (3), if the final column is not pivotal, all r pivots are in the columns of $\tilde{A}$ itself. Each of these columns corresponds to one of the variables. Hence there are exactly r pivotal variables and $n - r$ non–pivotal variables. When we do the back substitution, any non–pivotal variables will become parameters. If there are any parameters, there are infinitely many solutions. ∎

Example 1 Consider the following systems of equations, written as augmented matrices:

$$[A|\mathbf{a}] = \begin{bmatrix} 1 & 2 & 3 & | & 4 \\ 4 & 3 & 2 & | & 1 \end{bmatrix} \qquad [B|\mathbf{b}] = \begin{bmatrix} 1 & 2 & 3 & | & 4 \\ 4 & 3 & 2 & | & 1 \\ 5 & 5 & 5 & | & 6 \end{bmatrix} \qquad [C|\mathbf{c}] = \begin{bmatrix} 1 & 2 & 3 & | & 4 \\ 4 & 3 & 2 & | & 1 \\ 1 & 2 & 4 & | & 5 \end{bmatrix}$$

Which, if any, have no solutions? Unique solutions? Infinitely many solutions?

To answer these questions, row reduce. Starting with $[A|\mathbf{a}]$, one row operation gives us

$$\begin{bmatrix} 1 & 2 & 3 & | & 4 \\ 0 & -5 & -10 & | & -15 \end{bmatrix}.$$

There are two pivots, niether in the last column, and three variables. So there is a one paramter family of solutions to $A\mathbf{x} = \mathbf{a}$.

Next, $[B|\mathbf{b}]$ row reduces to

$$\begin{bmatrix} 1 & 2 & 3 & | & 4 \\ 0 & -5 & -10 & | & -15 \\ 0 & 0 & 0 & | & 1 \end{bmatrix}.$$

There is a pivot in the final column, so $B\mathbf{x} = \mathbf{b}$ has no solution.

Finally, $[C|\mathbf{c}]$ row reduces to

$$\begin{bmatrix} 1 & 2 & 3 & | & 4 \\ 0 & -5 & -10 & | & -15 \\ 0 & 0 & 1 & | & 1 \end{bmatrix}.$$

There is one pivot for each variable, none in the final column. So $C\mathbf{x} = \mathbf{c}$ has a unique solution.

Theorem 1 enables us to see the validity of a result we "borrowed" back in the first chapter. Recall that in our discussion of why only square matrices can have inverses, we asserted that whenever a matrix A has more columns than rows, $A\mathbf{x} = 0$ has a *non–zero* solution.

Theorem 2 (More Columns than Rows) *If A is an $m \times n$ matrix with $m < n$, then $A\mathbf{x} = 0$ has infinitely many solutions, and, in particular, non–zero solutions.*

Proof: The equation $A\mathbf{x} = 0$ *always* has at least one solution, namely $\mathbf{x} = 0$, so the last column cannot be pivotal, and there will be r pivotal variables, where r is the number of pivots after row reduction. By Theroem 1, $r \leq m$, and since $m < n$, we are in case (3) of the Theorem 1. ∎

Example 2 We can see Theorem 2 that the system given by $\begin{bmatrix} 1 & 2 & 3 & | & 4 \\ 4 & 3 & 2 & | & 1 \end{bmatrix}$ cannot have a unique solution. But Theorem 2 still leaves two possiblities open. While this ystem does have a one parameter family of solutions, as we saw in Example 1, the system

$$\begin{bmatrix} 1 & 2 & 3 & | & 4 \\ 1 & 2 & 3 & | & 1 \end{bmatrix}$$

doesn't have any solutions at all. In fact, it corresponds to the equations $x_1 + 2x_2 + 3x_3 = 4$ and $x_1 + 2x_2 + 3x_3 = 1$. There is no way $x_1 + 2x_2 + 3x_3$ can equal two different things at once. So there are two ways a matrix can fail to have unique solutions: It can have infinitely many, or it can have none. Theorem 2 doesn't tells us which case occurs. To decide that, row reduce, and use Theorem 1. Still, Theorem 2 does have its uses, as we've seen.

Theorem 2 summarizes about all that can be said about the solution set of $A\mathbf{x} = b$ without doing the row reduction of $[A|\mathbf{b}]$. It is tempting to think that one could say more just on the basis of comparing the numbers of variables and equations. For example, if A is an $n \times n$ matrix, then we have exactly as many equations as variables in $A\mathbf{x} = \mathbf{b}$. One might hope that this means that $A\mathbf{x} = \mathbf{b}$ will always have a unique solution. But this is wrong, as you see from $[B|\mathbf{b}]$ in Example 1. If you want to know how many solutions there are, you have to do the row reduction to find out where the pivots are. That brings us to a key question:

Row reduction ivolves arbitry choices. Does the number and placement of the pivots depend on these choices?

In Theorems 1 and 2, we are using row reduction to determine properties of a system $A\mathbf{x} = \mathbf{b}$ of linear equations, though in Theorem 2, row reduction only enters into the proof and not the statement. *The remarkable feature of this is that row reduction involves arbitrary choices.* Specifically, whenever it is necessary to swap rows, there are several ways to do it, and you could swap rows even when it isn't strictly speaking necessary. This will lead to different row reduced forms of $[A|\mathbf{b}]$. *How different can they be?*

We can see from Theorem 1 that which columns end up being pivotal, and in particular how many end up being pivotal cannot depend too much on the choices we make in doing the row reduction.

For example, suppose that one sequence of choices leads to a row reduced form $[\tilde{A}|\tilde{\mathbf{b}}]$ of $[A|\mathbf{b}]$ in which the final column is pivotal. By Theorem 1, we then know that $A\mathbf{x} = \mathbf{b}$ has no solutions. Since $A\mathbf{x} = \mathbf{b}$ either has solutions or it doesn't, as a matter of objective fact, it cannot be the case that any different sequence of choices in the row reduction of $[A|\mathbf{b}]$ could lead to a non–pivotal final column, for then Theorem 1 would say that $A\mathbf{x} = \mathbf{b}$ does have solutions.

• *So if one way of doing the row reduction leads to a pivotal final column, they all do, and if one does not, none of them do.*

Taking this line of reasoning a bit further leads to the following theorem:

Theorem 3 (Invariance of the Pivotal Columns) *Let $[A|\mathbf{b}]$ be an $m \times (n+1)$ augmented matrix. Then any two ways of row reducing it produce the same number of pivotal columns in the exact same places. Moreover, the number and position of pivotal columns among the first n columns is independent of what $\mathbf{b}$ is.*

Example 3 Consider the matrix

$$[A|\mathbf{b}] = \begin{bmatrix} 1 & 2 & 3 & | & 4 \\ 2 & 2 & 2 & | & 2 \end{bmatrix}.$$

If we subtract twice the first row from the second, we get

$$\begin{bmatrix} 1 & 2 & 3 & | & 4 \\ 0 & -2 & -4 & | & -6 \end{bmatrix}.$$

This is in row reduced form, and the first two columns are the pivotal ones.

We could also do the row reduction differently. Go back to $[A|\mathbf{b}]$, and divide the second row through by 2, and then swap the rows. This gives

$$\begin{bmatrix} 1 & 1 & 1 & | & 1 \\ 1 & 2 & 3 & | & 4 \end{bmatrix}.$$

Now subtracting the first row from the second, we get

$$\begin{bmatrix} 1 & 1 & 1 & | & 1 \\ 0 & 1 & 2 & | & 3 \end{bmatrix}.$$

This is in row reduced form. It is different from what we got before, but notice that still the pivotal columns are the same. *The theorem says that this always happens.*

Proof: The idea of the proof is to reduce the matter to something we know, namely that any two ways of doing the row reduction lead to the same results as far as the last column being pivotal is concerned. To apply this to the other columns, for each $k = 1, 2, \ldots, n$, let $\mathbf{v}_k$ denote the kth column of A, and for $k = 2, 3, \ldots, n$, let

$$A_{k-1} = [\mathbf{v}_1, \mathbf{v}_2, \ldots, \mathbf{v}_{k-1}].$$

This is simply the matrix obtained from A by crossing out everything but the first $k-1$ columns. Now consider the augmented matrix

$$[A_{k-1}|\mathbf{v}_k] \ .$$

Notice that this is the $m \times k$ matrix obtained from $[A|\mathbf{b}]$ by crossing out everything but the first k columns. *For this reason, any sequence of row operations that row reduces $[A|\mathbf{b}]$ also row reduces $[A_{k-1}|\mathbf{v}_k]$ for each $k = 2, 3, \ldots, n$.* Moreover, this row reduction of $[A_{k-1}|\mathbf{v}_k]$ is exactly what we would get if we row reduced $[A|\mathbf{b}]$ (with the same sequence of row operations), and then crossed out everything but the first k columns. This is because the row operations do not mix up the columns.

Now we are home free: For $k \geq 2$, the kth column in any row reduction of $[A|\mathbf{b}]$ will be pivotal if and only if the final column of $[A_{k-1}|\mathbf{v}_k]$ is pivotal for the *same* row reduction; i.e., the one using the same sequence of row operations. But we know that whether or not the final column of $[A_{k-1}|\mathbf{v}_k]$ is pivotal or not does not depend on how the row reduction is done. It is pivotal exactly when

$$A_{k-1}\mathbf{x} = \mathbf{v}_k \tag{5.2}$$

has no solution. (Here $\mathbf{x}$ is in R^{k-1}). So whether the kth column in the row reduction of $[A|\mathbf{b}]$ ends up being pivotal or not does not depend on how the row reduction is done. Finally the first column will be pivotal exactly when $\mathbf{v}_1 \neq 0$. This proves the first part.

To prove the second part, simply notice that for $k = 2, 3, \ldots, n$, whether or not the kth column will be pivotal or not depends only on whether (5.2) is solvable, and this has nothing to do with $\mathbf{b}$. Likewise, the condition for the first column to be pivotal has nothing to do with $\mathbf{b}$. ∎

Theorem 3 tells us that it makes sense to speak of "the pivotal columns of $[A|\mathbf{b}]$", instead of just the "pivotal columns of a row reduction of $[A|\mathbf{b}]$" because whether or not a column ends up being pivotal or not doesn't depend on how the row reduction is done. It just depends on $[A|\mathbf{b}]$ itself. Regarding A as a submatrix of $[A|\mathbf{b}]$, as in the proof, the same applies to A: No matter how you row reduce A, the same columns end up being pivotal. So which columns are pivotal and which are not has a definite meaning. From now on, we'll use the term "pivotal column" in this way, without reference to any particular row reduction. Therefore, Theorem 3 allows us to make the following definition:

Definition (Rank of a Matrix) The *rank r* of an $m \times n$ matrix A is the number of pivotal columns in A.

Theorem 1 tells us that the number and placement of pivots among the columns of $[A|\mathbf{b}]$ tells us how many solutions there are of $A\mathbf{x} = \mathbf{b}$. Since the rank tells us how many pivotal columns are among the the first n columns, we can restate part of Theorem 1 in terms of the rank as follows:

Theorem 4 (Existence, Uniqueness and Rank) *Let A be an $m \times n$ matrix with rank r. Then:*

(1) The solutions of $Ax = b$ are unique whenever they exist if and only if $r = n$.

(2) The equation $Ax = b$ is solvable for every b in R^m if and only if $r = m$.

(3) If A is a sqaure $n \times n$, then A is invertible if and only if $r = n$.

Corollary *If A is a square matrix that has a left or right inverse B, then A is invertible, and B is its inverse.*

Proof of the Corollary: If $AB = I$, then for any b, $A(Bb) = b$, so that $Ac = b$ always has a solution, namely Bb. Part (2) of Theorem 4 now implies that A has full rank and is invertible. Likewise, if $BA = I$, then $x = B(Ax)$ so $Ax = 0$ means $x = 0$. Part (1) of Theorem 4 now implies that A has full rank and is invertible. As we've seen in the last chapter, when A is invertible, any left or right inverse is the inverse. ■

The corrolary tells us something very nice: When looking for the inverse of a square matrix A, we just need to look for a matrix B with $AB = I$ (or $BA = I$). *Then $BA = I$ (or $AB = I$) follows for free.*

Proof of Theorem 4: By Theorem 1, $r \leq n$. If $r = n$, there are no parameters and solutions are unique when they exist. If $r < n$, there are paramters, and solutons are not unique when they exist.

If $r = m$, there are m pivotal columns in A itself. Since $[A|b]$ is $m \times n$, it can have at most m pivots, so there are none left for the last column. Hence the last column is never pivotal in this case, and $Ax = b$ always has a solution.

For the converse, suppose that $r < m$. Then at some point in the row reduction of $[A|0]$ we produce a row that is all zeros, since we get exactly one pivot for each non–zero row in the row reduction of $[A|0]$.

Let $[\tilde{A}|0]$ denote the augmented matrix at the stage of the row reduction where a row of all zeros has been produced. We may suppose, swapping two rows if need be that this is the mth row. Now let's modify $[\tilde{A}|0]$ by changing the final column to e_m. The mth row now corresponds to the impossible equation $0 \cdot x = 1$, and the modified augmented matrix $[\tilde{A}|e_m]$ corresponds to a system that has no solutions.

Now we can undo the row operations that led from $[A|0]$ to $[\tilde{A}|0]$ by using the inverses of these same row operations, which are also row operations, in the reverse order. If we apply this reversed sequence to $[\tilde{A}|e_m]$, the result is $[A|b]$ for some vector b. This is because the row operations act separately on each column, so since this sequence takes us back to A in the first n columns for $[\tilde{A}|0]$, it does so no matter what is in the final column. In particular it does so when the final column is e_m.

This does it: $[\tilde{A}|e_m]$ and $[A|b]$ are equivalent, since they are related by row operations. Then since $\tilde{A}x = e_m$ has no solutions, neither does $Ax = b$. This shows that if A does not have full row rank, there is a vector b for which $Ax = b$ has no solutions, and completes the proof of (2).

For (3), if A is $n \times n$ and its rank is n, by parts (1) and (2) $Ax = b$ has a unique solution for every b in R^n, so that A is invertible. On the other hand, if A is invertible, then $Ax = b$ has a unique solution for every b in R^n by *either* part (1) *or* (2), the rank of A is n. ■

The final part of Theorem 4 has many applications. Here is one:

Example 4 Let $P(s)$ be a polynomial of degree n in the variable s, so that

$$P(s) = a_0 + a_1 s + a_2 s^2 + \cdots + a_n s^n$$

for some coefficients $a_0, a_1, a_2, \ldots, a_n$. Suppose that we are given $n+1$ distinct points in the s, t plane:

$$(s_1, t_1), (s_2, t_2), \ldots, (s_{n+1}, t_{n+1}) \, .$$

The *polynomial fitting problem* is to choose the coefficients $a_0, a_1, a_2, \ldots, a_n$ so that the graph of $t = P(s)$ passes through each of these points. In other words, we want to choose the coefficients so that

$$P(s_i) = t_i \qquad \text{for} \quad i = 1, 2, \ldots, n+1 \, . \tag{5.3}$$

For this to happen, we must have $s_i \neq s_j$ whenever $i \neq j$. That is, no two of our points can lie on the same vertical line, because the graph of a function could never pass through both of these points. So we require that the s values are all different, but there is no restriction on the t values.

The system of equations (5.3) may not look linear at first, but remember that the variables we are trying to determine are the coefficients $a_0, a_1, a_2, \ldots, a_n$, and it is a linear system in these variables, even if $P(s)$ is a non–linear function of s. Indeed, let $\mathbf{x} = \begin{bmatrix} a_0 \\ a_1 \\ \vdots \\ a_n \end{bmatrix}$,

$$\mathbf{b} = \begin{bmatrix} t_1 \\ t_2 \\ \vdots \\ t_{n+1} \end{bmatrix}, \text{ and}$$

$$A = \begin{bmatrix} 1 & s_1 & s_1^2 & s_1^3 & \cdots & s_1^n \\ 1 & s_2 & s_2^2 & s_2^3 & \cdots & s_2^n \\ 1 & s_3 & s_3^2 & s_3^3 & \cdots & s_3^n \\ & \vdots & & & \vdots & \\ 1 & s_{n+1} & s_{n+1}^2 & s_{n+1}^3 & \cdots & s_{n+1}^n \end{bmatrix} \, .$$

Matrices of this form, in which each row consists of the zero through nth power of some set of numbers, are called *van der Monde* matrices. The name is Dutch for "from the moon", but you see that they come from the polynomial fitting problem since (5.3) is equivalent to $A\mathbf{x} = \mathbf{b}$.

We now apply Theorem 4 to show that the polynomial fitting problem always has a unique solution. To

Since A is square, Theorem 4 says that if $A\mathbf{x} = 0$ has only the zero solution, then A is invertible, and $A\mathbf{x} = \mathbf{b}$ always has a unique solution.

Hencew we focus our attention on $A\mathbf{x} = 0$, which in this case corresponds to

$$P(s_i) = 0 \quad \text{for} \quad i = 1, 2, \ldots, n+1 . \tag{5.4}$$

This would mean that each of the $n+1$ distinct numbers s_i are roots of $P(s)$. But a non–zero nth degree polynomial can have at most n distinct roots. (Use the Fundamental Theorem of algebra to factor $P(s)$). So the only way to achieve (5.4) is with $a_0 = a_1 = \ldots = a_n = 0$, which means $\mathbf{x} = 0$. That is, $A\mathbf{x} = 0$ has only the zero solution $\mathbf{x} = 0$. That does it:

• *Given any $n+1$ distinct numbers $s_1, s_2, \ldots, s_{n+1}$, and any $n+1$ numbers $t_1, t_2, \ldots, t_{n+1}$, there is one and only one way to choose the coefficients of an nth degree polynomial $P(s)$ so that the graph of $t = P(s)$ passes through each of the points (s_i, t_i), $i = 1, 2, \ldots, n$.*

Computing inverses

For the rest of this section, we focus on $n \times n$ matrices A with rank $r = n$. Theorem 4 says that such an A is invertible. How do we find A^{-1}?

We know A^{-1} exists. Suppose we write A^{-1} in terms of its columns as

$$A^{-1} = [\mathbf{v}_1, \mathbf{v}_2, \ldots, \mathbf{v}_n] .$$

If we can find each of the columns $\mathbf{v}_j$, we've found A^{-1}. To get an equation that $\mathbf{v}_j$ must satisfy, note that $AA^{-1} = I$ is the same as

$$[A\mathbf{v}_1, A\mathbf{v}_2, \ldots, A\mathbf{v}_n] = [\mathbf{e}_1, \mathbf{e}_2, \ldots, \mathbf{e}_n] .$$

That is, to find the jth column of A^{-1}, namely $\mathbf{v}_j$ all we need to do is to solve the equation

$$A\mathbf{v}_j = \mathbf{e}_j .$$

We can do this by row reduction! In fact, we can do it by row reduction alone, without any back substitution.

Example 5 Consider the Van der Monde matrix

$$A = \begin{bmatrix} 1 & 1 & 1 \\ 1 & 2 & 4 \\ 1 & 3 & 9 \end{bmatrix} .$$

This has rank 3, as you can check, and as we'll soon see. Let's solve $A\mathbf{v}_1 = \mathbf{e}_1$. The uaugmented matrix is

$$[A|\mathbf{e}_1] = \begin{bmatrix} 1 & 1 & 1 & | & 1 \\ 1 & 2 & 4 & | & 0 \\ 1 & 3 & 9 & | & 0 \end{bmatrix} .$$

Row reducing with no row swaps leads to

$$\begin{bmatrix} 1 & 1 & 1 & | & 1 \\ 0 & 1 & 3 & | & -1 \\ 0 & 0 & 2 & | & 1 \end{bmatrix} .$$

At this stage, we *could* proceed to find the unique solution by back substitution, which is the next step according to our general strategy. But in this case, we can keep going and obtain the unique solution by *further row reduction, and no back substitution.* Divide the last row through by 2 so that all of the pivots are 1:

$$\left[\begin{array}{ccc|c} 1 & 1 & 1 & 1 \\ 0 & 1 & 3 & -1 \\ 0 & 0 & 1 & 1/2 \end{array}\right].$$

now keep going with row operationsw, and "clean out" the entries above the pivots. First, subtract the second column for the first. This gives

$$\left[\begin{array}{ccc|c} 1 & 0 & -2 & 2 \\ 0 & 1 & 3 & -1 \\ 0 & 0 & 1 & 1/2 \end{array}\right].$$

Now subtract twice the third row from the second, and add twice the third row to the first. This cancels out everything above the last pivot and gives us

$$\left[\begin{array}{ccc|c} 1 & 0 & 0 & 3 \\ 0 & 1 & 0 & -5/2 \\ 0 & 0 & 1 & 1/2 \end{array}\right].$$

This corresponds to the system

$$I\mathbf{x} = \left[\begin{array}{c} 3 \\ -5/2 \\ 1/2 \end{array}\right]. \tag{5.5}$$

This equation solves itself! The unique solution is $\mathbf{v}_1$, and so the first column of A^{-1} is the vector on the right side of (5.5).

Notice that we solved this without back substitution – just by continuing until we had row reduced A to the identity matrix. Let's go through this schematicaly. We'll learn something interesting.

For *any* square matrix A, he augmented matrix corresponding to $A\mathbf{v}_j = \mathbf{e}_j$ is $[A|\mathbf{e}_j]$. If we row reduced it (and if A were 4×4, just to be concrete), we get the row reduced form:

$$\left[\begin{array}{cccc|c} \bullet & * & * & * & * \\ 0 & \bullet & * & * & * \\ 0 & 0 & \bullet & * & * \\ 0 & 0 & 0 & \bullet & * \end{array}\right],$$

where bullets denote non–zero entries, and asterisks denote entries that may or may not be zero. It looks like this becuase the rank equals the number of columns and rows, so we can only put the pivots along the diagonal, filling it up.

At this stage, we *could* proceed to find the unique solution by back substitution. But as in the example, we can keep going and obtain the unique solution by *further row reduction, and no back substitution*.

Multiply each row through by the inverse of its pivot. The pivot is non-zero, so you can do this. After this step, each pivots is 1. We don't need to use bullets to denote them any more – we know their values now. We have:

$$\begin{bmatrix} 1 & * & * & * & | & * \\ 0 & 1 & * & * & | & * \\ 0 & 0 & 1 & * & | & * \\ 0 & 0 & 0 & 1 & | & * \end{bmatrix}.$$

Now "clean out" the remaining entries in the pivotal columns. For example, whatever the first entry in the second column is, if we subtract this entry times the second row from the first, the result is

$$\begin{bmatrix} 1 & 0 & * & * & | & * \\ 0 & 1 & * & * & | & * \\ 0 & 0 & 1 & * & | & * \\ 0 & 0 & 0 & 1 & | & * \end{bmatrix}.$$

Continuing in this way, we arrive at a matrix of the form

$$\begin{bmatrix} 1 & 0 & 0 & 0 & | & * \\ 0 & 1 & 0 & 0 & | & * \\ 0 & 0 & 1 & 0 & | & * \\ 0 & 0 & 0 & 1 & | & * \end{bmatrix}.$$

If we let $\mathbf{c}$ denote the final column of this new matrix, it represents a very simple equation: $I\mathbf{x} = \mathbf{c}$. This equation solves itself; $\mathbf{x} = \mathbf{c}$ is the unique solution. Since we arrived at $I\mathbf{x} = \mathbf{c}$ starting from $A\mathbf{x} = \mathbf{e}_j$ by row reduction, the two equations are equivalent, and so $\mathbf{c}$ is the unique solution of $A\mathbf{x} = \mathbf{e}_j$. That is, $\mathbf{c} = \mathbf{v}_j$. We've got a method for finding $\mathbf{v}_j$ using row reduction alone!

Now we could do this for each of $\mathbf{v}_1, \mathbf{v}_2, \ldots, \mathbf{v}_n$, one at a time. But, we can be more efficient if we use the fact that the operations of row reduction act separately on each column of a matrix. Let's adjoin all n vectors $\mathbf{e}_j$ at once, forming the $n \times 2n$ matrix

$$[A|\mathbf{e}_1, \mathbf{e}_2, \ldots, \mathbf{e}_n]$$

which has the form

$$\begin{bmatrix} * & * & * & * & | & 1 & 0 & 0 & 0 \\ * & * & * & * & | & 0 & 1 & 0 & 0 \\ * & * & * & * & | & 0 & 0 & 1 & 0 \\ * & * & * & * & | & 0 & 0 & 0 & 1 \end{bmatrix}.$$

Since we are adjoining a copy of the identity matrix on the right, another way to describe this $n \times 2n$ matrix would be to write

$$[A|I] .$$

Row reducing this matrix to standard row reduced form yields

$$\begin{bmatrix} \bullet & * & * & * & | & * & * & * & * \\ 0 & \bullet & * & * & | & * & * & * & * \\ 0 & 0 & \bullet & * & | & * & * & * & * \\ 0 & 0 & 0 & \bullet & | & * & * & * & * \end{bmatrix} ,$$

provided A has full rank. (Otherwise, one or more of the entries marked by bullets would be zero.) In case A has full rank, we can proceed as above with more row operations to arrive at

$$\begin{bmatrix} 1 & 0 & 0 & 0 & | & * & * & * & * \\ 0 & 1 & 0 & 0 & | & * & * & * & * \\ 0 & 0 & 1 & 0 & | & * & * & * & * \\ 0 & 0 & 0 & 1 & | & * & * & * & * \end{bmatrix} , \tag{5.6}$$

The $(n+j)$th column of this matrix is the solution of $A\mathbf{x} = \mathbf{e}_j$, which is the jth column of A^{-1}. Therefore, the $n \times 2n$ matrix (5.6) is $[I|A^{-1}]$, and so the right half of this matrix is A^{-1}.

Example 6 Let's apply this method to compute the inverse of the Van der Monde matrix

$$A = \begin{bmatrix} 1 & 1 & 1 \\ 1 & 2 & 4 \\ 1 & 3 & 9 \end{bmatrix} .$$

We augment it with the identity, obtaining

$$\begin{bmatrix} 1 & 1 & 1 & | & 1 & 0 & 0 \\ 1 & 2 & 4 & | & 0 & 1 & 0 \\ 1 & 3 & 9 & | & 0 & 0 & 1 \end{bmatrix} .$$

Cleaning out the first column leads to

$$\begin{bmatrix} 1 & 1 & 1 & 1 & 0 & 0 \\ 0 & 1 & 3 & -1 & 1 & 0 \\ 0 & 2 & 8 & -1 & 0 & 1 \end{bmatrix} .$$

One more row operation takes us to row reduced form:

$$\begin{bmatrix} 1 & 1 & 1 & 1 & 0 & 0 \\ 0 & 1 & 3 & -1 & 1 & 0 \\ 0 & 0 & 2 & 1 & -2 & 1 \end{bmatrix} .$$

Dividing the final row through by 2 we get

$$\begin{bmatrix} 1 & 1 & 1 & 1 & 0 & 0 \\ 0 & 1 & 3 & -1 & 1 & 0 \\ 0 & 0 & 1 & 1/2 & -1 & 1/2 \end{bmatrix}.$$

Three more row operation clear out the upper right entries of the 3×3 part on the left, giving

$$\begin{bmatrix} 1 & 0 & 0 & 3 & -3 & 1 \\ 0 & 1 & 0 & -5/2 & 4 & -3/2 \\ 0 & 0 & 1 & 1/2 & -1 & 1/2 \end{bmatrix}.$$

This means that

$$A^{-1} = \frac{1}{2} \begin{bmatrix} 6 & -6 & 2 \\ -5 & 8 & -3 \\ 1 & -2 & 1 \end{bmatrix},$$

as you can easily check.

Section 6: The LU Factorization

Let A be an $m \times n$ matrix, and suppose that you need to solve $A\mathbf{x} = \mathbf{b}$ for many different vectors $\mathbf{b}$ in R^m, a situation that frequently arises in applications. If A happens to be an invertible square matrix, you could work out A^{-1}, and then deal with each new $\mathbf{b}$ by multiplying out $A^{-1}\mathbf{b}$. But what if A is not invertible, or not even square?

Applying our methods as they stand, you would have to row reduce $[A|\mathbf{b}]$, *starting from scratch*, for each different $\mathbf{b}$. This would quickly get repetitive. Indeed, since we row reduce from left to right, we can, and probably would, row reduce $[A|\mathbf{b}]$ using the exact same sequence of row operations to deal with the first n columns, no matter what $\mathbf{b}$ is. Always using this same sequence of row operations to deal with the first n columns, you would row reduce $[A|\mathbf{b}]$ to $[U|\mathbf{c}]$ where U would always be the same row reduced matrix, but where $\mathbf{c}$ would depend on what $\mathbf{b}$ was. Then you would go on from here, and finish the row reduction (there just one column to go), and find the solution set by back substitution, provided it is non-empty.

Doing things this way is certainly wasteful. We would keep spending a lot of effort recomputing U, which always turns out the same. All we need to compute when we are given a new $\mathbf{b}$ is how the new $\mathbf{c}$ depends on this new $\mathbf{b}$. Since row operations act column by column, if we have a record of the row operations we are using, we can use this record directly to transform $\mathbf{b}$ into $\mathbf{c}$ without wasting time recomputing U. So let's see how to keep a record of what the row operations are, and to use this to pass directly from $[A|\mathbf{b}]$ to $[U|\mathbf{c}]$ for each new $\mathbf{b}$. This is a very useful short–cut.

The key to doing this is to think of row operations as linear transformations. This makes sense because if we apply a row operation to a matrix A, it acts separately on each of the columns of A. For example, consider the operation of adding a times the ith row

to the jth row. If $\mathbf{v} = \begin{bmatrix} v_1 \\ \vdots \\ v_i \\ \vdots \\ v_j \\ \vdots \\ v_m \end{bmatrix}$ is *any* column of A, this row operation transforms $\mathbf{v}$ into

$\begin{bmatrix} v_1 \\ \vdots \\ v_i \\ \vdots \\ v_j + av_i \\ \vdots \\ v_m \end{bmatrix}$. Each entry of the output vector is a linear form in the entries of the input vector, so the transformation form R^m to R^m that assigns the output vector $\begin{bmatrix} v_1 \\ \vdots \\ v_i \\ \vdots \\ v_j + av_i \\ \vdots \\ v_m \end{bmatrix}$

to the input vector $\begin{bmatrix} v_1 \\ \vdots \\ v_i \\ \vdots \\ v_j \\ \vdots \\ v_m \end{bmatrix}$ is a linear transformation. Let R be the corresponding matrix.

Since applying the row operation to A is the same as applying this linear transformation to each of the columns of A, if $A = [\mathbf{v}_1, \mathbf{v}_2, \cdots, \mathbf{v}_n]$, then after the row transformation we have $[R\mathbf{v}_1, R\mathbf{v}_2, \cdots, R\mathbf{v}_n]$. But

$$[R\mathbf{v}_1, R\mathbf{v}_2, \cdots, R\mathbf{v}_n] = R[\mathbf{v}_1, \mathbf{v}_2, \cdots, \mathbf{v}_n] = RA \ . \tag{6.1}$$

That is, applying the row transformation to A is the same as multiplying on the left by R, and this is true no matter what A is. In particular, taking $A = I$, we see that $R = RI$ is the result of applying the row operation to I. For example, if the operation is to add twice the second row to the fourth row of a matrix with 5 rows, R would be given by

$$R = \begin{bmatrix} 1 & 0 & 0 & 0 & 0 \\ 0 & 1 & 0 & 0 & 0 \\ 0 & 0 & 1 & 0 & 0 \\ 0 & 2 & 0 & 1 & 0 \\ 0 & 0 & 0 & 0 & 1 \end{bmatrix} , \tag{6.2}$$

since that's what you get when you apply this operation to $I_{5\times 5}$. Each of the row operations can be undone by another row operation of the same type, so all of the matrices arising in this way are invertible. Indeed, to undo the operation we just described, you would *subtract* twice the second row from the fourth. This corresponds to the matrix

$$\begin{bmatrix} 1 & 0 & 0 & 0 & 0 \\ 0 & 1 & 0 & 0 & 0 \\ 0 & 0 & 1 & 0 & 0 \\ 0 & -2 & 0 & 1 & 0 \\ 0 & 0 & 0 & 0 & 1 \end{bmatrix} ,$$

which you can check is the inverse of (6.2). We have focused here on row operations of the "adding a multiple of one row to another" type, but after a bit of thought, it will be clear that each row operation is represented by a matrix R in this way, and R can be computed by applying the operation to the identity matrix, and R is always invertible, since any row operation can be undone by a row operation of the same type.

Now let A be any $m \times n$ matrix, and consider the augmented matrix $[A|I]$, where I denotes the $m \times m$ identity matrix. Let's begin row reducing this matrix left to right. Let R_1 be the $m \times m$ matrix corresponding to the first row operation that we use. Applying this row operation produces

$$[R_1 A | R_1 I] = [R_1 A | R_1]$$

since R acts separately on each of the columns of $[A|I]$, as in (6.1), and therefore separately on A and I. Let R_2 denote the second row operation we use. Applying this to what we have so far, namely $[R_1 A | R_1]$, results in $[R_2 R_1 A | R_2 R_1]$. Keep going in this way. After some number N of operations, we arrive at

$$[R_N \cdots R_2 R_1 A | R_N \cdots R_2 R_1]$$

where $U = R_N \cdots R_2 R_1 A$ is in row reduced form. (Since row reduced matrices "live in the upper right corner", U is the traditional symbol to use in this context). Let

$$R = R_N \cdots R_2 R_1 \ . \tag{6.3}$$

Then these N row operations $R_1, R_2, \ldots, R_N$ have transformed $[A|I]$ into $[U|R]$ where U is in row reduced form. The product R is the record that we are looking for: If $\mathbf{b}$ is any vector in R^m, and we apply our sequence of row operations $R_1, R_2, \ldots, R_N$, the result is

$$R_N \cdots R_2 R_1 [A|\mathbf{b}] = R[A|\mathbf{b}] = [RA | R\mathbf{b}] = [U | R\mathbf{b}] \ .$$

(Again, we are using the fact that when we multiply $[A|\mathbf{b}]$ on the left by R, R acts separately on each column, as in (6.1), and hence seperetely on A and $\mathbf{b}$). Therefore, with $\mathbf{c} = R\mathbf{b}$, $A\mathbf{x} = \mathbf{b}$ is equivalent to $U\mathbf{x} = \mathbf{c}$, and the latter equation is easy to deal with since U is in row reduced form.

The procedure for finding R is something familiar. Recall our method for computing inverses by row reduction: You write down the augmented matrix $[A|I]$, and apply row operations. The difference is that now A need not be square, and there is *even less work to do* since we stop as soon as we get a row reduced matrix U in the first n columns, instead of continuing until we get the $n \times n$ identity matrix in the first n columns. Let's summarize this new procedure for using row reduction to solve $A\mathbf{x} = \mathbf{b}$ in which $\mathbf{b}$ only enters after we've dealt with A:

The A First and b Later Procedure for Solving $A\mathbf{x} = \mathbf{b}$. *Let A be any $m \times n$ matrix, and write down $[A|I]$ where I is the $m \times m$ identity matrix. Apply row operation to transform this into $[U|R]$ where U is in row reduced form. Then for any $\mathbf{b}$ in R^m, $A\mathbf{x} = \mathbf{b}$ is equivalent to $U\mathbf{x} = \mathbf{c}$ where $\mathbf{c} = R\mathbf{b}$. Since U is in row reduced form, the latter equation is easily dealt with by back substitution.*

Example 1 Let $A = \begin{bmatrix} 1 & 1 & 1 \\ 2 & 4 & 3 \\ 3 & 7 & 6 \end{bmatrix}$, so that $[A|I] = \begin{bmatrix} 1 & 1 & 1 & 1 & 0 & 0 \\ 2 & 4 & 3 & 0 & 1 & 0 \\ 3 & 7 & 6 & 0 & 0 & 1 \end{bmatrix}$. Cleaning up the first column leads to, in two row operations,

$$\begin{bmatrix} 1 & 1 & 1 & 1 & 0 & 0 \\ 0 & 2 & 1 & -2 & 1 & 0 \\ 0 & 4 & 3 & -3 & 0 & 1 \end{bmatrix}.$$

Cleaning up the second column , in one more row operation, finishes the job:

$$\begin{bmatrix} 1 & 1 & 1 & 1 & 0 & 0 \\ 0 & 2 & 1 & -2 & 1 & 0 \\ 0 & 0 & 1 & 1 & -2 & 1 \end{bmatrix}. \tag{6.4}$$

Therefore, we have that

$$U = \begin{bmatrix} 1 & 1 & 1 \\ 0 & 2 & 1 \\ 0 & 0 & 1 \end{bmatrix} \quad \text{and} \quad R = \begin{bmatrix} 1 & 0 & 0 \\ -2 & 1 & 0 \\ 1 & -2 & 1 \end{bmatrix}.$$

Now having computed R and U, let's solve $A\mathbf{x} = \mathbf{b}$ for some $\mathbf{b}$ in R^3. For example, let's take $\mathbf{b} = \begin{bmatrix} 1 \\ 1 \\ 1 \end{bmatrix}$. Then $A\mathbf{x} = \mathbf{b}$ is equivalent to $U\mathbf{x} = \mathbf{c}$ where

$$\mathbf{c} = R\mathbf{b} = \begin{bmatrix} 1 & 0 & 0 \\ -2 & 1 & 0 \\ 1 & -2 & 1 \end{bmatrix} \begin{bmatrix} 1 \\ 1 \\ 1 \end{bmatrix} = \begin{bmatrix} 1 \\ -1 \\ 0 \end{bmatrix} \tag{6.5}$$

Next, $U\mathbf{x} = \mathbf{c}$ corresponds to the system of linear equations

$$\begin{aligned} x_1 + x_2 + x_3 &= 1 \\ 2x_2 + x_3 &= -1 \\ x_3 &= 0 . \end{aligned} \tag{6.6)}$$

Back substitution gives us $x_3 = 0$, $x_2 = -1/2$ and $x_1 = 3/2$, so that $\mathbf{x} = \frac{1}{2} \begin{bmatrix} 3 \\ -1 \\ 0 \end{bmatrix}$. As you can check, we do indeed have $A\mathbf{x} = \mathbf{b}$ for this $\mathbf{x}$.

Just for comparison, let's compute A^{-1}: If you keep on going from (6.4), in three more row operations you will find that

$$A^{-1} = \frac{1}{2} \begin{bmatrix} 3 & 1 & -1 \\ -3 & 3 & -1 \\ 2 & -4 & 2 \end{bmatrix} \,.$$

Notice that not only does it take more work to compute A^{-1} than to compute U and R, it takes more arithmetic operations to multiply out $A^{-1}\mathbf{b}$ than it does to *both* to multiply out $C\mathbf{b}$ *and* solve the system (6.6). This method is computationally effective even for nice, invertible square matrices.

The reason that it took fewer steps to both multiply out $R\mathbf{b} = \mathbf{c}$ and solve $U\mathbf{x} = \mathbf{c}$ than it took to multiply out $A^{-1}\mathbf{b}$ in this example is that A^{-1} is a "full" 3×3 matrix, with no special structure, but U is row reduced, and R has a very special structure, namely R is an example of a *unit lower triangular matrix*. Just looking at R, you can probably guess what the definition is going to be, but here it is:

Definition (Unit Lower Triangular Matrices) An $n \times n$ matrix L is called *lower triangular* in case $L_{i,j} = 0$ for $j > i$. L is called *unit lower triangular* it is lower triangular, and moreover, $L_{i,i} = 1$ for $i = 1, 2, \ldots, n$.

Unit lower triangular matrices look like

$$\begin{bmatrix} 1 & 0 & 0 & 0 & 0 \\ * & 1 & 0 & 0 & 0 \\ * & * & 1 & 0 & 0 \\ * & * & * & 1 & 0 \\ * & * & * & * & 1 \end{bmatrix} \,, \tag{6.7}$$

where, as usual, asterisks denote possibly non–zero entries.

Example 2 The simplest example of a unit lower triangular matrix is the identity. Another example is given by (6.2), which represents the operation of adding twice the second row to the fourth row in a matrix with five rows.

There is an important generalization of what we saw in Example 2, namely, the matrix R representing the operation of adding a times the ith row to the jth row with $i < j$ is always unit lower triangular. This is because R is just the identity matrix with the j, i entry changed from 0 to a. (Look at (6.2) and make sure you understand why it is the j, ith entry, and not the i, jth entry). Since the second index of this non–zero entry is less than the first, the matrix R is unit lower triangular.

This is half of the explanation for why the matrix R in Example 1 turned out to be unit lower triangular. Here is the other half: If L and K are two $n \times n$ unit lower triangular matrices, so is their product. To see this notice that

$$(LK)_{i,j} = \sum_{k=1}^{n} L_{i,k} K_{k,j} = \sum_{j \leq k \leq i} L_{i,k} K_{k,j} \,, \tag{6.8}$$

2-60

since $L_{i,k} = 0$ if $k > i$, and $K_{k,j} = 0$ if $j < k$. Since there is no k with $j \leq k \leq i$ if $i < j$, $(LK)_{i,j} = 0$ for $i < j$. Also from (6.8), thre is just one term, namely $k = i$, in the sum for $(LK)_{i,i}$. Therefore, $(LK)_{i,i} = L_{i,i}K_{i,i}, = 1$.

Now suppose we can row reduce a matrix A to row reduced form U without ever having to swap rows. This is quite often the case, and it happened in Example 1. Then the matrix R is the product

$$R = R_N R_{N-1} \cdots R_2 R_1$$

of matrices R_j, each of which represents the operation of subtracting a multiple of one row from another row *below* it. Therefore, each R_j is unit lower triangular, and so is R itself, being the product of unit lower triangular matrices. This explains not only why the R in Example 1 turned out to be unit lower triangular, but why this will frequently occur.

The fact that R will often be unit lower triangular is a good thing because solving equations involving unit lower triangular matrices is always easy. In fact, if L is any $n \times n$ unit lower triangular matrix, and $\mathbf{b}$ is any vector in R^n, then, as you see from (6.7), the first equation in the system corresponding too $LU = \mathbf{b}$ tells us what x_1 is. The second equation has the form $L_{2,1}x_1 + x_2 = b_2$, and since x_1 is now determined, this tells us what x_2 is. Continuing in this way, which is called *forward substitution*, one determines all of the variables, and sees that $L\mathbf{x} = \mathbf{b}$ always has a unique solution. Since this is true for *every* $\mathbf{b}$, the matrix L is invertible. We've learned quite a bit about unit lower triangular matrices in the preceeding paragraphs. Let's summarize it in a theorem:

Theorem 2 (Unit Lower Triangular Matrices)

(1) The product of any two $n \times n$ unit lower triangular matrices is again a unit lower triangular matrix.

(2) Every unit lower triangular matrix is invertible, and its inverse is again a unit lower triangular matrix.

(3) The matrix R representing the operation of subtracting a multiple of one row from another row below it is unit lower triangular.

Proof: We've already explained why all of these statements are true except that the inverse of a unit lower triangular matrix is unit lower triangular. To see this, consider how you would compute the inverse by row reduction of $[L|I]$. Since we can row reduce L to the identity by subtracting multiples of rows from rows beneath them, as is clear from (6.7), we can pass from $[L|I]$ to $[I|R]$ using only row operations of this type, with the result that R, which is L^{-1}, is unit lower triangular. ∎

In fact, not only are unit lower triangular matrices always invertible, but it is easy to compute the inverse even for relatively large matrices, using only subtraction and multiplication, and in particular, no division.

Example 3 Lets compute the inverse of

$$L = \begin{bmatrix} 1 & 0 & 0 & 0 \\ 2 & 1 & 0 & 0 \\ 3 & 5 & 1 & 0 \\ 4 & 6 & 7 & 1 \end{bmatrix}.$$

To do this, we write down

$$[L|I] = \begin{bmatrix} 1 & 0 & 0 & 0 & 1 & 0 & 0 & 0 \\ 2 & 1 & 0 & 0 & 0 & 1 & 0 & 0 \\ 3 & 5 & 1 & 0 & 0 & 0 & 1 & 0 \\ 4 & 6 & 7 & 1 & 0 & 0 & 0 & 1 \end{bmatrix}$$

and row reduce. Cleaning up the first column gives us

$$\begin{bmatrix} 1 & 0 & 0 & 0 & 1 & 0 & 0 & 0 \\ 0 & 1 & 0 & 0 & -2 & 1 & 0 & 0 \\ 0 & 5 & 1 & 0 & -3 & 0 & 1 & 0 \\ 0 & 6 & 7 & 1 & -4 & 0 & 0 & 1 \end{bmatrix}.$$

Cleaning up the second column gives us

$$\begin{bmatrix} 1 & 0 & 0 & 0 & 1 & 0 & 0 & 0 \\ 0 & 1 & 0 & 0 & -2 & 1 & 0 & 0 \\ 0 & 0 & 1 & 0 & 7 & -5 & 1 & 0 \\ 0 & 0 & 7 & 1 & 8 & -6 & 0 & 1 \end{bmatrix}.$$

Finally, cleaning up the third column gives us

$$\begin{bmatrix} 1 & 0 & 0 & 0 & 1 & 0 & 0 & 0 \\ 0 & 1 & 0 & 0 & -2 & 1 & 0 & 0 \\ 0 & 0 & 1 & 0 & 7 & -5 & 1 & 0 \\ 0 & 0 & 0 & 1 & -41 & 29 & -7 & 1 \end{bmatrix},$$

and we see that

$$L^{-1} = \begin{bmatrix} 1 & 0 & 0 & 0 \\ -2 & 1 & 0 & 0 \\ 7 & -5 & 1 & 0 \\ -41 & 29 & -7 & 1 \end{bmatrix}.$$

We've covered a lot of ground at this point, so let's summarize: If one is going to use row reduction to solve linear systems of equations, the "A first, b later" procedure is a very good way to go about it, and the matrices R and U with $RA = U$ that it produces are nice matrices: U is row reduced, and R is always invertible – often even unit lower triangular. Because of this, even when A is invertible, not only is it *less work* to compute R and U than it is to compute A^{-1}, it is *no more work* to multiply out $Rb = c$, and then to solve $Ux = c$ than it is to multiply out $A^{-1}b$. Better yet, the procedure works whether or not A invertible, or even square. Finally, it is all based on a very simple idea: keeping a record of the row operations needed to bring A into row reduced form. The matrix R is this record.

Therefore, in computer programs for doing linear algebra, you will find that what is built in for doing row reduction are methods for computing R and U given A.

Actually though, they don't give R, but, in most cases, R^{-1}. This may seem perverse at first, and in some sense it is, since $\mathbf{c} = R\mathbf{b}$ is what we want to compute, and if we are given R^{-1}, we have to *solve* $R^{-1}\mathbf{c} = \mathbf{b}$ to find $\mathbf{c}$. So why do they do this? There is a point of view from which this makes very good sense, and since it is helpful in many other contexts, and since most computer programs for dealing with linear algebra are built with the assumption that the user understands this point of view, it will be worth our while to examine it here.

Up to this point, when we've been solving systems of linear equations $A\mathbf{x} = \mathbf{b}$ by row reduction and back substitution, we haven't made much use of the fact that A represents a linear transformation f_A. Writing $A\mathbf{x} = \mathbf{b}$ as $f_A(\mathbf{x}) = \mathbf{b}$ suggests a different way of thinking about the system. It suggests the possiblity of writing f_A as the composition product of two or more "simple" transformations.

To see why this helps, consider something very familiar. Consider the single variable equation:

$$e^{x^2} = 2 . \tag{6.9}$$

You would solve this by first taking the natural logarithm of both sides, getting

$$x^2 = \ln(2) .$$

You would then take the square root of both sides, getting

$$x = \pm\sqrt{\ln(2)} .$$

The point is that the function $f(x) = e^{x^2}$ can be written as

$$f = g \circ h \quad \text{where} \quad g(y) = e^y \quad \text{and} \quad h(x) = x^2 .$$

the functions g and h are simple, familiar functions, and their inverses are included on any scientific calculator, so it is easy to deal with them one at a time. This "divide and conquer" strategy of factorization into easy pieces is so natural, you would probably just do it to solve (6.9) without explicitly thinking about it as involving a factorization. (The equation (6.9) is non-linear, but linear equations in one variable are too simple to be interesting, and the point we are making now about factorizations is quite general, and doesn't have anything particular to do with linearity).

The "A first, $\mathbf{b}$ later" version of row reduction gives us an invertible matrix R and a row reduced matrix U such that

$$RA = U . \tag{6.10}$$

Since R is invertible, if we let L denote R^{-1}, and multiply through on the left by L, we will have

$$A = LU . \tag{6.11}$$

If R was unit lower triangular, then by Theorem 2, L is also unit lower triangular. Therefore, we see that the "A first, $\mathbf{b}$ later" version of row reduction gives us a factorization of

A into a product of two nice matrices. When L is unit lower triangular, this factorization is called the LU factorization of A.

If you have an LU factorization of A, it is easy to solve $Ax = b$ for any b. Using the factorization, $Ax = b$ is the same as $L(Ux) = b$. So if we let

$$y = Ux , \qquad (6.12)$$

this becomes $Ly = b$. Since L is unit lower triangular, it is easy to solve this to find y. Once we have y, we can then solve (6.12) for x.

Example 4 Let A be the same as in Example 1, so that $A = \begin{bmatrix} 1 & 1 & 1 \\ 2 & 4 & 3 \\ 3 & 7 & 6 \end{bmatrix}$, and let's compute its LU decomposition.

We have already found R and U with $RA = U$ in Example 1. (Finding them now would be the first step if we hadn't). Next, we invert R to find L. This works out easily, even more easily than in Example 3, with the result that

$$L = \begin{bmatrix} 1 & 0 & 0 \\ 2 & 1 & 0 \\ 3 & 2 & 1 \end{bmatrix} \quad \text{and} \quad U = \begin{bmatrix} 1 & 1 & 1 \\ 0 & 2 & 1 \\ 0 & 0 & 1 \end{bmatrix} ,$$

and we have the factorization $A = LU$.

Now let's solve $Ax = b$ for $b = \begin{bmatrix} 1 \\ 1 \\ 1 \end{bmatrix}$ as in Example 1, but using the LU factorization this time.

Then the system corresponding to $Ly = b$ is

$$x_1 = 1$$
$$2x_1 + x_2 = 1 \qquad (6.13)$$
$$3x_1 + 2x_2 + x_3 = 1$$

Solving this, we find that the solution of $Ly = b$ is $y = \begin{bmatrix} 1 \\ -1 \\ 0 \end{bmatrix}$.

Solving $Ux = y$ is something we've already done in Example 1. (What we called c there is called y here). The equation $Ux = y$ corresponds to the system (6.6), and so $x = \frac{1}{2} \begin{bmatrix} 3 \\ -1 \\ 0 \end{bmatrix}$, just as we found in Example 1.

As you see, there is a close relation between Examples 1 and 4: In each of these we solved $A\mathbf{x} = \mathbf{b}$ for $A = \begin{bmatrix} 1 & 1 & 1 \\ 2 & 4 & 3 \\ 3 & 7 & 6 \end{bmatrix}$ and $\mathbf{b} = \begin{bmatrix} 1 \\ 1 \\ 1 \end{bmatrix}$. In Example 1, we did this from the point of view of the "A first, $\mathbf{b}$ later" version of row reduction, and in Example 4 using the LU factorization of A. The similarity between the two examples brings out the fact that the LU decomposition is just another way of talking about "A first, $\mathbf{b}$ later" version of row reduction. It is more complicated, since in this approach the "simplification through factorization" idea plays an important role, as well as the idea of "keeping a record of the row operations in a matrix R", which is the key idea behind the "A first, $\mathbf{b}$ later" procedure.

It is worth discussing this procedure in the context of the "simplification through factorization" idea because this idea is a main theme of linear algebra. In the following chapters we will see several such factorizations.

One thing we've left out of the discussion so far is what R, and hence L, will look like if we do have to swap rows in the row reduction of A.

Consider the operation of swapping rows i and j. Let P be the corresponding matrix. This is just the identity matrix, but with rows i and j interchanged. This matrix P is an example of a *permutation matrix*. A permutation matrix is a matrix that is obtained from the identity matrix by rearranging the order of the rows. By definition, we also include the original order, so that the identity matrix itself is a permutation. The matrix P is called a *pair permutation matrix* if it just has two rows of the identity interchanged.

Example 5 The matrix

$$P = \begin{bmatrix} 0 & 1 & 0 \\ 0 & 0 & 1 \\ 1 & 0 & 0 \end{bmatrix}$$

is the permutation martrix so that if A is a 3×3 matrix with $A = \begin{bmatrix} \mathbf{a}_1 \\ \mathbf{a}_2 \\ \mathbf{a}_3 \end{bmatrix}$, then

$$PA = \begin{bmatrix} \mathbf{a}_2 \\ \mathbf{a}_3 \\ \mathbf{a}_1 \end{bmatrix} .$$

To see this recall that the ith row of PA is a linear combinations of the rows of A with coefficients taken from the ith row of P. Since each row of P has just one non–zero entry, these linear combinations are easy to work out. And taking $A = I$, it is clear which row gets sent where.

Now notice that P has the form $P = [\mathbf{e}_3, \mathbf{e}_1, \mathbf{e}_2]$. Since the columns of P are orthonotmal, P is an isometry. It follows that

$$P^t P = I . \tag{6.14}$$

In fact, since P is square, P^t is the inverse of P, and not just a left inverse. (You should check explicity that $P^t P = I$, just to see it).

What we saw in this example is true in general. Any permuation matrix just has the standard basis vectors as its rows and columns, just in a different order. Hence it is always an isometery, and so whenever P is a permutation matrix, P^t is its inverse.

Now suppose we are row reducing $[A|I]$ to find R and U, and at some stage we need to swap two rows. Let P be the permutation matrix that does this. Form the matrix PA, and keep a record of P as well, and start over. If you can now row reduce $[PA|I]$ to the form $[U|R]$ without swapping rows, R will be unit lower triangular, and we have

$$R(PA) = U$$

with R unit lower triangular. Letting $L = R^{-1}$ as usual, we have

$$PA = LU$$

where P is a permutation matrix, L is unit lower triangular, and U is in row reduced form.

If you are so unlucky as to need to swap rows again in the row reduction of $[PA|I]$, let P_1 be the first pair permutation you used – keep records! – and let P_2 be the second. Let P be their product, which does both row swaps. Now start over with $[PA|I]$. Every time you swap rows, you get at least one more row "in the right place", so the worst possible case is that you would have to do this m times for an $m \times n$ matrix A. We have arrived at the following conclusion:

Theorem 3 (LU Factorization) *Every $m \times n$ matrix A can be written in the form*

$$PA = LU$$

where P is an $m \times m$ permutation matrix, L is a unit lower triangular matrix, and U is in row reduced form

If we have A written in this form, it is easy to solve $A = \mathbf{b}$: Multiply on the left by P to obtain $PA\mathbf{x} = P\mathbf{b}$. Let $\mathbf{c} = P\mathbf{b}$, and write this in the form $LU = \mathbf{c}$, which we solve in two easy steps, as before, in Example 4. Indd̄ed, if $PA = LU$, then

$$A\mathbf{x} = \mathbf{b} \iff PA\mathbf{x} = P\mathbf{b}$$

$$\iff LU\mathbf{x} = P\mathbf{b}$$

$$\iff U\mathbf{x} = L^{-1}P\mathbf{b}$$

Therefore:

• *Solving $U\mathbf{x} = L^{-1}(P\mathbf{b})$ is equivalent to solving $A\mathbf{x} = \mathbf{b}$, and it is always easy solve $U\mathbf{x} = L^{-1}(P\mathbf{b})$ since U is in row reduced form.*

If you swap rows only when it is strictly necessary, you aren't likely to do much of it in typical problems. Therefore, we won't worry too much here about all the "starting over". It isn't the most efficient thing to do, but unless you are going to compute a lot of $PA = LU$ factorizations by hand, or write a program to do so, neither of which is contemplated here, we won't go into the alternatives.

Problems for Chapter Two

Problems for Section 1

2.1.1 Find a one–to–one paramterization of the solution set in R^3 of the equation

$$2x_1 + 3x_2 - x_3 = 1 .$$

Also, write down the corresponding parameterization matrix.

2.1.2 Find a one–to–one paramterization of the solution set in R^3 of the equation

$$x_1 - 2x_2 + x_3 = 2 .$$

Also, write down the corresponding parameterization matrix.

2.1.3 Find a one–to–one paramterization of the solution set in R^4 of the equation

$$x_1 - 3x_2 + x_3 + x_4 = 0 .$$

Also, write down the corresponding parameterization matrix.

2.1.4 Find a one–to–one paramterization of the solution set in R^4 of the equation

$$3x_2 + x_3 = 0 .$$

Also, write down the corresponding parameterization matrix.

2.1.5 Consider the lines parameterized by

$$\begin{bmatrix} 1 \\ 0 \\ 1 \end{bmatrix} + s \begin{bmatrix} 2 \\ 3 \\ 1 \end{bmatrix} \quad \text{and} \quad \begin{bmatrix} -1 \\ -3 \\ 0 \end{bmatrix} + t \begin{bmatrix} 4 \\ 6 \\ 2 \end{bmatrix}$$

as s and t vary. Is this two different parameterizations of the same line, or are the lines different?

2.1.6 Consider the lines parameterized by

$$\begin{bmatrix} 1 \\ 2 \\ 2 \end{bmatrix} + s \begin{bmatrix} -1 \\ 1 \\ -3 \end{bmatrix} \quad \text{and} \quad \begin{bmatrix} 0 \\ 3 \\ -1 \end{bmatrix} + t \begin{bmatrix} 2 \\ -2 \\ 6 \end{bmatrix}$$

as s and t vary. Is this two different parameterizations of the same line, or are the lines different?

Problems for Section 2

2.2.1 Find the distance from $\mathbf{y} = \begin{bmatrix} 1 \\ 1 \\ 1 \end{bmatrix}$ to the plane given by

$$2x_1 + 3x_2 - x_3 = 1 .$$

Also find the vector in this plane that is closest to $\mathbf{y}$.

2.2.2 Find the distance from $\mathbf{y} = \begin{bmatrix} 1 \\ 0 \\ 2 \end{bmatrix}$ to the plane given by

$$x_1 - 2x_2 + x_3 = 2 .$$

Also find the vector in this plane that is closest to $\mathbf{y}$.

2.2.3 Find the orthogonal projection matrix P_S onto the solution set S of $\mathbf{a} \cdot \mathbf{x} = 0$ where $\mathbf{a} = \begin{bmatrix} 1 \\ 0 \\ 2 \end{bmatrix}$. Which vector in S is closest to $\mathbf{y} = \begin{bmatrix} 1 \\ 1 \\ 1 \end{bmatrix}$?

2.2.4 Find the orthogonal projection matrix P_S onto the solution set S of $\mathbf{a} \cdot \mathbf{x} = 0$ where $\mathbf{a} = \begin{bmatrix} 1 \\ -1 \\ 2 \end{bmatrix}$. Which vector in S is closest to $\mathbf{y} = \begin{bmatrix} 1 \\ 1 \\ 1 \end{bmatrix}$?

2.2.5 Consider the plane in $I\!\!R^3$ that passes through the three points

$$(1, -2, 3) \qquad (2, 0, 1) \qquad \text{and} \qquad (3, 3, 3) .$$

(a) Find a one–to–one parameterization of this plane.

(b) Find a linear equation whose solution set is this plane.

(c) Is the point (-1,1,3) on this plane?

2.2.6 Consider the plane in $I\!\!R^3$ that passes through the three points

$$(1, 1, 1) \qquad (2, 3, 4) \qquad \text{and} \qquad (4, 3, 2) .$$

(a) Find a one–to–one parameterization of this plane.

(b) Find a linear equation whose solution set is this plane.

(c) Is the point (1,1,1) on this plane?

Problems for Section 3

2.3.1 Give a one–to–one parameterization of the solution set of the following the system of equations:

$$x_1 + 2x_2 + x_3 = 0$$
$$x_2 + 2x_3 = 2$$

2.3.2 Give a one–to–one parameterization of the solution set of the following the system of equations:

$$2x_1 + x_2 - x_3 = 1$$
$$x_3 - 2x_2 = 1$$

2.3.3 Give a one–to–one parameterization of the solution set of the following the system of equations:

$$x_1 + 2x_2 + x_3 = 0$$
$$x_2 + 2x_3 = 2$$
$$x_3 = 3$$

2.3.4 Give a one–to–one parameterization of the solution set of the following the system of equations:

$$2x_1 + x_2 - x_3 = 1$$
$$x_3 - 2x_2 = 1$$
$$x_3 = 1$$

Problems for Section 4

2.4.1 Find the intersection of the lines given by

$$x_1 + 2x_2 = 5$$

and

$$2x_1 + x_2 = 5 \ .$$

2.4.2 Find the intersection of the lines given by

$$x_1 + 2x_2 = 5$$

and

$$2x_1 + 2x_2 = 5 \ .$$

2.4.3 Consider the system of equations

$$x + 2y + z = b$$

$$2x + y + 2z = 2$$

$$3x + 3y + az = 3$$

(a) For which values of a and b, if any, does this system have a unique solution? Give the solution for any such values of a and b.

(b) For which values of a and b, if any, does this system have no solution?

(c) For which values of a and b, if any, does this system have infinitely many solutions?

2.4.4 Consider the system of equations

$$x + z = 1$$

$$2x + ay + z = 1$$

$$2x - y = b$$

(a) For which values of a and b, if any, does this system have a unique solution? Give the solution for any such values of a and b.

(b) For which values of a and b, if any, does this system have no solution?

(c) For which values of a and b, if any, does this system have infinitely many solutions?

2.4.5 Consider the system of equations

$$x - 2y + az = 2$$

$$x + y + z = 0$$

$$3y + z = 2$$

(a) For which values of a and b, if any, does this system have a unique solution? Give the solution for any such values of a and b.

(b) For which values of a and b, if any, does this system have no solution?

(c) For which values of a and b, if any, does this system have infinitely many solutions?

2.4.6 Consider the system of equations

$$2x + ay - z = 1$$

$$x + by + -\frac{1}{2}z = 2$$

$$x + 2y + 3z = 0$$

(a) For which values of a and b, if any, does this system have a unique solution? Give the solution for any such values of a and b.

(b) For which values of a and b, if any, does this system have no solution?

(c) For which values of a and b, if any, does this system have infinitely many solutions?

2.4.7 Consider the system of equations

$$x - 2y + -z = 3$$
$$2x + y + 3z = a$$
$$3x + ay - z = 5$$

(a) For which values of a, if any, does this system have a unique solution? Give the solution for any such values of a and b.

(b) For which values of a, if any, does this system have no solution?

(c) For which values of a, if any, does this system have infinitely many solutions?

2.4.8 Consider the following systems:

$$(a) \qquad 2x - 3y - w = 2$$
$$y + z + w = 2$$
$$-x + 2y + z + w = 1$$
$$3x - 2y + w = 0$$

$$(b) \qquad x + 2y - z = 2$$
$$2x - y + 2z = 1$$
$$x + 2y = 0$$

Which of these systems has a unique solution – neither, both, just (a) or just (b)? In each case where there is a unique solution, find it.

2.4.9 Analyze the following systems by reducing them. Determine whether there is no solution, one solution, or are infinitely many solutions. If there is a unique solution, say what it is.

a) $2x - y = 0$

$\qquad -x + 2y - z = 1$

$\qquad x - y + 3z - w = 0$

$\qquad y + 2w = 0$

b) $2x + 2y + z = 0$

$\qquad x + z = 1$

$\qquad x + 2y = 5$

Problems for Section 5

2.5.1 Compute the inverse of

$$B = \begin{bmatrix} 1 & 2 & 4 \\ 2 & 4 & 1 \\ 4 & 1 & 2 \end{bmatrix}$$

2.5.2 Compute the inverse of

$$B = \begin{bmatrix} 1 & 2 & 3 & 4 \\ 0 & 1 & 2 & 3 \\ 0 & 0 & 1 & 2 \\ 0 & 0 & 0 & 1 \end{bmatrix}$$

2.5.3 Compute the inverse of

$$B = \begin{bmatrix} 1 & 1 & -1 \\ 1 & -1 & 2 \\ 1 & 2 & 0 \end{bmatrix}$$

2.5.4 Compute the inverse of

$$B = \begin{bmatrix} 1 & 2 & -1 \\ 2 & -1 & 2 \\ 1 & 2 & 0 \end{bmatrix}$$

2.5.6 Let C be a 2 by 2 matrix such that

$$C \begin{bmatrix} 1 \\ 2 \end{bmatrix} = \begin{bmatrix} 2 \\ 1 \end{bmatrix} \quad \text{and} \quad C^2 \begin{bmatrix} 1 \\ 2 \end{bmatrix} = \begin{bmatrix} -1 \\ 1 \end{bmatrix}$$

What is C? (I.e., give the entries.)

2.5.7 Let C be a 3×3 matrix such that

$$C \begin{bmatrix} 1 \\ 0 \\ 1 \end{bmatrix} = \begin{bmatrix} 2 \\ 1 \\ 1 \end{bmatrix} \quad C \begin{bmatrix} 2 \\ 1 \\ 3 \end{bmatrix} = \begin{bmatrix} -1 \\ 1 \\ -2 \end{bmatrix} \quad \text{and} \quad C \begin{bmatrix} 3 \\ 0 \\ 1 \end{bmatrix} = \begin{bmatrix} 0 \\ -1 \\ 1 \end{bmatrix} .$$

Find the inverse of C.

Problems for Section 6

2.6.1 For the matrix A, where

$$A = \begin{bmatrix} 1 & 2 & 4 \\ 2 & 5 & 1 \\ 1 & 1 & 1 \end{bmatrix}$$

find an invertible matrix R and a row reduced matrix U so that $RA = U$. Also, compute $R\mathbf{b}$ where $\mathbf{b} = \begin{bmatrix} 1 \\ 1 \\ 1 \end{bmatrix}$, and solve the equation $U\mathbf{x} = R\mathbf{b}$. Finally, solve $A\mathbf{x} = \mathbf{b}$.

2.6.2 For the matrix A, where

$$A = \begin{bmatrix} 1 & 2 & -1 \\ 2 & -1 & 2 \\ 1 & 2 & 0 \end{bmatrix}$$

find an invertible matrix R and a row reduced matrix U so that $RA = U$. Also, compute $R\mathbf{b}$ where $\mathbf{b} = \begin{bmatrix} 1 \\ 0 \\ 2 \end{bmatrix}$, and solve the equation $U\mathbf{x} = R\mathbf{b}$. Finally, solve $A\mathbf{x} = \mathbf{b}$.

2.6.3 For the matrix A, where

$$A = \begin{bmatrix} 1 & 2 & 4 \\ 2 & 5 & 1 \\ 1 & 1 & 1 \end{bmatrix}$$

find a unit lower triangular matrix L and a row reduced matrix U so that $A = LU$. Also, solve $L\mathbf{y} = \mathbf{b}$ where $\mathbf{b} = \begin{bmatrix} 1 \\ 1 \\ 1 \end{bmatrix}$, and solve the equation $U\mathbf{x} = \mathbf{y}$ for this $\mathbf{y}$. Finally, solve $A\mathbf{x} = \mathbf{b}$.

2.6.4 For the matrix A, where

$$A = \begin{bmatrix} 1 & 2 & -1 \\ 2 & -1 & 2 \\ 1 & 2 & 0 \end{bmatrix}$$

find a unit lower triangular matrix L and a row reduced matrix U so that $A = LU$. Also, solve $L\mathbf{y} = \mathbf{b}$ where $\mathbf{b} = \begin{bmatrix} 1 \\ 0 \\ 2 \end{bmatrix}$, and solve the equation $U\mathbf{x} = \mathbf{y}$ for this $\mathbf{y}$. Finally, solve $A\mathbf{x} = \mathbf{b}$.

2.6.5 For the matrix A, where

$$A = \begin{bmatrix} 1 & 2 & 4 \\ 2 & 4 & 1 \\ 4 & 1 & 2 \end{bmatrix}$$

find a permutation matrix P, a unit lower triangular matrix R and a row reduced matrix U so that $PA = LU$. Also, compute $R\mathbf{b}$ where $\mathbf{b} = \begin{bmatrix} 1 \\ 1 \\ 1 \end{bmatrix}$, and solve the equation $U\mathbf{x} = R\mathbf{b}$.

Finally, solve $A\mathbf{x} = \mathbf{b}$.

2.6.6 For the matrix A, where

$$A = \begin{bmatrix} 1 & 2 & 4 \\ -1 & -2 & 1 \\ -2 & -1 & 2 \end{bmatrix}$$

find a permutation matrix P, a unit lower triangular matrix R and a row reduced matrix U so that $PA = LU$. Also, compute $R\mathbf{b}$ where $\mathbf{b} = \begin{bmatrix} 1 \\ 1 \\ 1 \end{bmatrix}$, and solve the equation $U\mathbf{x} = R\mathbf{b}$.

Finally, solve $A\mathbf{x} = \mathbf{b}$.

2.6.7 (a) Explain why whenever P is a permutation matrix, then P is length preserving; i.e., $|P\mathbf{x}| = |\mathbf{x}|$ for all $\mathbf{x}$.

(b) Explain why whenever P is a permutation matrix, then P is invertible, and $P^{-1} = P^t$.

2.6.8 Let R be the matrix $R = \begin{bmatrix} 1 & 0 & 0 & 0 & 0 \\ 0 & 1 & 0 & 0 & 0 \\ 3 & 0 & 1 & 0 & 0 \\ 0 & 0 & 0 & 1 & 0 \\ 0 & 0 & 0 & 2 & 1 \end{bmatrix}$.

(a) Let A be any other matrix with 5 rows. Explain in words what multiplying A on the left by R does to the rows of A.

(b) Explain in words what you would do to undo the effect of multiplying A on the left by R.

(c) Without doing any computation, write down the inverse of R.

(d) Note that $(R^t A)^t = A^t((R^t)^t) = A^t R$. Let C be any matrix with 5 columns. Explain in words what multiplying C on the right by R does to the columns of C.

(e) How would your answers to (a), (b) and (c) change if the definition of R is changed to $\begin{bmatrix} 1 & 0 & 0 & 0 & 0 \\ 0 & 1 & 0 & 0 & 0 \\ 3 & 0 & 1 & 0 & 0 \\ 0 & 0 & 2 & 1 & 0 \\ 0 & 0 & 0 & 0 & 1 \end{bmatrix}$? What feature of this small change is responsible for the difference?

2.6.9 Let R be the matrix $R = \begin{bmatrix} 1 & 0 & 0 & 0 & 0 \\ 0 & 1 & 0 & 0 & 0 \\ 0 & 2 & 1 & 0 & 0 \\ 0 & 0 & 0 & 1 & 0 \\ 0 & 0 & 0 & 2 & 1 \end{bmatrix}$.

(a) Let A be any other matrix with 5 rows. Explain in words what multiplying A on the left by R does to the rows of A.

(b) Explain in words what you would do to undo the effect of multiplying A on the left by R.

(c) Without doing any computation, write down the inverse of R.

(d) Note that $(R^t A)^t = A^t((R^t)^t) = A^t R$. Let C be any matrix with 5 columns. Explain in words what multiplying C on the right by R does to the columns of C.

(e) How would your answers to (a), (b) and (c) change if the definition of R is changed to $\begin{bmatrix} 1 & 0 & 0 & 0 & 0 \\ 2 & 1 & 0 & 0 & 0 \\ 0 & 2 & 1 & 0 & 0 \\ 0 & 0 & 2 & 1 & 0 \\ 0 & 0 & 0 & 2 & 1 \end{bmatrix}$? What feature of this small change is responsible for the difference?

2.6.10 Suppose that A is an $m \times n$ matrix, and we have a factorization $A = QU$ where U is in row reduced form, and Q is a length preserving matrix. Explain why $\mathbf{x}$ is a solution of $A\mathbf{x} = \mathbf{b}$ if and only if $U\mathbf{x} = Q^t\mathbf{b}$. Note that the last equation is easy to deal with since U is in row reduced form. (The thing we are taking advantage of in this factorization is that length preserving matrices have a let inverse that requires no work to write down. In the next chapter we will develop a way of producing factorizations of this type.)

Overview of Chapter Three

In Chapter 2 we focused on developing a general method for determining the solution set of a linear system $A\mathbf{x} = \mathbf{b}$. There, A was a given $m \times n$ matrix, $\mathbf{b}$ was a given vector in $I\!R^m$, and the object was to find all vectors $\mathbf{x}$ such that $A\mathbf{x} = \mathbf{b}$.

Here we are concerned with a different issue. *We want to determine the set of all vectors* $\mathbf{b}$ *for which* $A\mathbf{x} = \mathbf{b}$ *has a solution.* This set is called the *image* of A, because it is the image of $I\!R^n$ in $I\!R^m$ under the linear transformation corresponding to A.

The relevance of this question to the problem of solving $A\mathbf{x} = \mathbf{b}$ lies in three facts, the first of which is just the definition of the image restated:

(1) $A\mathbf{x} = \mathbf{b}$ has a solution if and only if $\mathbf{b}$ lies in the image of A.

(2) When the image of A is not all of $I\!R^m$, it is only a thin slice of it, as we will see. For example, if A is 3×3, it might be that the image of A is a plane. (This is the picture we have in mind when we use the phrase "thin slice").

(3) Very often in practice, the vector of "given" data $\mathbf{b}$ is obtained through laboratory experiments. Even a slight amount of "noise" corrupting the data will usually knock $\mathbf{b}$ out of the image of A when it is just a slice through $I\!R^m$.

For example, if the image of A is a plane in $I\!R^3$, and you take an exact data vector $\mathbf{b}$ in this plane, but then add some "random noise" to each entry, you will almost surely move $\mathbf{b}$ off of the plane, and render the equation $A\mathbf{x} = \mathbf{b}$ unsolvable.

There is a very good way to deal with this problem, which is what this chapter is about. In learning to deal with it, we will also learn a considerable amount about linear transformations, geometry and matrices that is generally useful, and in our view, a number of beautiful ideas.

Section 1 of this chapter begins with a careful look at an example. It comes from electric circuit theory. (We assume nothing about circuit theory here; the mathematical issues are close to the surface in this problem, which is why we chose it.) The circuit in question is the one introduced in Example 2 of the very first section of Chapter One. The problem is to solve a system of equations that describes the "voltages to currents" transformation for this circuit in order to deduce the voltages from measurements of the currents. There are three currents, denoted I_1, I_2 and I_3, and a physical law says that

$$I_1 + I_2 + I_3 = 0 . \tag{0.1}$$

The equation (0.1) describes a plane in $I\!R^3$, and if the data does not lie in this plane, the equations relating the currents and the voltages have no solution.

With even slighly noisy data, there will be a conflict between laboratory data and (0.1), a law of Nature. Something has got to yield, and it isn't going to be Nature. We've got to adjust the data to "wash out the noise". The basic idea is that we want to find a vector $\mathbf{c}$ that is in the image, but is as close to $\mathbf{b}$ as possible. We consider this to be a "corrected" version of $\mathbf{b}$, and we then solve the corrected system $A\mathbf{x} = \mathbf{c}$ instead.

There is another way of looking at this. When we are trying to solve $A\mathbf{x} = \mathbf{b}$, we are looking for a vector $\mathbf{x}$ that makes $|A\mathbf{x} - \mathbf{b}| = 0$. It is usually easier to work with the squared distance since this gets rid of some square roots, so we can restate the problem considered in Chapter 2 as follows:

- *How do we find all of the vectors $\mathbf{x}$ that make $|A\mathbf{x} - \mathbf{b}|^2 = 0$?*

Our new goal, if the solution set of $A\mathbf{x} = \mathbf{b}$ is empty, will be to answer the question

- *How do we find all of the vectors $\mathbf{x}$ that make $|A\mathbf{x} - \mathbf{b}|^2$ as small as possible?*

Stating the two questions this way emphasizes their similarity. Moreover, we'll see that the second question can be answered by solving $A\mathbf{x} = \mathbf{c}$, where $\mathbf{c}$ is the "corrected" version of $\mathbf{b}$. Solutions of this second problem are called "least squares solutions".

The problem of finding least squares solutions is essentially geometric in character. Since it is a question about distances, geometry must play a significant role in any method for answering it. For this reason, geometry plays a much more central role in this chapter than the last one. This is not a burden: The geometry we will learn has many other applications, and brings a great deal of insight into the nature of the problems of linear algebra. Besides, it really is beautiful.

Section 1: Subspaces and the Images of Linear Transformations

Suppose we are given an $m \times n$ matrix A. In the previous chapter, we developed a general method for finding the solution set of $A\mathbf{x} = \mathbf{b}$ for any $\mathbf{b}$ in $\mathbb{R}^m$. One of our main concerns in this chapter is to describe the set of vectors $\mathbf{b}$ in $\mathbb{R}^m$ for which $A\mathbf{x} = \mathbf{b}$ has a solution. Let's contrast the question we answered in the previous chapter to the one we are raising now:

- *Given an $m \times n$ matrix A and a vector $\mathbf{b}$ in $\mathbb{R}^m$, what are all of the solutions of the equation $A\mathbf{x} = \mathbf{b}$?*

- *Given an $m \times n$ matrix A, what are all of the vectors $\mathbf{b}$ in $\mathbb{R}^m$ for which the equation $A\mathbf{x} = \mathbf{b}$ has a solution?*

At first sight, you might well think that the second question is less interesting, in practical terms at least, than the first one. In fact, one might well think that with a method for answering the first question in hand, we've got the subject all wrapped up. Yet the book goes on. Why is this?

There is more to linear algebra than solving equations like $A\mathbf{x} = \mathbf{b}$, despite the centrality of this issue, as we'll see. But even if that were our only concern, the question of determining the set of those vectors $\mathbf{b}$ in $\mathbb{R}^m$ for which $A\mathbf{x} = \mathbf{b}$ has a solution is a matter of practical importance.

Here is why: Consider an example from early on in the book. In Example 2 from the very first section we considered a simple electric circuit with two voltages V_1 and V_2 and three currents I_1, I_2 and I_3. These were related by*

$$\frac{1}{3R} \begin{bmatrix} -2 & -1 \\ 1 & 2 \\ 1 & -1 \end{bmatrix} \begin{bmatrix} V_1 \\ V_2 \end{bmatrix} = \begin{bmatrix} I_1 \\ I_2 \\ I_3 \end{bmatrix} . \tag{1.1}$$

Suppose that you have measured the three currents I_1, I_2 and I_3 in a laboratory experiment, and have obtained the data

$$I_1 = 1.02 \text{ amps} \quad , \quad I_2 = -2.04 \text{ amps} \quad \text{and} \quad I_3 = 0.99 \text{ amps} . \tag{1.2}$$

If you then wanted to know what were the voltages V_1 and V_2 that produced these currents, you might try to solve

$$\frac{1}{3R} \begin{bmatrix} -2 & -1 \\ 1 & 2 \\ 1 & -1 \end{bmatrix} \begin{bmatrix} V_1 \\ V_2 \end{bmatrix} = \begin{bmatrix} 1.02 \\ -2.04 \\ 0.99 \end{bmatrix} . \tag{1.3}$$

Unfortunately, this equation has no solutions so you can't possibly solve it to find V_1 and V_2. *What went wrong?*

---

* We didn't use the matrix notation back in the very first section, but now that you know it, you can easily translate the formulas from the Example 2 there.

There is nothing wrong with (1.1), which is a statement of physical law. In particular, as explained in Example 2 of the very first section, (1.1) incorporates

$$I_1 + I_2 + I_3 = 0 , \qquad (1.4)$$

one of Kirchhoff's rules. This expresses the conservation of electrical charge, which is an exact and absolute law of nature. Any solution of (1.1) will respect this law. Indeed, let

$\mathbf{v} = \begin{bmatrix} V_1 \\ V_2 \end{bmatrix}$, and let $A = \begin{bmatrix} -2 & -1 \\ 1 & 2 \\ 1 & -1 \end{bmatrix}$. Then we can rewrite (1.1) as

$$\begin{bmatrix} I_1 \\ I_2 \\ I_3 \end{bmatrix} = \frac{1}{3R} A\mathbf{v} . \qquad (1.5)$$

Now define $\mathbf{a}$ by $\mathbf{a} = \begin{bmatrix} 1 \\ 1 \\ 1 \end{bmatrix}$ so that

$$I_1 + I_2 + I_3 = \mathbf{a} \cdot \begin{bmatrix} I_1 \\ I_2 \\ I_3 \end{bmatrix} . \qquad (1.6)$$

Then combining (1.5) and (1.6), no matter what the voltage vector $\mathbf{v}$ is,

$$I_1 + I_2 + I_3 = \frac{1}{3R} \mathbf{a} \cdot (A\mathbf{v})$$
$$= \frac{1}{3R} (A^t \mathbf{a}) \cdot \mathbf{v}$$
$$= 0$$

since

$$A^t \mathbf{a} = 0 ,$$

as you can easily check. Please do this now; it amounts to checking that $\mathbf{a}$ is orthogonal to each of the rows of A^t, and therefore, to each of the columns of A.

• *The fact that $\mathbf{a}$ is orthogonal to each of the columns of A ensures that every vector in the image of A is orthogonal to $\mathbf{a}$, and hence that the image of A lies in the plane given by $\mathbf{a} \cdot \mathbf{x} = 0$.*

But for the laboratory data (1.2),

$$\mathbf{a} \cdot \begin{bmatrix} I_1 \\ I_2 \\ I_3 \end{bmatrix} = \mathbf{a} \cdot \begin{bmatrix} 1.02 \\ -2.04 \\ 0.99 \end{bmatrix} = -0.03 . \qquad (1.7)$$

Hence (1.4) is not satisfied, and so (1.3) has no solution.

Now we can see what went wrong. The condition (1.4) is an *exact* physical law, but laboratory data is *essentially never* exact. There is almost always some experimental error, some "noise" in the data. The vector $\begin{bmatrix} 1.02 \\ -2.04 \\ 0.99 \end{bmatrix}$ isn't the "right" right hand side of (1.3), just an approximate measurement of it.

We could go back and try to measure the currents more carefully, but there will always be some noise, so we'd better just figure out how to deal with it. There is a natural thing to do:

- *Replace the vector* $\begin{bmatrix} 1.02 \\ -2.04 \\ 0.99 \end{bmatrix}$ *by the closest vector to it that does satisfy (1.3), or equivalently, (1.4).*

Now (1.4) is the equation of a plane, and we know how to find the closest point in a plane to any given vector. As explained in Theorem 2.2.1 and the following examples, the point in the plane given by (1.4) that is closest to $\begin{bmatrix} 1.02 \\ -2.04 \\ 0.99 \end{bmatrix}$ is, since $|\mathbf{a}|^2 = 3$,

$$\begin{bmatrix} 1.02 \\ -2.04 \\ 0.99 \end{bmatrix} - \frac{1}{3}\left(\begin{bmatrix} 1.02 \\ -2.04 \\ 0.99 \end{bmatrix} \cdot \mathbf{a}\right)\mathbf{a} = \begin{bmatrix} 1.02 \\ -2.04 \\ 0.99 \end{bmatrix} + \begin{bmatrix} 0.01 \\ 0.01 \\ 0.01 \end{bmatrix} = \begin{bmatrix} 1.03 \\ -2.03 \\ 1.00 \end{bmatrix} .$$

The right hand side is the physically possible vector of currents that is closest to the vector of currents that were measured in the laboratory. To find the voltages, use this "corrected" experimental data as the right hand side in (1.1), and solve for V_1 and V_2. This time, there is no problem, and we find the unique solution*

$$V_1 = -0.01 \quad \text{and} \quad V_2 = -1.01 .$$

Notice what went on here. We had an equation $A\mathbf{x} = \mathbf{b}$ to solve that wasn't solvable because the right hand side $\mathbf{b}$ consisted of "noisy data". We corrected this by changing the right hand side from the given vector $\mathbf{b}$ into the nearest vector $\mathbf{c}$ for which $A\mathbf{x} = \mathbf{c}$

---

* These are the voltages that reproduce the measured currents as closely as possible, as explained above. That is as far as the mathematics itself takes us. However, if you had reason to believe that the voltages were set very accurately to integer values, you would go on from the least squares solution to conclude that $V_1 = 0$ and $V_2 = -1$. Our concern here will be with the mathematics that gets us to the least squares solution. Still, it is worth noting, if only in a footnote, that there is non–mathematic issue of interpretation that depends on the context of the problem.

does have a solution. This "washes out" the noise in our laboratory data, and allows us to proceed with the computation.

In this particular case, the set of "right hand sides" satisfying the condition (1.4), constituted a plane, and since we know how to find the closest vector in a plane, we had no problem computing the "corrected right hand side". In general though, if we are going to compute the closest vector in the set of vectors for which $A\mathbf{x} = \mathbf{b}$ has a solution, we need to know something about what this set is. *This is the practical motivation for investigating the nature of this set:* Knowledge of it will tell us how to "correct" faulty data that makes equations unsolvable. Let's now begin the systematic investigation:

Definition (Image of a Matrix A) Let A be an $m \times n$ matrix. The *image* of A is the set of vectors $\mathbf{b}$ in $I\!\!R^m$ that are of the form $\mathbf{b} = A\mathbf{x}$ for some $\mathbf{x}$ in $I\!\!R^n$. This subset of $I\!\!R^m$ is denoted by Img(A)

Our goal is to be able to answer the following question:

- *Given an $m \times n$ matrix A and a vector $\mathbf{b}$ in $I\!\!R^m$, is there a vector $\mathbf{c}$ in the* Img(A) *that is closest to $\mathbf{b}$, and if so, what is it?*

Before going further with the theory, let's investigate Img(A) itself in an example. We need to know what Img(A) is before we can go looking for closest vectors in it!

Example 1: Consider the matrix $A = \begin{bmatrix} 1 & -1 & 0 \\ 2 & 0 & 2 \\ 0 & 1 & 1 \end{bmatrix}$. Let $\mathbf{b} = \begin{bmatrix} x \\ y \\ z \end{bmatrix}$ denote an arbitrary

element of $I\!\!R^3$. To determine Img(A), form the augmented matrix corresponding to the equation $A\mathbf{x} = \mathbf{b}$. This is

$$[A|\mathbf{b}] = \begin{bmatrix} 1 & -1 & 0 & | & x \\ 2 & 0 & 2 & | & y \\ 0 & 1 & 1 & | & z \end{bmatrix} ,$$

and row reducing it to upper triangular form we find

$$\begin{bmatrix} 1 & -1 & 0 & | & x \\ 0 & 2 & 2 & | & y - 2x \\ 0 & 0 & 0 & | & z - (y/2) + x \end{bmatrix} .$$

Evidently there will be a solution if and only if the final entry in the last row is zero. That is, if

$$2x - y + 2z = 0 .$$

This is the equation of a plane through the origin in $I\!\!R^3$:

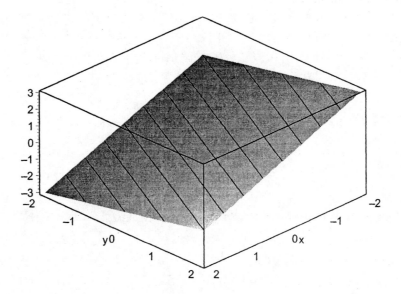

The lines drawn on the plane are lines of constant height z. Notice the different scale used for the z axes to adjust the proportions of the graph. This plane (imagine the patch extended indefinitely) is the set of points $\mathbf{b}$ in $I\!\!R^3$ for which the equation $A\mathbf{x} = \mathbf{b}$ has a solution. As you see, this is only a "slice" of $I\!\!R^3$ in this case!

In Example 1, $\text{Img}(A)$ turned out to be the solution set of a single linear equation. More generally, it will turn out to be the solution set of a system of linear equations. (In this example, there was just one equation in the system). The connection with solution sets is useful, since we know quite a lot about parameterizing solution sets. Let's do another example in more variables.

Example 2: Consider the matrix $A = \begin{bmatrix} 1 & 2 & 1 & 2 \\ 2 & 3 & 2 & 3 \\ 1 & 1 & 1 & 1 \\ 0 & 1 & 0 & 1 \end{bmatrix}$. Let $\mathbf{b} = \begin{bmatrix} b_1 \\ b_2 \\ b_3 \\ b_4 \end{bmatrix}$ denote an arbitrary

element of $I\!\!R^4$. (With four or more variables, it pays to use subscripts...) To determine $\text{Img}(A)$, form the augmented matrix corresponding to the equation $A\mathbf{x} = \mathbf{b}$. This is

$$[A|\mathbf{b}] = \begin{bmatrix} 1 & 2 & 1 & 2 & | & b_1 \\ 2 & 3 & 2 & 3 & | & b_2 \\ 1 & 1 & 1 & 1 & | & b_3 \\ 0 & 1 & 0 & 1 & | & b_4 \end{bmatrix}.$$

Row reducing this with no row swaps, we get

$$\left[\begin{array}{cccc|c} 1 & 2 & 1 & 2 & b_1 \\ 0 & -1 & 0 & -1 & b_2 - 2b_1 \\ 1 & 0 & 0 & 0 & b_3 - b_1 + b_1 \\ 0 & 0 & 0 & 0 & b_4 + b_2 - 2b_1 \end{array}\right] .$$

Hence there is a solution of $A\mathbf{x} = \mathbf{b}$ if and only if

$$b_1 - b_2 + b_3 = 0$$

$$-2b_1 + b_2 + b_4 = 0$$

is satisfied. This system can be writen in matrix form as $B\mathbf{b} = 0$ where

$$B = \left[\begin{array}{cccc} 1 & -1 & 1 & 0 \\ -2 & 1 & 0 & 1 \end{array}\right] .$$

Therefore, $\mathbf{b}$ belongs to $\mathrm{Img}(A)$ if and only if $\mathbf{b}$ belongs to the solution set of $B\mathbf{y} = 0$. Let's parameterize the solution set of $B\mathbf{y} = 0$, and thus get a parameterization of $\mathrm{Img}(A)$. To do this, we row reduce $[B|0]$, obtaining in one step

$$\left[\begin{array}{cccc|c} 1 & -1 & 1 & 0 & 0 \\ 0 & -1 & 2 & 1 & 0 \end{array}\right] .$$

The pivotal variables are y_1 and y_2. Putting $t_1 = y_3$ and $t_2 = y_4$ we have

$$y_2 = 2t_1 + t_2 \qquad \text{and} \qquad y_1 = (2t_1 + t_2) - t_1 = t_1 + t_2 .$$

Hence $\mathbf{y}$ is in the solution set when

$$\left[\begin{array}{c} y_1 \\ y_2 \\ y_3 \\ y_4 \end{array}\right] = \left[\begin{array}{c} t_1 + t_2 \\ 2t_1 + t_2 \\ t_1 \\ t_2 \end{array}\right] = t_1 \left[\begin{array}{c} 1 \\ 2 \\ 1 \\ 0 \end{array}\right] + t_2 \left[\begin{array}{c} 1 \\ 1 \\ 0 \\ 1 \end{array}\right] .$$

This is a one–to–one parameterization of $\mathrm{Img}(A)$.

If you ponder the matter a bit, you will see from this example that for any $m \times n$ matrix A, $\mathrm{Img}(A)$ will be the solution set of a homogenous equation $B\mathbf{y} = 0$ for some matrix B with m columns – one for each variable in R^m. (The equation in Example 1 can be written in this form for an 1×3 matrix B instead of the dot product form if we so choose). Moreover, you can always find B by row reduction. You can then get a one–to–one parameterization of $\mathrm{Img}(A)$ by row reducing $[B|0]$ and using the methods of the last chapter.

All of this is very straightforward, but geometric considerations will help us to find a better* parameterization more directly. We will still be using this connection between images and solution sets of homogenous equations, so let's not forget it. We will just be doing it in a more geometric way. Let's think a bit about the general geometric properties of images.

The geometric properties of images

Here is one general statement we can make about images:

- If A is any $m \times n$ matrix, $\operatorname{Img}(A)$ contains the origin in R^m for the simple reason that $A\mathbf{x} = 0$ always has at least the zero solution.

Is there any other general statement we can make about images of linear transformations? Yes, there is. First recall that a subset of $I\!R^m$ is called *affine* in case it contains the line through every two points in it. A plane in $I\!R^3$ has this property: If $\mathbf{x}_0$ and $\mathbf{x}_1$ are any two points in a plane, then every point on the line through them is also in the plane. That is, for every t, $(1-t)\mathbf{x}_0 + t\mathbf{x}_1$ lies in the plane. Here is the general fact

- If A is any $m \times n$ matrix, $\operatorname{Img}(A)$ is an affine subset of R^m.

Indeed, if $\mathbf{b}_1$ and $\mathbf{b}_2$ belong to $\operatorname{Img}(A)$, then by definition there are $\mathbf{x}_1$ and $\mathbf{x}_2$ with $A\mathbf{x}_1 = \mathbf{b}_1$ and $A\mathbf{x}_2 = \mathbf{b}_2$. Then for any number t,

$$A((1-t)\mathbf{x}_1 + t\mathbf{x}_2) = (1-t)A\mathbf{x}_1 + tA\mathbf{x}_2 = (1-t)\mathbf{b}_1 + t\mathbf{b}_2 .$$

Since the left hand side is in $\operatorname{Img}(A)$, it follows that $(1-t)\mathbf{b}_1 + t\mathbf{b}_2$ is in $\operatorname{Img}(A)$ for every t. Hence the line through $\mathbf{b}_1$ and $\mathbf{b}_2$ is contained in $\operatorname{Img}(A)$, which is what it means for $\operatorname{Img}(A)$ to be affine.

The algebraic properties of images

Affine sets that contain the origin have a special algebraic property: are *They are closed under taking linear combinations.*

That is, let S be an affine set that contains the origin. Then for any $\mathbf{v}_1$ and $\mathbf{v}_2$ in S, and any numbers t_1 and t_2, the linear combination

$$t_1\mathbf{v}_1 + t_2\mathbf{v}_2 \tag{1.8}$$

also belongs to S. The fact that S is affine means that

$$(1-t)\mathbf{v}_1 + t\mathbf{v}_2 \tag{1.9}$$

is in S for each t. This is less general than (1.8) since in (1.9) the coefficients add up to 1, while in (1.8) they are unconstrained.

---

* In this context, a "better" parameterization is one that helps us calculate which vector $\mathbf{c}$ in $\operatorname{Img}(A)$ is closest to some given $\mathbf{b}$ in $I\!R^n$. How much work this is will depend on the parameterization of $\operatorname{Img}(A)$ that we use to do the calcuations.

But if S contains 0, then it also contains $2t_1\mathbf{v}_1$ since this is on the line through 0 and $\mathbf{v}_1$. Likewise, S contains $2t_2\mathbf{v}_2$. It therefore contains $t_1\mathbf{v}_1 + t_2\mathbf{v}_2$ which is the midpoint of the line through $2\tau_1\mathbf{v}_1$ and $2t_2\mathbf{v}_2$. So when S is affine and it contains the origin, it contains all linear combinations of every pair of elements in it. Morover, for any three vectors $\mathbf{v}_1$, $\mathbf{v}_2$ and $\mathbf{v}_3$,

$$t_1\mathbf{v}_1 + t_2\mathbf{v}_2 + t_3\mathbf{v}_3 = (t_1\mathbf{v}_1 + t_2\mathbf{v}_2) + t_3\mathbf{v}_3$$

is a linear combination of two vectors in S, namely $t_1\mathbf{v}_1 + t_2\mathbf{v}_2$ and $\mathbf{v}_3$. So it too belongs to S. Continuing in this way, you see that if $\mathbf{v}_1, \mathbf{v}_2, \ldots, \mathbf{v}_k$ are any k vectors in S, and $t_1, t_2, \ldots, t_k$ are any numbers, then the linear combination $t_1\mathbf{v}_1 + t_2\mathbf{v}_2 + \cdots + t_k\mathbf{v}_k$ belongs to S. This brings us to the definition of a subspace of $I\!R^m$:

Definition: (Subspace) A set S of vectors in $I\!R^m$ is called a *subspace* of $I\!R^m$ in case any linear combination

$$t_1\mathbf{v}_1 + t_2\mathbf{v}_2 + \cdots + t_k\mathbf{v}_k \tag{1.10}$$

of vectors $\mathbf{v}_1, \mathbf{v}_2, \ldots, \mathbf{v}_k$ in S also belongs to S.

Theorem 1 (Affine Sets and Subspaces) *A subset S of $I\!R^m$ is a subspace if and only if it is an affine subset containing the origin* 0.

Proof: We have already seen in the discussion leading to the definition that when S is affine and contains the origin, then it is a subspace. On the other hand, if S is a subspace, take $k = 0$, $t_1 = 0$, and any $\mathbf{v}_1$ in S to conclude that $0 = t_1\mathbf{v}_1$ is in S. Next, for any t and any $\mathbf{v}_1$ and $\mathbf{v}_2$ in S, the linear combination $(1 - t)\mathbf{v}_1 + t\mathbf{v}_2$ is in S, so S is affine. ∎

The point of the theorem is although the definition of a subspace is algebraic, expressed in terms of linear combinations (1.10), there is an equivalent geometric way to think about subspaces. Since you have a good intuition into when a set in $I\!R^m$ contains the line through every two points, you have a good intuition into when a sets in $I\!R^m$ is affine.

Example 3 Lines and planes in $I\!R^3$ are affine sets, so lines and planes through the origin are subspaces of $I\!R^3$. Of course $I\!R^3$ itself is affine and contains the origin, so it is a subspace, but in a trivial sort of way. Also, the set $\{0\}$ consisting of the origin alone is a subspace. (Sets consisting of a single vector $\mathbf{x}$ are affine in that $(1 - t)\mathbf{x} + t\mathbf{x} = \mathbf{x}$, which is in the set for all t). A subspace of $I\!R^m$ is called *nontrivial* if it is neither $I\!R^m$ itself or the zero subspace. We will see in a while that there are no other examples: Every nontrivial subspace in $I\!R^3$ is either a line through the origin, or a plane through the origin.

Important note about notation: We are going to use 0 to denote the zero subspace; that is, the subspace consisting of just the zero vector 0. Now the symbol 0 has three meanings: The *number* 0, the *zero vector* in $I\!R^n$, any n, and the *zero subspace* of $I\!R^n$, any n. This may seem dangerous, but it is standard practice, and the context will always make the meaning of the symbol 0 evident.

Example 4 Let A be any $m \times n$ matrix. Then $\text{Img}(A)$ is a subspace, as is the solution set of $A\mathbf{x} = 0$. We've observed above that $\text{Img}(A)$ is affine and that it always contains the origin, so we know by Theorem 1 that it is a subspace. Alternatively, (1.10) can be checked directly. If each $\mathbf{v}_j$ is in $\text{Img}(A)$, there is an $\mathbf{x}_j$ with $A\mathbf{x}_j = \mathbf{v}_j$. Then

$$A(t_1\mathbf{x}_1 + t_2\mathbf{x}_2 + \cdots + t_k\mathbf{x}_k) = t_1 A\mathbf{x}_1 + t_2 A\mathbf{x}_2 + \cdots + t_k A\mathbf{x}_k = t_1\mathbf{v}_1 + t_2\mathbf{v}_2 + \cdots + t_k\mathbf{v}_k .$$

This equality shows that the right side is in the image of A, and hence $\text{Img}(A)$ is closed under taking linear combinations.

In the same way, if $A\mathbf{v}_j = 0$ for each j, then

$$A(t_1\mathbf{v}_1 + t_2\mathbf{v}_2 + \cdots + t_k\mathbf{v}_k) = t_1 A\mathbf{v}_1 + t_2 A\mathbf{v}_2 + \cdots + t_k A\mathbf{v}_k = 0$$

and this displays $t_1\mathbf{v}_1 + t_2\mathbf{v}_2 + \cdots + t_k\mathbf{v}_k$ as belonging to the solution set of $A\mathbf{x} = 0$. Hence this solution set is a subspace. Furthermore, as noted following Example 2, $\text{Img}(A)$ is the solution set of some equation $B\mathbf{y} = 0$, so the second part of this example could have been deduced from the first.

However, the solution set of $A\mathbf{x} = \mathbf{b}$ is a subspace only in the special case that $\mathbf{b} = 0$. Indeed, since $A0 = 0$, the solution set of $A\mathbf{x} = \mathbf{b}$ contains the origin 0 if and only if $\mathbf{b} = 0$.

Example 4 brings us to the following definition:

Definition: (Kernel) Let A be an $m \times n$ matrix. The *kernel* of A is the solution set of the equation $A\mathbf{x} = 0$. It is denoted by $\text{Ker}(A)$.

(Presumably the name is meant to indicate that kernels are really important, "at the heart of the matter". Often the more descriptive term *nullspace* is used in place of kernel.)

Kernels and images are the key examples of subspaces, and the reason we are talking about subspaces in the first place. Later we will see that every subspace is both an image and a kernel.

Example 5: Let A and B be the 2×2 matrices

$$A = \begin{bmatrix} 1 & 2 \\ 2 & 4 \end{bmatrix} \quad \text{and} \quad B = \begin{bmatrix} 1 & 3 \\ 2 & 4 \end{bmatrix} .$$

What are the images and kernels of A and B?

The augmented matrix $[A|\mathbf{b}] = \begin{bmatrix} 1 & 2 & | & b_1 \\ 2 & 4 & | & b_2 \end{bmatrix}$ row reduces to

$$\begin{bmatrix} 1 & 2 & | & b_1 \\ 0 & 0 & | & b_2 - 2b_1 \end{bmatrix} .$$

Hence $\mathbf{b}$ is in $\text{Img}(A)$ if and only if $b_2 - 2b_1 = 0$. That is, if $\mathbf{b}$ lies on the line $y = 2x$ in the x, y plane. Thus $\text{Img}(A)$ is the line through the origin with slope 2.

To find $\text{Ker}(A)$, parameterize the solution set of $A\mathbf{x} = 0$. The augmented matrix $[A|0]$ row reduces to $\begin{bmatrix} 1 & 2 & | & 0 \\ 0 & 0 & | & 0 \end{bmatrix}$ which represents the single equation $x + 2y = 0$. This is the equation of a line in the plane. Hence in this case $\text{Ker}(A)$ is the line in $I\!R^2$ passing through 0 with slope $-1/2$.

On the other hand, $[B|\mathbf{b}] = \begin{bmatrix} 1 & 3 & | & b_1 \\ 2 & 4 & | & b_2 \end{bmatrix}$ row reduces to

$$\begin{bmatrix} 1 & 3 & | & b_1 \\ 0 & -2 & | & b_2 - 2b_1 \end{bmatrix} .$$

There is no pivot in the final column, and there is a pivot in each of the first two. Hence $B\mathbf{x} = \mathbf{b}$ has a unique solution for every $\mathbf{b}$. This means $\text{Img}(B) = I\!R^2$ and $\text{Ker}(B) = 0$.

Now let's take stock of where we are. Given an $m \times n$ matrix A, we've introduced two subspaces associated to A: the kernel of A, $\text{Ker}(A)$, which is a subspace of $I\!R^n$, and the image of A, $\text{Img}(A)$, which is a subspace of $I\!R^m$. In Example 2, we've seen how we can get a one–to–one parameterization of $\text{Img}(A)$ by identifying it with $\text{Ker}(B)$, where B is found by row reducing $[A|\mathbf{b}]$. Then a row reduction of $[B|0]$ gives the desired parameterization.

This works, and it isn't too bad. We've done it in Example 2. But it involves two row reduction. First, one to find B, and then another to parameterize $\text{Ker}(B)$. If we only want the system of equations describing the image, then only one row reduction is involved, and this is a good way to proceed. In particular, if the $\text{Img}(A)$ is a plane in $I\!R^3$, and you want the equation of the plane, then the method demonstrated in Example 1 is very efficient.

But if it is a parameterization of $\text{Img}(A)$ that we are after, we can be more direct if we work with the columns of A instead of the rows of A. The next theorem, simple as it is, is the key to this.

Theorem 2: (Image of A and Columns of A) *Let A be an $m \times n$ matrix. A vector* $\mathbf{b}$ *in $I\!R^m$ is in* Img(A) *if and only if* $\mathbf{b}$ *is a linear combination of the columns of A.*

Proof: Let the columns of A be $\{\mathbf{v}_1, \mathbf{v}_2, \ldots, \mathbf{v}_n\}$. Then

$$A \begin{bmatrix} t_1 \\ t_2 \\ \vdots \\ t_n \end{bmatrix} = \sum_{j=1}^{n} t_j \mathbf{v}_j . \tag{1.11}$$

■

A subspace S of $I\!R^n$ can be formed out of any collection $\{\mathbf{v}_1, \mathbf{v}_2, \ldots, \mathbf{v}_k\}$ of vectors in $I\!R^n$ by taking S to be the set of *all* linear combinations of the $\{\mathbf{v}_1, \mathbf{v}_2, \ldots, \mathbf{v}_k\}$. Since a linear combination of linear combinations is just another linear combination, this set of all linear combinations is a subspace.

Definition: (Span of a Set) Let $\{\mathbf{v}_1, \mathbf{v}_2, \ldots, \mathbf{v}_k\}$ be any set of vectors in $I\!R^n$. The *span* of $\{\mathbf{v}_1, \mathbf{v}_2, \ldots, \mathbf{v}_k\}$ is the set of all linear combinations of the $\{\mathbf{v}_1, \mathbf{v}_2, \ldots, \mathbf{v}_k\}$. It is denoted by

$$\text{Sp}(\mathbf{v}_1, \mathbf{v}_2, \ldots, \mathbf{v}_k) . \tag{1.12}$$

Theorem 2 says that $\text{Img}(A)$ is the span of the columns of A. For this reason $\text{Img}(A)$ is often called the *column space* of A. It also says that every vector in $\text{Img}(A)$ can be written in the form

$$t_1\mathbf{v}_1 + t_2\mathbf{v}_2 + \cdots + t_n\mathbf{v}_n$$

so this gives us a parameterization of $\text{Img}(A)$. However, it is in general not a one–to–one parameterization.

Example 6 Let A be the matrix from Example 1, which has the columns

$$\mathbf{v}_1 = \begin{bmatrix} 1 \\ 2 \\ 0 \end{bmatrix} \qquad \mathbf{v}_2 = \begin{bmatrix} -1 \\ 0 \\ 1 \end{bmatrix} \qquad \mathbf{v}_3 = \begin{bmatrix} 0 \\ 2 \\ 1 \end{bmatrix}.$$

Notice that

$$\begin{bmatrix} 1 \\ 4 \\ 1 \end{bmatrix} = \mathbf{v}_1 + \mathbf{v}_3 \qquad \text{and} \qquad \begin{bmatrix} 1 \\ 4 \\ 1 \end{bmatrix} = 3\mathbf{v}_1 + 2\mathbf{v}_2 - \mathbf{v}_3.$$

Thus, the same vector is described by the two different triples of parameters $(t_1, t_2, t_3) = (1, 0, 1)$ and $(t_1, t_2, t_3) = (3, 1, -1)$. Since different sets of parameters correspond to the same vector, one can't tell whether or not two vectors are equal by comparing parameter values. If the parameter values are the same, then certainly you've got equality. But if they aren't, maybe and maybe not. "Maybe and maybe not" is bad news in mathematics; In the next section we're going to try to do better.

Section 2: Linear Independence and Bases

We saw in the previous section that if $A = [\mathbf{v}_1, \mathbf{v}_2, \ldots, \mathbf{v}_k]$ is any $m \times k$ matrix, then every vector $\mathbf{b}$ in $\text{Img}(A)$ can be written as a linear combination of the columns of A:

$$x_1\mathbf{v}_1 + x_2\mathbf{v}_2 + \cdots + x_k\mathbf{v}_k = \mathbf{b} . \tag{2.1}$$

We also saw there in Example 4 that although we can regard $x_1, x_2, \ldots, x_k$ as parameters for $\text{Img}(A)$, this does not always give us a one–to–one parameterization of $\text{Img}(A)$.

When does it give us a one–to–one parameterization? Since (2.1) is equivalent to

$$A \begin{bmatrix} x_1 \\ x_2 \\ \vdots \\ x_k \end{bmatrix} = \mathbf{b} , \tag{2.2}$$

we see that the values of the parameters $x_1, x_2, \ldots, x_n$ are uniquely determined by $\mathbf{b}$ if and only if the solutions of (2.2) are unique. This in turn is the case if and only if $A\mathbf{x} = 0$ has only the zero solution, or, in other words

$$\text{Ker}(A) = 0 \tag{2.3}$$

Writing (2.3) back out in terms of the columns of A, this means that

$$x_1\mathbf{v}_1 + x_2\mathbf{v}_2 + \cdots + x_k\mathbf{v}_k = 0 \tag{2.4}$$

if and only if $x_j = 0$ for each $j = 1, 2, \ldots, n$. This leads to the following definition:

Definition: (Linear Independence) A set of vectors $\{\mathbf{v}_1, \mathbf{v}_2, \ldots, \mathbf{v}_k\}$ in $I\!\!R^m$ is *linearly independent* in case

$$x_1\mathbf{v}_1 + x_2\mathbf{v}_2 + \cdots + x_k\mathbf{v}_k = 0 \quad \Rightarrow \quad x_j = 0 \quad \text{for each} \quad j . \tag{2.5}$$

Theorem 1: (Independence and Uniqueness) *For any set of vectors* $\{\mathbf{v}_1, \mathbf{v}_2, \ldots, \mathbf{v}_k\}$ *in* $I\!\!R^n$*, the following are equivalent:*

(1) $\{\mathbf{v}_1, \mathbf{v}_2, \ldots, \mathbf{v}_k\}$ *is linearly independent.*

(2) $(x_1\mathbf{v}_1 + x_2\mathbf{v}_2 + \cdots + x_k\mathbf{v}_k) = (y_1\mathbf{v}_1 + y_2\mathbf{v}_2 + \cdots + y_k\mathbf{v}_k) \quad \Rightarrow \quad x_j = y_j$ *for each* j.

(3) $\text{Ker}([\mathbf{v}_1, \mathbf{v}_2, \ldots, \mathbf{v}_k]) = 0$.

(4) The rank of the $n \times k$ *matrix* $[\mathbf{v}_1, \mathbf{v}_2, \ldots, \mathbf{v}_k]$ *is* k.

For the most part, the validity of these statements has already been explained in the opening discussion. But it won't hurt to go through the logic again.

Proof: The proof works by showing that each of these statements is equivalent to *(4)*. We know from Theorem 2.5.4 that *(4)* holds if and only if $[\mathbf{v}_1, \mathbf{v}_2, \ldots, \mathbf{v}_k]\mathbf{x} = 0$ has only the zero solution and this is just another way of expressing *(3)*. So *(3)* and *(4)* are equivalent.

Next, notice that with $A = [\mathbf{v}_1, \mathbf{v}_2, \ldots, \mathbf{v}_k]$, *(2)* is just a longer way of writing

$$A\mathbf{x} = A\mathbf{y} \Rightarrow \mathbf{x} = \mathbf{y} \ .$$

So *(2)* just says that solutions of $A\mathbf{x} = \mathbf{b}$ are unique whenever they exist. Again by Theorem 2.5.4, we know that this is the case if and only if the rank of A is k, the number of columns. Hence *(2)* and *(4)* are equivalent.

Finally, (2.5) can be rewritten in terms of A as $A\mathbf{x} = 0 \Rightarrow \mathbf{x} = 0$, which means that $A\mathbf{x} = 0$ has only the zero solution. Again by Theorem 2.5.4, this is the case if and only if the rank of A is k, the number of columns. Thus *(1)* and *(4)* are equivalent. ∎

Example 1 Let $\mathbf{x}_0 + t_1\mathbf{v}_1 + t_2\mathbf{v}_2 + \cdots + t_k\mathbf{v}_k$ be a parameterization of the solution set of some linear system of equations $A\mathbf{x} = \mathbf{b}$. We know from the last chapter that this parameterization will always be one–to–one, so that

$$\mathbf{x}_0 + t_1\mathbf{v}_1 + t_2\mathbf{v}_2 + \cdots + t_k\mathbf{v}_k = \mathbf{x}_0 + s_1\mathbf{v}_1 + s_2\mathbf{v}_2 + \cdots + s_k\mathbf{v}_k \qquad (2.6)$$

implies that $t_j = s_j$ for each j. Cancelling of the $\mathbf{x}_0$ from both sides of (2.6), we see that $\{\mathbf{v}_1, \mathbf{v}_2, \ldots, \mathbf{v}_k\}$ is linearly independent. So you already have a considerable experience working with linearly independent sets of vectors:

• *The sets $\{\mathbf{v}_1, \mathbf{v}_2, \ldots, \mathbf{v}_k\}$ – not including the $\mathbf{x}_0$ – that we used to parameterize solution sets of linear systems are always linearly independent.*

The linear independence of these sets is a direct reflection of the fact that the parameterizations obtained using row reduction are always one–to–one.

Theorem 1 provides a useful means of checking for independence, since we know how to check, by using row reduction, whether or not $A\mathbf{x} = 0$ has non–zero solutions or not.

Example 2 Let

$$\mathbf{v}_1 = \begin{bmatrix} 1 \\ 2 \\ 0 \end{bmatrix} \qquad \mathbf{v}_2 = \begin{bmatrix} -1 \\ 0 \\ 1 \end{bmatrix} \qquad \mathbf{v}_3 = \begin{bmatrix} 0 \\ 2 \\ 1 \end{bmatrix} \ .$$

Is $\{\mathbf{v}_1, \mathbf{v}_2, \mathbf{v}_3\}$ a linearly independent set of vectors? To check, form $A = [\mathbf{v}_1, \mathbf{v}_2, \mathbf{v}_3] = \begin{bmatrix} 1 & -1 & 0 \\ 2 & 0 & 2 \\ 0 & 1 & 1 \end{bmatrix}$. Row reducing this yields $\begin{bmatrix} 1 & -1 & 0 \\ 0 & 2 & 2 \\ 0 & 0 & 0 \end{bmatrix}$. Evidently, the rank of A is 2, not 3, so that there is a non–pivotal variable, and $\{\mathbf{v}_1, \mathbf{v}_2, \mathbf{v}_3\}$ is not a linearly independent set of vectors. On the other hand, $[\mathbf{v}_1, \mathbf{v}_2] = \begin{bmatrix} 1 & -1 \\ 2 & 0 \\ 0 & 1 \end{bmatrix}$ row reduces to $\begin{bmatrix} 1 & -1 \\ 0 & 2 \\ 0 & 0 \end{bmatrix}$, which has rank 2. This time, that's full column rank – there are just 2 columns – so that $\{\mathbf{v}_1, \mathbf{v}_2\}$ is a linearly independent set of vectors.

Before moving on, we point out that it wouldn't be right to summarize the result of Example 2 as being "$\mathbf{v}_1$ and $\mathbf{v}_2$ are independent, but $\mathbf{v}_3$ is not".

- *Independence is not a property of individual vectors, it is a property of sets of vectors, so it makes no sense to talk about an individual vector being independent or not.*

Indeed, there is nothing at all "wrong" with $\mathbf{v}_3$ itself in this example; you can, and should, check that both $\{\mathbf{v}_1, \mathbf{v}_3\}$ and $\{\mathbf{v}_2, \mathbf{v}_3\}$ are independent. Be sure you understand this point. A lot of "first time around" confusion comes from trying to think about linear independence as a property of individual vectors.

There is another way to look at independence, in terms of *solvability* of systems involving the $k \times n$ matrix formed using the vectors $\mathbf{v}_1, \mathbf{v}_2, \ldots, \mathbf{v}_k$ as *rows*.

Theorem 2: (Independence and Solvability) *For any set of vectors* $\{\mathbf{v}_1, \mathbf{v}_2, \ldots, \mathbf{v}_k\}$ *in* $I\!\!R^n$, *the following are equivalent:*

(1) $\{\mathbf{v}_1, \mathbf{v}_2, \ldots, \mathbf{v}_k\}$ *is linearly independent.*

(2) $\mathrm{Img}\left(\begin{bmatrix} \mathbf{v}_1 \\ \mathbf{v}_2 \\ \vdots \\ \mathbf{v}_k \end{bmatrix}\right) = I\!\!R^k.$

(3) The rank of the matrix $\begin{bmatrix} \mathbf{v}_1 \\ \mathbf{v}_2 \\ \vdots \\ \mathbf{v}_k \end{bmatrix}$ *is k; i.e., this matrix has full row rank.*

Proof: First of all, *(2)* and *(3)* are equivalent by Theorem 2.5.4. It remains to show that either *(2)* or *(3)* imply *(1)*, and that *(1)* implies either of *(2)* or *(3)*.

We now show that *(2)* implies *(1)*. Let $A = \begin{bmatrix} \mathbf{v}_1 \\ \mathbf{v}_2 \\ \vdots \\ \mathbf{v}_k \end{bmatrix}$, and note that $A^t = [\mathbf{v}_1, \mathbf{v}_2, \ldots, \mathbf{v}_k]$.

To prove linear independence whenever *(2)*, it suffices to show that then $\mathrm{Ker}(A^t) = 0$, because A^t is the matrix considered in Theorem 1.

Suppose *(2)* holds so that $\mathrm{Img}(A) = I\!\!R^k$. Let $\mathbf{b}$ be any vector in $\mathrm{Ker}(A^t)$. Then $\mathbf{b}$ belongs to $I\!\!R^k$ and by *(2)*, $A\mathbf{x} = \mathbf{b}$ has a solution $\mathbf{x}$. Multiplying both sides by A^t, we get $A^t A \mathbf{x} = A^t \mathbf{b} = 0$. Taking the dot product of both sides with $\mathbf{x}$, we get $|A\mathbf{x}|^2 = \mathbf{x} \cdot A^t A \mathbf{x} = 0$. Hence $\mathbf{b} = A\mathbf{x} = 0$, and so $\mathrm{Ker}(A^t) = 0$. By Theorem 1, *(1)* holds whenever *(2)* does.

Next, we show that *(1)* implies *(3)*. This is because when $\{\mathbf{v}_1, \mathbf{v}_2, \ldots, \mathbf{v}_k\}$ is linearly independent, we can never get a row of all zeros when we are row reducing $[A|0]$. This is because at every stage of the row reduction, each row is a linear combination of the rows of the original matrix $[A|0]$. If one of these linear combinations was all zeros, we would have $a_1 \mathbf{v}_1 + a_2 \mathbf{v}_2 + \cdots + a_k \mathbf{v}_k = 0$ for some numbers $a_1, a_2, \ldots, a_k$. Hence we never get a row of all zeros, and when we row reduce $[A|0]$ we get a pivot in every row. This means that the rank of A is k, the number of rows, and hence *(3)* holds. $\blacksquare$

Theorem 2 gives us another useful check for the independence of k vectors in R^n: Form

the matrix $\begin{bmatrix} \mathbf{v}_1 \\ \mathbf{v}_2 \\ \vdots \\ \mathbf{v}_k \end{bmatrix}$ and row reduce it to determine whether or not it has full row rank. As

we will explain shortly, we will only be interested in the case $k \leq n$. Therefore, it will

often be easier to row reduce $\begin{bmatrix} \mathbf{v}_1 \\ \mathbf{v}_2 \\ \vdots \\ \mathbf{v}_k \end{bmatrix}$ than $[\mathbf{v}_1, \mathbf{v}_2, \ldots, \mathbf{v}_k]$ since the former has fewer rows

when $k < n$.

Example 3 Consider the vector $\mathbf{v}_1$ and $\mathbf{v}_3$ from Example 2. We claimed that $\{\mathbf{v}_1, \mathbf{v}_3\}$ is

linearly independent. To check, form $A = \begin{bmatrix} \mathbf{v}_1 \\ \mathbf{v}_3 \end{bmatrix} = \begin{bmatrix} 1 & 2 & 0 \\ 0 & 2 & 1 \end{bmatrix}$. This matrix is already row

reduced, and clearly has rank 2, so that $\{\mathbf{v}_1, \mathbf{v}_3\}$ is indeed an independent set of vectors. The matrix $[\mathbf{v}_1, \mathbf{v}_2]$ is however not in row reduced form, so if we used Theorem 1, we'd have to do a little more work. There is not much of a difference here, and in any case, a set of two vectors is linearly independent if and only if the two vectors are not proportional – think about this – so dealing with two vectors isn't going to be hard no matter how you go about it. But keep in mind that the amount of work you have to do will in general depend on whether you choose to use Theorem 1 of Theorem 2.

Theorem 2 also helps explain the term "independence". Theorem 2 says that no matter how the numbers b_j are chosen, each of the equations

$$\mathbf{v}_j \cdot \mathbf{x} = b_j \qquad , \qquad j = 1, 2, \ldots, k$$

are "independent equations", in the sense that you can adjust the values of the different b_j independently, and no matter how you do it, the system will be solvable.

Example 4 Consider the system of equations

$$x + 2y = 3$$
$$2x + 4y = 6$$

and notice that the second equation is just the first one multiplied through by 2. It doesn't tell us anything new. That is, it doesn't give us any information about the solution set that is *independent* of what the first equation tells us. Moreover, while this system has a solution, namely $x = y = 1$, we cannot *independently* adjust the right hand sides of the equations and still have a solution. For instance, if we change the right hand side of just the first equation so that we have

$$x + 2y = 4$$
$$2x + 4y = 6 \ ,$$

there isn't any solution since when $x + 2y = 4$, we must have $2x + 4y = 8$, not 6. How you have to change the right hand side of the second equation *depends* on how you change the right hand side in the first equation. The dependence of these equations in this sense corresponds exactly to the fact that $\left\{ \begin{bmatrix} 1 \\ 2 \end{bmatrix}, \begin{bmatrix} 2 \\ 4 \end{bmatrix} \right\}$ is not independent. This is the sort of thing people had in mind when they tagged (2.5) with the name "independence"; Theorem 2 provides the connection.

Let's take stock of where we are: If $\{\mathbf{v}_1, \mathbf{v}_2, \ldots, \mathbf{v}_k\}$ is independent, then a parameterization

$$t_1 \mathbf{v}_1 + t_2 \mathbf{v}_2 + \cdots + t_k \mathbf{v}_k$$

based on these vectors never represents the same vector with more than one set of parameters. But we also want to be sure there is at least one:

Definition: (Spanning Set) A collection of vectors $\{\mathbf{v}_1, \mathbf{v}_2, \ldots, \mathbf{v}_k\}$ in a subspace S of $I\!R^n$ *spans* S in case every vector in S can be written as a linear combination

$$\mathbf{w} = t_1 \mathbf{v}_1 + t_2 \mathbf{v}_2 + \cdots + t_k \mathbf{v}_k \tag{2.7}$$

in at least one way. In this case, $\{\mathbf{v}_1, \mathbf{v}_2, \ldots, \mathbf{v}_k\}$ is a *spanning set* for S.

• *By(2.7) and the basic definition of matrix–vector multiplication, or as well by Theorem 1.5.1, the span of* $\{\mathbf{v}_1, \mathbf{v}_2, \ldots, \mathbf{v}_k\}$ *is the image of the matrix* $[\mathbf{v}_1, \mathbf{v}_2, \ldots, \mathbf{v}_k]$. *That is:*

$$\text{Img}([\mathbf{v}_1, \mathbf{v}_2, \ldots, \mathbf{v}_k]) \quad \text{is the span of} \quad \{\mathbf{v}_1, \mathbf{v}_2, \ldots, \mathbf{v}_k\} \tag{2.8}$$

Example 5 The vectors $\{\mathbf{e}_1, \mathbf{e}_2, \mathbf{e}_3\}$ in $I\!R^3$ span $I\!R^3$, since certainly any vector $\mathbf{w}$ in $I\!R^3$ can be expressed as a linear combination of them. We've been using this fact all along; the only thing new here is another way to talk about it.

We can get other examples of spanning sets from (2.8). Let $A = \begin{bmatrix} 1 & 2 & 1 & 2 \\ 2 & 3 & 2 & 3 \\ 1 & 1 & 1 & 1 \\ 0 & 1 & 0 & 1 \end{bmatrix}$,

the matrix considered in Example 2 of the Section 1. We found there that $\text{Img}(A)$ is the subspace S of $I\!R^4$ paramterized by

$$t_1 \begin{bmatrix} 1 \\ 2 \\ 1 \\ 0 \end{bmatrix} + t_2 \begin{bmatrix} 1 \\ 1 \\ 0 \\ 1 \end{bmatrix}. \tag{2.9}$$

Let $\{\mathbf{v}_1, \mathbf{v}_2, \mathbf{v}_3, \mathbf{v}_4\}$ be the four columns of A. Then according to (2.8), the span of $\{\mathbf{v}_1, \mathbf{v}_2, \mathbf{v}_3, \mathbf{v}_4\}$ is the subspace parameterized in (2.9). Now let $\mathbf{u}_1$ and $\mathbf{u}_2$ denote the two vectors in (2.9) so this this expression becomes $t_1 \mathbf{u}_1 + t_2 \mathbf{u}_2$. Evidently, every vector in S is also a linear combination of $\mathbf{u}_1$ and $\mathbf{u}_2$, so $\{\mathbf{u}_1, \mathbf{u}_2\}$ is another spanning set for S. Using the notation from (1.12),

$$\mathrm{Sp}(\mathbf{v}_1, \mathbf{v}_2, \mathbf{v}_3, \mathbf{v}_4) = \mathrm{Sp}(\mathbf{u}_1, \mathbf{u}_2) \ .$$

This is an example of two different sets of vectors, containing different numbers of vectors, that span the same subspace S of R^4.

Putting "span" and "linearly independent" together brings us to the central definition of this section:

Defintion: (Basis) A collection of vectors $\{\mathbf{v}_1, \mathbf{v}_2, \ldots, \mathbf{v}_k\}$ in a subspace S of $I\!R^n$ is a *basis* for S in case it is linearly independent and it spans S.

Example 6 The standard basis vectors

$$\{\mathbf{e}_1, \mathbf{e}_2 \ldots, \mathbf{e}_n\}$$

are in fact a basis for $I\!R^n$ since

$$\begin{bmatrix} x_1 \\ x_2 \\ \vdots \\ x_n \end{bmatrix} = x_1 \mathbf{e}_1 + x_2 \mathbf{e}_2 + \ldots + \ldots x_n \mathbf{e}_n \ .$$

This shows at once that we can write any $\mathbf{x}$ in $I\!R^n$ as a linear combination of standard basis elements, and that the linear combination is the zero vector if and only if each $x_i = 0$. So this collection is independent, and it spans $I\!R^n$. This is what it means to be a basis in the proper technical sense, which is the only way we will use the word from here on out.

Example 7 Let A be the 4×4 matrix in Example 2 of the previous section. There we found that

$$t_1 \begin{bmatrix} 1 \\ 2 \\ 1 \\ 0 \end{bmatrix} + t_2 \begin{bmatrix} 1 \\ 1 \\ 0 \\ 1 \end{bmatrix}$$

was a one–to–one parameterization of $\mathrm{Img}(A)$. The fact that every vector in $\mathrm{Img}(A)$ can be written as such a linear combination means that

$$\left\{ \begin{bmatrix} 1 \\ 2 \\ 1 \\ 0 \end{bmatrix}, \begin{bmatrix} 1 \\ 1 \\ 0 \\ 1 \end{bmatrix} \right\} \tag{2.10}$$

spans $\mathrm{Img}(A)$, and the fact that this parameterization is one–to–one means, by Theorem 1, that this set of vectors is linearly independent. Hence (2.10) is a basis for $\mathrm{Img}(A)$.

If you look back at Example 2 of the previous section, you will see that we actually found the parameterization of Img(A) by working out a parameterization of Ker(B) for some other matrix B. The connection between bases and one–to–one parameterizations that we used in Example 7 is a useful general fact, and is worth recording as a Theorem.

Theorem 3 (Bases and One–to–one Paramterizations) *Let S be a subspace of $I\!R^n$ Then*

$$t_1\mathbf{v}_1 + t_2\mathbf{v}_2 + \cdots + t_k\mathbf{v}_k \tag{2.11}$$

is a one–to–one parameterization of S if and only if $\{\mathbf{v}_1, \mathbf{v}_2, \ldots, \mathbf{v}_k\}$ is a basis for S.

Proof: Suppose that (2.11) is a parameterization of S. Then every $\mathbf{v}$ in S can be written in the form (2.11), so $\{\mathbf{v}_1, \mathbf{v}_2, \ldots, \mathbf{v}_k\}$ spans S. Since the parameterization is one–to–one, Theorem 1 says that $\{\mathbf{v}_1, \mathbf{v}_2, \ldots, \mathbf{v}_k\}$ is linearly independent. Hence it is a basis for S.

Conversely, if $\{\mathbf{v}_1, \mathbf{v}_2, \ldots, \mathbf{v}_k\}$ spans S, then every $\mathbf{v}$ in S can be written in the form (2.11), so (2.11) is a parameterization of S. Moreover, since $\{\mathbf{v}_1, \mathbf{v}_2, \ldots, \mathbf{v}_k\}$ is independent, there cannot be two different ways to write $\mathbf{v}$ as a linear combination of the $\{\mathbf{v}_1, \mathbf{v}_2, \ldots, \mathbf{v}_k\}$. Hence the parameterization is one–to–one. ∎

According to Theorem 3, at least when S is Ker(A) for some matrix A, finding a basis for S is essentially same thing as finding a one–to–one parameterization of the solution set of $A\mathbf{x} = 0$, and this is something with which we have a lot of experience now.

Coordinates with respect to a basis

To conclude this section, fix any subspace S of $I\!R^n$, and suppose that $\{\mathbf{v}_1, \mathbf{v}_2, \ldots, \mathbf{v}_k\}$ is a basis for S. Then each vector $\mathbf{w}$ in S has an expression as a linear combination

$$\mathbf{w} = \sum_{j=1}^{k} x_j \mathbf{v}_j \tag{2.12}$$

for exactly one set of numbers $x_1, x_2, \ldots, x_k$. *How do we find these numbers?*

Organize the numbers $x_1, x_2, \ldots, x_k$ into the vector $\mathbf{x} = \begin{bmatrix} x_1 \\ x_2 \\ \vdots \\ x_k \end{bmatrix}$ in $I\!R^k$, and form the

matrix $A = [\mathbf{v}_1, \mathbf{v}_2, \ldots, \mathbf{v}_k]$ Then (2.12) is equivalent to the equation

$$\mathbf{w} = A\mathbf{x} . \tag{2.13}$$

which can now be solved, by row reduction, to find $\mathbf{x}$. For every $\mathbf{w}$ in S, this procedure leads to a uniquely determined $\mathbf{x}$ in $I\!R^k$.

Example 8 Let A be the 4×4 matrix $A = \begin{bmatrix} 1 & 2 & 1 & 2 \\ 2 & 3 & 2 & 3 \\ 1 & 1 & 1 & 1 \\ 0 & 1 & 0 & 1 \end{bmatrix}$ considered in Example 4 of

the previous section and Example 7 of this section. Then

$$A = \begin{bmatrix} 1 & 2 & 1 & 2 \\ 2 & 3 & 2 & 3 \\ 1 & 1 & 1 & 1 \\ 0 & 1 & 0 & 1 \end{bmatrix} \begin{bmatrix} 1 \\ 1 \\ 1 \\ 1 \end{bmatrix} = \begin{bmatrix} 6 \\ 10 \\ 4 \\ 2 \end{bmatrix}.$$

Evidently, the vector on the right is in $\mathrm{Img}(A)$. Since, as we saw in Example 7, (2.10) is a basis for $\mathrm{Img}(A)$, we must have

$$\begin{bmatrix} 6 \\ 10 \\ 4 \\ 2 \end{bmatrix} = x_1 \begin{bmatrix} 1 \\ 2 \\ 1 \\ 0 \end{bmatrix} + x_2 \begin{bmatrix} 1 \\ 1 \\ 0 \\ 1 \end{bmatrix} \tag{2.14}$$

for some numbers x_1 and x_2. To find these numbers, rewrite this as the matrix equation

$$\begin{bmatrix} 1 & 1 \\ 2 & 1 \\ 1 & 0 \\ 0 & 1 \end{bmatrix} \begin{bmatrix} x_1 \\ x_2 \end{bmatrix} = \begin{bmatrix} 6 \\ 10 \\ 4 \\ 2 \end{bmatrix}.$$

We don't even have to row reduce this to solve it: The last row corresponds to the equation $x_2 = 2$, and the third row corresponds to the equation $x_1 = 4$. And indeed, you can check that if you take $x_1 = 4$ and $x_2 = 2$, then (2.14) holds. We didn't need to do the full row reduction because we know in advance, by Theorem 3, that there is a unique solution. As soon as we have considered enough equations in the system to determine the x_j, we can stop.

The numbers $x_1 = 4$ and $x_2 = 2$ that we found in Example 8 are called the *coordinates* of $\begin{bmatrix} 6 \\ 10 \\ 4 \\ 2 \end{bmatrix}$ *with respect to the basis (2.10).*

You see that coordinates and the parameters in a one–to–one parameterization are essentially the same thing. If S is a subspace of $I\!R^n$, and $\{\mathbf{v}_1, \mathbf{v}_2, \ldots, \mathbf{v}_k\}$ is a basis for S, then we have the *parameterization function*

$$\begin{bmatrix} t_1 \\ t_2 \\ \vdots \\ t_k \end{bmatrix} \mapsto \sum_{j=1}^{k} t_j \mathbf{v}_j \tag{2.15}$$

which is a one–to–one function from $\mathbb{R}^k$ onto S.

Since it is one–to–one and onto, it is invertible, and so there is an inverse function

$$\mathbf{w} \mapsto \begin{bmatrix} x_1 \\ x_2 \\ \vdots \\ x_k \end{bmatrix} \tag{2.16}$$

taking vectors $\mathbf{w}$ in S to vectors $\mathbf{x}$ in $\mathbb{R}^k$ such that

$$\sum_{j=1}^{k} x_j \mathbf{v}_j = \mathbf{w} . \tag{2.17}$$

It is common practice to refer to the inverse function as the coordinate transformation for the basis $\{\mathbf{v}_1, \mathbf{v}_2, \ldots, \mathbf{v}_k\}$ and the numbers x_j in (2.16) and (2.17) as the *coordinates of* $\mathbf{w}$ *with respect to the basis* $\{\mathbf{v}_1, \mathbf{v}_2, \ldots, \mathbf{v}_k\}$.

Now if we have two different bases for a subsapce S, we have two different parameterizations, and hence two different sets of coordinates. How are they related to one another? We take up this question in the next section.

Section 3: Dimension

The first thing we show in this section is that any two bases for a given subspace S of $\mathbb{R}^n$ have the same number of elements.

Theorem 1: (All Bases are the Same Size) *Suppose that S is a subspace of $\mathbb{R}^n$. No linearly independent set of vectors in S contains more elements than any spanning set in S. In particular, any basis for S contains exactly the same number vectors as any other.*

Proof: Suppose that $\{\mathbf{v}_1, \mathbf{v}_2, \ldots, \mathbf{v}_m\}$ spans S, and that $\{\mathbf{u}_1, \mathbf{u}_2, \ldots, \mathbf{u}_\ell\}$ is linearly independent. We will show that $\ell \leq m$.

Let V denote this $n \times m$ matrix $[\mathbf{v}_1, \mathbf{v}_2, \ldots, \mathbf{v}_m]$. Since $\{\mathbf{v}_1, \mathbf{v}_2, \ldots, \mathbf{v}_m\}$ spans S, we have from (2.8) that every vector in S belongs to $\text{Img}(V)$. In particular, for each j, $\mathbf{u}_j$ belongs to S, and hence to $\text{Img}(V)$. This means there is a vector $\mathbf{r}_j$ in $\mathbb{R}^m$ with $V\mathbf{r}_j = \mathbf{u}_j$. Now let U be the $n \times \ell$ matrix $[\mathbf{u}_1, \mathbf{u}_2, \ldots, \mathbf{u}_\ell]$, and let R be the $m \times \ell$ matrix $[\mathbf{r}_1, \mathbf{r}_2, \ldots, \mathbf{r}_\ell]$. Then by Theorem 1.5.2,

$$VR = [V\mathbf{r}_1, V\mathbf{r}_2, \ldots, V\mathbf{r}_\ell] = [\mathbf{u}_1, \mathbf{u}_2, \ldots, \mathbf{u}_\ell] = U . \tag{3.1}$$

Now suppose $\mathbf{x}$ is any vector in $\mathbb{R}^\ell$ with $R\mathbf{x} = 0$. Then $V(R\mathbf{x}) = 0$ too, but by (3.1),

$$V(R\mathbf{x}) = (VR)\mathbf{x} = U\mathbf{x} .$$

Therefore

$$R\mathbf{x} = 0 \Rightarrow U\mathbf{x} = 0 .$$

But since $\{\mathbf{u}_1, \mathbf{u}_2, \ldots, \mathbf{u}_\ell\}$ is linearly independent, Theorem 3.2.1 says that $\text{Ker}(U) = 0$. Hence the only solution of $R\mathbf{x} = 0$ is the zero vector, and so $\text{Ker}(R) = 0$.

Since R is an $m \times \ell$ matrix, this shows that $\ell \leq m$. By Theorem 2.5.2, for any matrix A with more columns than rows, $A\mathbf{x} = 0$ always has a non zero solution. That doesn't happen for R, so it can't have more columns than rows.

This shows that no spanning set contains fewer vectors than any linearly independent set. Since a basis is both spanning and linearly independent, the second part follows immediately. ∎

It follows right away from Theorem 1 that no set of more than n vectors in $\mathbb{R}^n$ can be independent since $\{\mathbf{e}_1, \mathbf{e}_2, \ldots, \mathbf{e}_n\}$ is a spanning set for R^n with n elements. This leads to the fact that *every subspace S has a basis*, which is included in the next theorem.

Theorem 2: (Constructing Bases) *Let $\{\mathbf{v}_1, \mathbf{v}_2, \ldots, \mathbf{v}_k\}$ be a set of vectors in a non-zero subspace S of $\mathbb{R}^n$.*

(1) Any set $\{\mathbf{v}_1, \mathbf{v}_2, \ldots, \mathbf{v}_k\}$ of vectors in S that spans S contains a basis. That is, any spanning set can be reduced to obtain a basis.

(2) Any set $\{\mathbf{v}_1, \mathbf{v}_2, \ldots, \mathbf{v}_k\}$ of vectors in S that is linearly independent is contained in a basis $\{\mathbf{u}_1, \mathbf{u}_2, \ldots, \mathbf{u}_m\}$, $m \geq k$, with $\mathbf{u}_j = \mathbf{v}_j$ for $j \leq k$.

(3) Every non-zero subspace of $\mathbb{R}^n$ has a basis.

Proof: Consider (1). If the vectors are linearly independent, we already have a basis, and there is nothing to do. If not, we have

$$x_1 \mathbf{v}_1 + x_2 \mathbf{v}_2 + \ldots + x_k \mathbf{v}_k = 0$$

and not all of the x_j's are zero. Renumbering the vectors if need be, we can assume that $x_k \neq 0$ and then

$$\mathbf{v}_k = -\frac{x_1}{x_k} \mathbf{v}_1 - \frac{x_2}{x_k} \mathbf{v}_2 - \ldots - \frac{x_{k-1}}{x_k} \mathbf{v}_{k-1}$$

and so $\mathbf{v}_k$ can be eliminated from any linear combination, and so

$$\mathrm{Sp}(\mathbf{v}_1, \mathbf{v}_2, \ldots, \mathbf{v}_{k-1}) = \mathrm{Sp}(\mathbf{v}_1, \mathbf{v}_2, \ldots, \mathbf{v}_k) = S .$$

Thus the reduced set still spans. Now keep reducing until what remains is independent. This clearly happens, at worst, by the time we've reduced to a single non-zero vector.

For (2), if $\mathrm{Sp}(\mathbf{v}_1, \mathbf{v}_2, \ldots, \mathbf{v}_k) = S$ our set spans and so it is already a basis. If not, define $\mathbf{v}_{k+1}$ to be some vector in S that is not in $\mathrm{Sp}(\mathbf{v}_1, \mathbf{v}_2, \ldots, \mathbf{v}_k)$. If the enlarged set $\{\mathbf{v}_1, \mathbf{v}_2, \ldots, \mathbf{v}_{k+1}\}$ were not linearly independent, we would have

$$x_1 \mathbf{v}_1 + x_2 \mathbf{v}_2 + \ldots + x_{k+1} \mathbf{v}_{k+1} = 0 \qquad (3.2)$$

without all of the $x_j = 0$ for each j. But if $x_{k+1} \neq 0$, we could divide by x_{k+1} in this equation, and solve for $\mathbf{v}_{k+1}$ as a linear combination of $\{\mathbf{v}_1, \mathbf{v}_2, \ldots, \mathbf{v}_k\}$. This can't be since, by choice, $\mathbf{v}_{k+1}$ does not belong to $\mathrm{Sp}(\mathbf{v}_1, \mathbf{v}_2, \ldots, \mathbf{v}_k)$. So $x_{k+1} = 0$, and this leaves us with

$$x_1 \mathbf{v}_1 + x_2 \mathbf{v}_2 + \ldots + x_k \mathbf{v}_k = 0 .$$

But by the independence of $\{\mathbf{v}_1, \mathbf{v}_2, \ldots, \mathbf{v}_k\}$, it can't be that any $x_j \neq 0$. Thus all of the x_j's in (3.2) are zero, and the augmented set $\{\mathbf{v}_1, \mathbf{v}_2, \ldots, \mathbf{v}_{k+1}\}$ is linearly independent. This process must stop by the time $k = n$, or else we would produce a set of $n+1$ linearly independent vectors in $I\!\!R^n$, which is impossible. So for some $k \leq n$, $\{\mathbf{v}_1, \mathbf{v}_2, \ldots, \mathbf{v}_k\}$ spans S. *(3)* follows from *(2)*: Start with any non–zero vector $\mathbf{v}_1$, and keep adding vectors as above until you get a basis. This terminates in $n-1$ steps at most. ∎

Now that we know that every subspace S has at least one basis, and that any two bases for a subspace S of $I\!\!R^n$ have the same number of elements, we can make the following definition:

Definition (Dimension) The number of elements in a basis for a subspace S of $I\!\!R^n$ is called the *dimension* of S, and is denoted by $\dim(S)$.

Theorem 3 *Let S be a subspace of $I\!\!R^n$, and let $d = \dim(S)$. Then*

(1) Any set of d linearly independent vectors in S is a basis of S.

(2) Any set of d vectors in S that spans S is a basis of S.

Proof: Let $\{\mathbf{v}_1, \mathbf{v}_2, \ldots, \mathbf{v}_d\}$ be a set of d vectors in S that is linearly independent. If it does not span S, then according to Theorem 2, we can add to $\{\mathbf{v}_1, \mathbf{v}_2, \ldots, \mathbf{v}_d\}$ vectors in S to obtain a basis with more than d elements. This cannot be, according to Theorem 1. So $\{\mathbf{v}_1, \mathbf{v}_2, \ldots, \mathbf{v}_d\}$ must also span S, and hence be a basis.

In the same vein, let $\{\mathbf{v}_1, \mathbf{v}_2, \ldots, \mathbf{v}_d\}$ be a set of d vectors that spans S. Then according to Theorem 2, if it is not a basis, we can delete vectors from it to obtain a basis. But this would produce a basis with fewer than d elements, which Theorem 1 says is impossible. So it must be the case that $\{\mathbf{v}_1, \mathbf{v}_2, \ldots, \mathbf{v}_d\}$ is already a basis. ∎

We've waited a while for examples in this section, but we really need all three theorems and the definition of dimension to give good examples. Some ideas are only powerful in combination with one another!

Example 1 Let S be the plane through the origin in $I\!R^3$ given by $x_1 + x_2 + x_3 = 0$. Then x_2 and x_3 are non–pivotal, so we set $x_2 = t_1$ and $x_3 = t_2$. Then solving for the pivotal variable, $x_1 = -x_2 - x_3 = -t_1 - t_2$. Then any vector in the plane S is given by

$$
\begin{bmatrix} x_1 \\ x_2 \\ x_3 \end{bmatrix} = \begin{bmatrix} -t_1 - t_2 \\ t_1 \\ t_2 \end{bmatrix} = t_1 \begin{bmatrix} -1 \\ 1 \\ 0 \end{bmatrix} + t_2 \begin{bmatrix} -1 \\ 0 \\ 1 \end{bmatrix} .
$$

This is a one–to–one parameterization of S, and hence $\{\mathbf{v}_1, \mathbf{v}_2\}$ is a basis for S where

$$
\mathbf{v}_1 = \begin{bmatrix} -1 \\ 1 \\ 0 \end{bmatrix} \quad \text{and} \quad \mathbf{v}_2 = \begin{bmatrix} -1 \\ 0 \\ 1 \end{bmatrix} . \tag{3.3}
$$

Since we've found a basis of S with two elements, $\dim(S) = 2$. Of course the dimension of a plane should be 2! If it weren't, there would have been something wrong with our definition.

Now let's find a *second* basis of S. A very useful feature of the standard basis vectors in $I\!R^n$ is that they are orthonormal. Can we find an orthonormal basis for S? Yes, we can. There is a procedure for doing this for *any subspace*. We'll get to that later, but it is simple enough to do it now for this 2 dimensional space.

First of all, define

$$
\mathbf{u}_1 = \frac{1}{|\mathbf{v}_1|}\mathbf{v}_1 = \frac{1}{\sqrt{2}} \begin{bmatrix} -1 \\ 1 \\ 0 \end{bmatrix} .
$$

This is a unit vector; that is, $|\mathbf{u}_1| = 1$. Notice that $\mathbf{u}_1$ belongs to S since S is a subspace, and $\mathbf{u}$ is a multiple of $\mathbf{v}_1$, a vector in S. Now define a vector

$$
\mathbf{w} = \mathbf{v}_2 - (\mathbf{v}_2 \cdot \mathbf{u}_1)\mathbf{u}_1 = \begin{bmatrix} -1 \\ 0 \\ 1 \end{bmatrix} - \frac{1}{2} \begin{bmatrix} -1 \\ 1 \\ 0 \end{bmatrix} = \frac{1}{2} \begin{bmatrix} -1 \\ -1 \\ 2 \end{bmatrix} .
$$

The point of this definition is that $\mathbf{w}$ is orthogonal to $\mathbf{u}_1$. To see this, compute

$$\mathbf{u}_1 \cdot \mathbf{w} = \mathbf{u}_1 \cdot (\mathbf{v}_2 - (\mathbf{v}_2 \cdot \mathbf{u}_1)\mathbf{u}_1)$$

$$= \mathbf{u} \cdot \mathbf{v}_2 - (\mathbf{v}_2 \cdot \mathbf{u}_1)|\mathbf{u}_1|^2$$

$$= \mathbf{u} \cdot \mathbf{v}_2 - \mathbf{u} \cdot \mathbf{v}_2 = 0$$

Also $\mathbf{w}$ belongs to S since it is a linear combination of two vectors in S, namely $\mathbf{v}_2$ and $\mathbf{u}_1$. It still belongs to S if we divide by its length to make it a unit vector. Therefore define

$$\mathbf{u}_2 = \frac{1}{|\mathbf{w}|}\mathbf{w} = \frac{1}{\sqrt{6}}\begin{bmatrix} -1 \\ -1 \\ 2 \end{bmatrix} .$$

Now $\{\mathbf{u}_1, \mathbf{u}_2\}$ is pair of orthonormal vectors in S.

The fact that $\{\mathbf{u}_1, \mathbf{u}_2\}$ is pair of orthonormal vectors means it is linearly independent. Indeed, suppose

$$t_1\mathbf{u}_1 + t_2\mathbf{u}_2 = 0 .$$

Then

$$0 = (t_1\mathbf{u}_1 + t_2\mathbf{u}_2) \cdot (t_1\mathbf{u}_1 + t_2\mathbf{u}_2)$$

$$= t_1^2 + t_2^2$$

by the Pythagorean Theorem. Therefore $t_1 = t_2 = 0$. This proves the linear independence. Since we know that $\dim(S) = 2$, Theorem 3 says that $\{\mathbf{u}_1, \mathbf{u}_2\}$ is also a basis for S.

Although the vectors in the second basis $\{\mathbf{u}_1, \mathbf{u}_2\}$ are more complicated than the vectors in the original basis $\{\mathbf{v}_1, \mathbf{v}_2\}$, they have an advantage when we are asking geometric questions. For example, if we express a vector $\mathbf{w}$ in S in terms of $\{\mathbf{u}_1, \mathbf{u}_2\}$; that is,

$$\mathbf{w} = t_1\mathbf{u}_1 + t_2\mathbf{u}_2 ,$$

we know from the Pythagorean Theorem that $|\mathbf{w}|^2 = t_1^2 + t_2^2$. Computing lengths is easy if you use coordinates corresponding to an orthonormal basis.

Theorem 1 tells us that no matter which basis we use to parameterize a subspace, we will need the same number of parameters to do it. This number is $\dim(S)$. Any one–to–one parameterization of the plane in Example 1 (or any other plane) will use two parameters.

Example 2 If A is any $m \times n$ matrix, the dimension of $\mathrm{Ker}(A)$ is $n - r$ where r is the rank of A. This is because we know from chapter 2 that if we row reduce $[A|0]$, and find r pivots, then this is, by definition, the rank of A, and there will be $n - r$ non–pivotal variables, so that $\mathrm{Ker}(A)$ will have a one–to–one parameterization of the form $t_1\mathbf{v}_1 + t_2\mathbf{v}_2 + \cdots + t_{-r}\mathbf{v}_{n-r}$, and $\{\mathbf{v}_1, \mathbf{v}_2, \ldots, \mathbf{v}_{n-r}\}$ is a basis for $\mathrm{Ker}(A)$.

Dimension is a useful concept in part because it provides an easy way to decide when two subspaces S and $\tilde{S}$ are equal. Recall that sets, and in particular subspaces, are equal in case they have the same members which means that S is a subset of $\tilde{S}$, and $\tilde{S}$ is a subset of S. That is, using the standard symbols, $S \subset \tilde{S}$ and $\tilde{S} \subset S$. So one way to show that two subspaces are the same is to show that both

$$S \subset \tilde{S} \quad \text{and} \quad \tilde{S} \subset S .$$

Sometimes one of these is easier to show than the other, and dimension provides a way around that:

$$\tilde{S} \subset S \quad \text{and} \quad \dim(\tilde{S}) = \dim(S) \quad \Rightarrow \quad \tilde{S} = S . \tag{3.4}$$

In fact, any basis for $\tilde{S}$ is a set of $\dim(\tilde{S}) = \dim(S)$ linearly independent vectors in S since $\tilde{S} \subset S$. By Theorem 3, it is a basis for S as well as $\tilde{S}$. Since every element in S is a linear combination of the basis vectors, which also belong to $\tilde{S}$, every vector in S belongs to $\tilde{S}$. In other words, $S \subset \tilde{S}$. Since we already knew $\tilde{S} \subset S$, we now know $\tilde{S} = S$.

The proof of Theorem 4 below provides an important example of this kind of reasoning.

Definition (Orthogonal Complement) If S is any subspace of $I\!\!R^n$, the *orthogonal complement of S* is the set of all vectors $\mathbf{v}$ such that $\mathbf{v} \cdot \mathbf{w}$ for every $\mathbf{w}$ in S. It is denoted by $S^{\perp}$.

Theorem 4 (Dimension and Orthogonal Complements) *Let S be any subspace of $I\!\!R^n$. Then*

(1) The orthogonal complement $S^{\perp}$ is a subspace of $I\!\!R^n$. In particular, if $\{\mathbf{v}_1, \mathbf{v}_2, \ldots \mathbf{v}_d\}$ is any basis for S, and A is the $d \times n$ matrix whose ith row is $\mathbf{v}_i$, then

$$S^{\perp} = \operatorname{Ker}(A) . \tag{3.5}$$

(2) The dimensions of S and $S^{\perp}$ sum to n. That is,

$$\dim(S) + \dim(S^{\perp}) = n \tag{3.6}$$

(3) Taking the orthogonal complement twice gives you back S. That is,

$$(S^{\perp})^{\perp} = S . \tag{3.7}$$

Let's look at an example before we go into the proof.

Example 3: Let S be the plane through the origin in Example 1. We know that the vectors $\{\mathbf{v}_1, \mathbf{v}_2\}$ in (3.3) are a basis, so $S^{\perp}$ is the kernel of A where

$$A = \begin{bmatrix} -1 & 1 & 0 \\ -1 & 0 & 1 \end{bmatrix} .$$

One easily finds that $\text{Ker}(A)$ is spanned by

$$\mathbf{a} = \begin{bmatrix} 1 \\ 1 \\ 1 \end{bmatrix} .$$

which is no surprise since the equation of S is $x_1 + x + 2 + x_3 = 0$, which can be written as $\mathbf{a} \cdot \mathbf{x} = 0$. Thus $S^\perp$ is the line through the origin in the direction of $\mathbf{a}$. This line is just the normal line to the plane S. None of this depends on the particular coefficents of our equation, as long as it is nontrivial. In general, if S is any plane through the origin in $I\!\!R^3$, $S^\perp$ is the normal line of this plane.

Now suppose that $\mathbf{v}$ is orthogonal to $\mathbf{w}$ for every $\mathbf{w}$ in $S^\perp$. In this case, $\mathbf{w}$ is in $S^\perp$ if and only if it is a multiple of $\mathbf{a}$, so $\mathbf{v}$ belongs to $(S^\perp)^\perp$ if and only if $\mathbf{a} \cdot \mathbf{v} = 0$. But this is just the equation of the plane S. So we see that (3.7) holds in this example.

Notice that (3.6) is also satisfied in this example: $\dim(S) = 2$ and $\dim(S^\perp) = 1$, and $2 + 1 = 3$.

Having seen an example, and gained some geometric feeling for the relation between S and $S^\perp$, let's prove the theorem in general.

Proof: Let $\{\mathbf{w}_1, \mathbf{w}_2, \ldots, \mathbf{w}_d\}$ be a basis for S so that $d = \dim(S)$. Let A be the matrix

$$A = \begin{bmatrix} \mathbf{w}_1 \\ \mathbf{w}_2 \\ \vdots \\ \mathbf{w}_d \end{bmatrix}, \text{ and note that } \mathbf{v} \text{ belongs to } \text{Ker}(A) \text{ if and only if } A\mathbf{v} = \begin{bmatrix} \mathbf{w}_1 \cdot \mathbf{v} \\ \mathbf{w}_2 \cdot \mathbf{v} \\ \vdots \\ \mathbf{w}_d \cdot \mathbf{v} \end{bmatrix} = 0. \text{ Thus,}$$

$A\mathbf{v}$ is the zero vector if and only if $\mathbf{v}$ is orthogonal to each of the rows of A, and hence to any linear combination of them. Since the rows are a basis for S, we see that

$$S^\perp = \text{Ker}(A) . \tag{3.8}$$

This shows *(1)*, since Ker(A) is a subspace. Moreover, since the rows of A are linearly independent, being members of a basis, A has full row rank by Theorem 2 of the previous section. Hence the rank of A is $d = \dim(S)$, and so the dimension of $\text{Ker}(A)$ is $n - \dim(S)$ by Example 2. This together with (3.8) proves *(2)*.

Applying *(2)* with S replaced by $S^\perp$, we see that $\dim(S^\perp) + \dim((S^\perp)^\perp) = n$. Since we already know that $\dim(S^\perp) = n - \dim(S)$, this tells us that

$$\dim((S^\perp)^\perp) = n - (n - \dim(S)) = \dim(S)$$

so that S and $(S^\perp)^\perp$ have the same dimension.

Now to prove *(3)*, simply observe that $\mathbf{y}$ belongs to $(S^\perp)^\perp$ if and only if $\mathbf{y} \cdot \mathbf{v} = 0$ for every $\mathbf{v}$ in $S^\perp$. But this is true by the definition of $S^\perp$ if $\mathbf{y}$ belongs to S, so

$$S \subset (S^\perp)^\perp .$$

Since they have the same dimension, *(3)* is proved using (3.4). ■

If you try to prove $(S^\perp)^\perp = S$ by directly showing that $(S^\perp)^\perp \subset S$, you will find that this is not at all as easy as directly showing $S \subset (S^\perp)^\perp$, and you will see the value of the indirect argument using dimension.

At this point you may be wondering why $S^\perp$ is called the "orthogonal complement" of S. It is clear what the role of orthogonality is, but in what sense is $S^\perp$ *complementary* to S?

This is an important question, and let's begin by saying what the answer is not. $S^\perp$ *is not complementary to S in the set theoretic sense*. Recall the set theoretic complement of any subset S of $I\!\!R^n$ is just the set S^c of vectors that don't belong to S. For one thing, all subspaces contain the zero vector, so the set theoretic intersection of S and $S^\perp$ is not empty, while the set theoretic intersection of S and S^c is empty, by definition.

Example 4 Let S be the subspace in $I\!\!R^2$ spanned by the vector $\mathbf{w} = \begin{bmatrix} 1 \\ 1 \end{bmatrix}$. This is the line in $I\!\!R^2$ consisting of all multiples of $\mathbf{w}$. Since S is one dimensional, $\dim(S^\perp) = 2 - 1 = 1$, so $S^\perp$ is also one dimensional. It is easy to see that $\mathbf{v} = \begin{bmatrix} 1 \\ -1 \end{bmatrix}$ satisfies $\mathbf{v} \cdot (a\mathbf{w}) = 0$ for any a, so $\mathbf{v}$ belongs to $S^\perp$. Since $S^\perp$ is one dimensional, this single vector is a basis for $S^\perp$, and so $S^\perp$ is the line consisting of all multiples of $\mathbf{v}$.

Thus, in this example, S and $S^\perp$ are just two lines in $I\!\!R^2$. Together, they don't cover much of $I\!\!R^2$ at all, while, by definition, S and S^c cover the whole plane. However, every vector in $I\!\!R^2$ can be written as a sum of two vectors, one on each of these lines. Moreover, this can be done in exactly one way. It is an *algebraic* sort of complementarity that we are talking about here. We will return to this in the section on orthogonal projections in this chapter.

Given a subspace S and a vector $\mathbf{w}$, how do you check to se if $\mathbf{w}$ belongs to $S^\perp$? That's easy if you know a basis for S, or even a set of vectors whose span is S.

Theorem 4 *Let $\{\mathbf{v}_1, \mathbf{v}_2, \dots, \mathbf{v}_k\}$ be a set of vectors that spans a subspace S of $I\!\!R^n$. Then a vector $\mathbf{w}$ in $I\!\!R^n$ belongs to $S^\perp$ if and only if $\mathbf{w} \cdot \mathbf{v}_j = 0$ for each $j = 1, 2, \dots, n$.*

Proof Suppose that $\mathbf{w}$ belongs to $S^\perp$. Then $\mathbf{w}$ is orthogonal to every vector in S, and in particular, to each of the $\mathbf{v}_j$. On the other hand, suppose that $\mathbf{w} \cdot \mathbf{v}_j = 0$ for each $j = 1, 2, \dots, n$. If $\mathbf{v}$ is any vector in S, then because $\{\mathbf{v}_1, \mathbf{v}_2, \dots, \mathbf{v}_k\}$ spans S, $\mathbf{v}$ can be written as a linear combination $\mathbf{v} = t_1\mathbf{v}_1 + t_2\mathbf{v}_2 + \cdots + t_k\mathbf{v}_n$. Then

$$\mathbf{w} \cdot \mathbf{v} = \mathbf{w} \cdot (t_1\mathbf{v}_1 + t_2\mathbf{v}_2 + \cdots + t_k\mathbf{v}_n)$$

$$= t_1(\mathbf{w} \cdot \mathbf{v}_1) + t_2(\mathbf{w} \cdot \mathbf{v}_2) + \cdots + t_k(\mathbf{w} \cdot \mathbf{v}_n) = 0 .$$

Hence $\mathbf{w}$ is orthogonal to every $\mathbf{v}$ in S, and so $\mathbf{w}$ belongs to $S^\perp$. ∎

Example 5 Let S be the line though the origin in $I\!\!R^3$ in the dierection of $\mathbf{a} = \begin{bmatrix} 1 \\ 1 \\ 1 \end{bmatrix}$.

Every vector on this line is a multiple of $\mathbf{a}$, so the set $\{\mathbf{a}\}$ spans S. (It is also linearly independent, as is any set consisting of a single non–zero vector, so it is a basis for S which is one dimensional, as a line should be). Now let's ask: for which values of t, if any, does the vector $\begin{bmatrix} t \\ 1 \\ 2 \end{bmatrix}$ belong to $S^\perp$. By Theorem 4, this vector belongs to $S^\perp$ exactly when

$$0 = \mathbf{a} \cdot \begin{bmatrix} t \\ 1 \\ 2 \end{bmatrix} = t + 1 + 2 \ .$$

Evidently $t = -3$ is the unique solution.

There is another way of looking at this example: We know that $S^\perp$ is the plane orthogonal to the line along $\mathbf{a}$, so $S^\perp$ is just the solution set of $\mathbf{a} \cdot \mathbf{x} = 0$. Hence, our vector belongs to $S^\perp$ exactly when it satisfies this equation. In the case of planes and lines in $I\!\!R^3$, Theorem 4 is telling us something very familiar. The point of Theorem 4 is that it holds in any dimension.

It is useful to restate Theorem 4 in matrix terms. Let $\{\mathbf{v}_1, \mathbf{v}_2, \ldots, \mathbf{v}_n\}$ be any collection of n vectors in $I\!\!R^m$, and let $S = \mathrm{Sp}(\mathbf{v}_1, \mathbf{v}_2, \ldots, \mathbf{v}_n)$ be their span. We will certainly be able to apply Theorem 4 to S, since we have a spanning set for free!

Let A be the $m \times n$ matrix $A = [\mathbf{v}_1, \mathbf{v}_2, \ldots, \mathbf{v}_n]$. Then by (2.8),

$$\mathrm{Img}(A) = \mathrm{Sp}(\mathbf{v}_1, \mathbf{v}_2, \ldots, \mathbf{v}_n) = S \ . \tag{3.9}$$

Given any $\mathbf{w}$ in $I\!\!R^n$, you could check for membership in S by trying to solve $A\mathbf{x} = \mathbf{w}$. If you succeed, $\mathbf{w}$ belongs to S, and otherwise not.

Now consider $S^\perp$. Theorem 4 says that a vector $\mathbf{w}$ in $I\!\!R^m$ belongs to $S^\perp$ if and only if $\mathbf{v}_j \cdot \mathbf{w} = 0$ for each j. We can write this in terms of A^t as follows: Since $A^t = \begin{bmatrix} \mathbf{v}_1 \\ \mathbf{v}_2 \\ \vdots \\ \mathbf{v}_n \end{bmatrix}$, we have from Theorem 1.5.1 that

$$A^t \mathbf{w} = \begin{bmatrix} \mathbf{v}_1 \cdot \mathbf{v} \\ \mathbf{v}_2 \cdot \mathbf{w} \\ \vdots \\ \mathbf{v}_n \cdot \mathbf{w} \end{bmatrix} \ .$$

Evidently, $\mathbf{v}_j \cdot \mathbf{w} = 0$ for each j if and only if $\mathbf{w}$ is in the kernel of A^t. That is,

$$\mathrm{Ker}(A^t) = S^\perp \ . \tag{3.10}$$

Together, (3.9) and (3.10) say that $\mathrm{Ker}(A^t) = (\mathrm{Img}(A))^\perp$. Then, by (3.7), we have $(\mathrm{Ker}(A^t))^\perp = \mathrm{Img}(A)$.

Our starting point was the set of vectors $\{\mathbf{v}_1, \mathbf{v}_2, \ldots, \mathbf{v}_n\}$ in $I\!\!R^m$. But every $m \times n$ matrix A can be written in the form $[\mathbf{v}_1, \mathbf{v}_2, \ldots, \mathbf{v}_n]$, and so the conclusions $\text{Ker}(A^t) = (\text{Img}(A))^\perp$ and $(\text{Ker}(A^t))^\perp = \text{Img}(A)$ hold for all matrices A. This proves the following theorem:

Theorem 5 (Images and Kernels) *Let A be an $m \times n$ matrix. Then*

$$\text{Img}(A) = \left(\text{Ker}(A^t)\right)^\perp \tag{3.11}$$

and so $A\mathbf{x} = \mathbf{b}$ has a solution if and only if $\mathbf{b}$ is orthogonal to every solution of $A^t\mathbf{y} = 0$.

Theorem 5 is a fundamental result. It does several things or us. First of all, if we know the solution set of $A^t\mathbf{y} = 0$, then Theorem 5 gives us a convenient check on whether or not $A\mathbf{x} = \mathbf{b}$ is solvable. In fact, it often happens that $\text{Ker}(A^t)$ has a low dimension, even one. Suppose that this is the case so that $\text{Ker}(A^t)$ is the line consisting of al multiples of some vector $\mathbf{v}$. Then $\mathbf{b}$ belongs to $\text{Img}(A) = (\text{Ker}(A^t))^\perp$ if and only if $\mathbf{b}$ is orthogonal to every multiple of $\mathbf{v}$, which of course is the case if and only if $\mathbf{b} \cdot \mathbf{v} = 0$. Thus in this case, the necessary and sufficient condition for solvability of $A\mathbf{x} = \mathbf{b}$ is $\mathbf{b} \cdot \mathbf{v} = 0$.

Example 6 Consider the $n \times n$ matrix

$$A = \begin{bmatrix} 1 & 0 & 0 & 0 & \cdots & 0 & -1 \\ -1 & 1 & 0 & 0 & \cdots & 0 & 0 \\ 0 & -1 & 1 & 0 & \cdots & 0 & 0 \\ 0 & 0 & -1 & 1 & \cdots & 0 & 0 \\ & & \vdots & & & \vdots & \\ 0 & 0 & 0 & 0 & \cdots & -1 & 1 \end{bmatrix}.$$

That is, for each $i = 1, 2, \ldots, n$, $(A\mathbf{x})_i = x_i - x_{i-1}$ where x_0 is interpreted as x_n. A bit of thought should convince you that A^t does something pretty similar: $(A^t\mathbf{x})_i = x_i - x_{i+1}$ where x_{n+1} is interpreted as x_1.

Then clearly,

$$|A^t\mathbf{x}|^2 = \sum_{i=1}^{n} (x_i - x_{i+1})^2$$

so that $A\mathbf{x} = 0$ if and only if $x_i = x_i + 1$ for each i. This means that $\mathbf{x}$ is in the kernel of A^t if and only if x_i is independent of i, which means that $\mathbf{x}$ is a multiple of $\mathbf{v} = \begin{bmatrix} 1 \\ 1 \\ 1 \\ \vdots \\ 1 \end{bmatrix}$.

Hence $\text{Ker}(A^t)$ is the line consisting of all multiples of $\mathbf{v}$, and $\text{Img}(A)$ is its orthogonal complement. Thus, $A\mathbf{x} = \mathbf{b}$ is solvable if and only if $\mathbf{b} \cdot \mathbf{v} = 0$. *This is very easy to check.*

For example, if $n = 5$ just to be specific, $A\mathbf{x} = \mathbf{b}$ is solvable for $\mathbf{b} = \begin{bmatrix} 1 \\ 1 \\ 1 \\ 1 \\ -4 \end{bmatrix}$, but not for

$$\mathbf{b} = \begin{bmatrix} 1 \\ 1 \\ 1 \\ 1 \\ -3 \end{bmatrix}.$$

The matrices A and A^t are not artificial; you see that they perform "finite differences", and they come up in many numerical treatments of differential equations. Arguments of the type we have just used, checking for solvability by checking for orthogonality, are call *Fredholm arguments*, and are important in integral equations (the subject that Fredholm himself was investigating) as well as differential and partial differential equations.

Section 4: Finding Bases For the Image of a Linear Transformation

Let A be an $m \times n$ matrix. We know that $\text{Img}(A)$ is the span of the columns of A, and hence the rows of A^t. We saw in Chapter 2 that if we apply any row operation to A^t we get a new matrix $\tilde{A}^t$ such that every row of A^t is a linear combination or rows of A^t, and every row of A^t is a linear combination of the rows of $\tilde{A}^t$. We can now express this more simply: The rows of A^t and $\tilde{A}^t$ span the same subspace of $I\!R^n$. Now row reduce A^t to row reduced form. At every stage, the span of the rows is unchanged, so at the end, when we get to a row reduced matrix U, the span of the rows of U is still $\text{Img}(A)$. Better yet:

- *The non–zero rows of any row reduced form U of A^t are a basis for $\text{Img}(A)$.*

To see why this is the case, we just need to see why the rows of U are independent since we have already explained why they span $\text{Img}(A)$. If we cross out any rows of U that are all zeros, we are left with an $r \times m$ matrix V with a pivot in every row. This matrix will look something like

$$\begin{bmatrix} \bullet & * & * & * & * & * \\ 0 & \bullet & * & * & * & * \\ 0 & 0 & 0 & \bullet & * & * \\ 0 & 0 & 0 & 0 & \bullet & * \end{bmatrix},$$

where the bullets denote pivotal entries as usual. Now suppose that some linear combination of the rows is zero. It is clear that the multiple of the first row must be zero, since nothing in the other rows can cancel the pivotal entry in the first row. But then by the same reasoning, the multiple of the second row must be zero, and so on, to the conclusion that all of the multiples must be zero. This means that the rows are independent.

Alternatively, since the rank of V is r, the number of rows, $\text{Img}(V) = I\!R^r$, and so the rows of V are independent by Theorem 3.2.2. (The second argument is shorter, but it is important to "see" why the non–zero rows of a row reduced matrix are linearly independent).

We now have a method for finding a basis for $\text{Img}(A)$: *row reduce A^t and discard zero rows.*

Example 1 Consider $A = \begin{bmatrix} 2 & 2 & -1 & 1 \\ 0 & -3 & -1 & -4 \\ 1 & 1 & 0 & 1 \\ -1 & 2 & 2 & 4 \end{bmatrix}$ so that $A^t = \begin{bmatrix} 2 & 0 & 1 & -1 \\ 2 & -3 & 1 & 2 \\ -1 & -1 & 0 & 2 \\ 1 & -4 & 1 & 4 \end{bmatrix}$. One

then row reduces A^t, in just a few steps to $\begin{bmatrix} 2 & 0 & 1 & -1 \\ 0 & -1 & 0 & 1 \\ 0 & 0 & 1 & 1 \\ 0 & 0 & 0 & 0 \end{bmatrix}$. Therefore, $\{\mathbf{v}_1, \mathbf{v}_2, \mathbf{v}_3\}$ is

a basis for $\text{Img}(A)$, where

$$\mathbf{v}_1 = \begin{bmatrix} 2 \\ 0 \\ 1 \\ -1 \end{bmatrix}, \quad \mathbf{v}_2 = \begin{bmatrix} 0 \\ -1 \\ 0 \\ 1 \end{bmatrix} \quad \text{and} \quad \mathbf{v}_3 = \begin{bmatrix} 0 \\ 0 \\ 1 \\ 1 \end{bmatrix}. \tag{4.1}$$

- *Further row reduction can lead to an even simpler basis.*

With a few more row operations, we can "clean out" each pivotal column so that the jth pivotal column is $\mathbf{e}_j$. This fully row reduced form of A is called the *echelon form* of A.

In this case it is $\begin{bmatrix} 1 & 0 & 0 & -1 \\ 0 & 1 & 0 & -1 \\ 0 & 0 & 1 & 1 \\ 0 & 0 & 0 & 0 \end{bmatrix}$. Therefore, $\{\mathbf{u}_1, \mathbf{u}_2, \mathbf{u}_3\}$ is a basis for $\text{Img}(A)$, where

$$\mathbf{u}_1 = \begin{bmatrix} 1 \\ 0 \\ 0 \\ -1 \end{bmatrix}, \quad \mathbf{u}_2 = \begin{bmatrix} 0 \\ 1 \\ 0 \\ -1 \end{bmatrix} \quad \text{and} \quad \mathbf{u}_3 = \begin{bmatrix} 0 \\ 0 \\ 1 \\ 1 \end{bmatrix}. \tag{4.2}$$

The approach that we have just taken is a straightforward approach relying on our expertise at row reduction. It tends to yield bases consisting of nice vectors with plenty of zeros in them. This is useful! If you are taking the trouble to calculate a basis, presumably you intend to do some further computations with it. The simpler your basis vectors are, the better they are.

If you are trying to find a basis for S, it can help to know $\dim(S)$. For example, if you know that $\dim(S) = 2$, and you have found a pair of vectors that spans S, then this must be a basis. This follows from Theorem 2 of the previous section, since any spanning set that isn't a basis can be reduced to form a basis. But we can't reduce our pair of vectors to get a basis, since every basis of S must have two vectors.

In the exact same way, any pair of linearly independent vectors in S must be a basis for S. Otherwise you could add in more vectors to get a basis, but you've already got 2, so there is no room to add anything. Here is the general statement:

- *If S is an r dimensional subspace of $\mathbb{R}^n$, then any set of r vectors in S that spans S is a basis for S, and any linearly independent set of r vectors in S is a basis for S.*

We are mostly interested in the case in which $S = \text{Ker}(A)$ or when $S = \text{Img}(A)$ for some matrix A. It turns out that we can determine $\dim(S)$ by just computing the rank of A. The following theorem explains how to do this.

Theorem 1 (The Dimension Formula) *Let A be an $m \times n$ matrix, and let r be the number of pivotal columns. Then*

$$\dim\left(\text{Img}(A)\right) = r$$
$$\dim\left(\text{Img}(A^t)\right) = r$$
$$\dim\left(\text{Ker}(A)\right) = n - r \tag{4.3}$$
$$\dim\left(\text{Ker}(A^t)\right) = m - r \ .$$

In particular,

$$\dim\left(\text{Img}(A)\right) + \dim\left(\text{Ker}(A)\right) = n \ . \tag{4.4}$$

The formula (4.4) is called the *dimension formula*.

Proof: We know that a one–to–one parameterization of Ker(A) will involve $n - r$ parameters, and so dim (Ker(A)) $= n - r$. By Theorem 3.3.5

$$\text{Img}(A^t) = (\text{Ker}(A))^\perp \ .$$

By Theorem 3.3.3, the dimensions of these spaces add up to n, so dim (Img(A^t)) $= r$. That gives us two of our four dimension formulas!

Now let s be the rank of A^t. Then the exact same analysis, with A^t in place of A, tells us that dim (Ker(A^t)) $= m - s$ and dim (Img(A)) $= s$.

We'll complete the proof by showing that $s = r$. We do this by showing that

$$\dim (\text{Img}(A)) = \dim \left(\text{Img}(A^t)\right) \ .$$

Let $\{\mathbf{u}_1, \mathbf{u}_2, \ldots, \mathbf{u}_r\}$ be a basis for Img(A^t). We'll show that $\{A\mathbf{u}_1, A\mathbf{u}_2, \ldots, A\mathbf{u}_r\}$ is a basis for Img(A). Once we have done this, we will know that dim (Img(A)) $= r$ too, and the proof will be complete.

First, clearly each $A\mathbf{u}_j$ belongs to Img(A). Therefore, with $S = \text{Sp}(A\mathbf{u}_1, A\mathbf{u}_2, \ldots, A\mathbf{u}_r)$, $S \subset \text{Img}(A)$, which means that

$$\dim(S) \le \dim (\text{Img}(A)) = s \ . \tag{4.5}$$

Next we claim that $\{A\mathbf{u}_1, A\mathbf{u}_2, \ldots, A\mathbf{u}_r\}$ is linearly independent. Since this set spans S by definition, this would mean it is a basis, and hence dim(S) $= r$. Combining this with (4.5), we have $r \le s$. Repeating the argument with A^t and A interchanged, we'd get that $s \le r$. Altogether we'd know $r = s$, and the proof would be complete. So all we have to do is to show that $\{A\mathbf{u}_1, A\mathbf{u}_2, \ldots, A\mathbf{u}_r\}$ is linearly independent.

Suppose there are numbers $a_1, a_2, \ldots a_r$ so that

$$\sum_{j=1}^{r} a_j(A\mathbf{u}_j) = 0 \ . \tag{4.6}$$

By linearity, (4.6) implies that $A\left(\sum_{j=1}^r a_j \mathbf{u}_j\right) = 0$ which means exactly that $\sum_{j=1}^r a_j \mathbf{u}_j$ is in Ker(A). But since $\{\mathbf{u}_1, \mathbf{u}_2, \ldots, \mathbf{u}_r\}$ is a basis for Img(A^t), $\sum_{j=1}^r a_j \mathbf{u}_j$ is in Img(A^t). Then by Theroem 3.3.5 applied with A^t in place of A, $\sum_{j=1}^r a_j \mathbf{u}_j$ is orthogonal to itself, and hence is the zero vector. Since $\{\mathbf{u}_1, \mathbf{u}_2, \ldots, \mathbf{u}_r\}$ is linearly independent, each $a_j = 0$ in (4.6). Therefore, $\{A\mathbf{u}_1, A\mathbf{u}_2, \ldots, A\mathbf{u}_r\}$ is linearly independent. ∎

Example 2 Let's compute the dimensions of Img(A), Img(A^t), Ker(A) and Ker(A^t) for the 3×5 matrix

$$A = \begin{bmatrix} 2 & 0 & 1 & -1 & 1 \\ 0 & -6 & 0 & 6 & 0 \\ 0 & 0 & 3 & 3 & 1 \end{bmatrix} \ ,$$

and use the results to find bases for these subspaces. This matrix happens to already be in row reduced form, and we can see that $r = 3$. Hence by Theorem 3, $\dim(\text{Img}(A)) = \dim(\text{Img}(A^t)) = 3$. Also, we have $\dim(\text{Ker}(A)) = 5 - 3 = 2$, and $\dim(\text{Ker}(A^t)) = 3 - 3 = 0$.

Now let's go on to find bases for $\text{Img}(A)$ and $\text{Img}(A^t)$, using what we know about dimensions. Since $\dim(\text{Ker}(A^t)) = 0$, $\text{Ker}(A^t) = 0$. Hence the three columns of A^t are linearly independent, and are also a spanning set by Theorem 3.1.2. Hence these three columns themselves are a basis for $\text{Img}(A^t)$.

Next note that since $\text{Img}(A)$ is a three dimensional subspace of $I\!R^3$, we have $\text{Img}(A) = I\!R^3$. There is no point in doing any work to compute a basis for $I\!R^3$; just use the standard basis $\{\mathbf{e}_1, \mathbf{e}_2, \mathbf{e}_3\}$. Dimension counting can save a lot of computation.

As we have seen, there are several ways to go about computing bases for a subspace, and some of the bases are nicer than others. If we are going to do much computation with a basis at all, it will be worth some effort to get a basis in which the calculations are as simple as possible. Here is one very nice sort of basis:

Definition (Orthonormal Basis) Let S be a subspace of $I\!R^n$. A basis $\{\mathbf{u}_1, \mathbf{u}_2, \ldots, \mathbf{u}_k\}$ for S is an *orthonormal basis* in case

$$\mathbf{u}_i \cdot \mathbf{u}_j = \begin{cases} 1 & \text{if } i = j \\ 0 & \text{if } i \neq j \end{cases} . \tag{4.7}$$

Note, in particular, that this means that $|\mathbf{u}_j| = 1$ for each j.

Here is a nice feature of orthonormal bases:

Theorem 2: (Coordinates for Orthonormal bases) *Any set $\{\mathbf{u}_1, \mathbf{u}_2, \ldots, \mathbf{u}_k\}$ of k orthonormal vectors in $I\!R^n$ is independent. Consequently any set $\{\mathbf{u}_1, \mathbf{u}_2, \ldots, \mathbf{u}_k\}$ of k orthonormal vectors in a k-dimensional subspace S of $I\!R^n$ is a basis for S. If*

$$\mathbf{w} = \sum_{j=1}^{k} x_j \mathbf{u}_j \tag{4.8}$$

then the jth coordinate of $\mathbf{w}$ with respect to $\{\mathbf{u}_1, \mathbf{u}_2, \ldots, \mathbf{u}_k\}$, x_j, is given by

$$x_j = \mathbf{w} \cdot \mathbf{u}_j \tag{4.9}$$

and

$$|\mathbf{w}|^2 = \sum_{j=1}^{k} x_j^2 . \tag{4.10}$$

Proof: Consider any linear combination $\mathbf{w}$ of $\{\mathbf{u}_1, \mathbf{u}_2, \ldots, \mathbf{u}_k\}$ as in (4.8). Then

$$|\mathbf{w}|^2 = \mathbf{w} \cdot \mathbf{w} = \left(\sum_{i=1}^{k} x_i \mathbf{u}_i \right) \cdot \left(\sum_{j=1}^{k} x_j \mathbf{u}_j \right) = \sum_{i,j=1}^{k} x_i x_j (\mathbf{u}_i \cdot \mathbf{u}_j) . \tag{4.11}$$

But Using (4.7) in the right side of (4.11) we have (4.10), and (4.10) shows that the linear combination $\mathbf{w}$ in (4.8) is the zero vector only when each $x_j = 0$, and this proves the independence. If the dimension of S is k, then $\{\mathbf{u}_1, \mathbf{u}_2, \ldots, \mathbf{u}_k\}$ must also span S, and hence it is a basis. Finally, taking the dot product of $\mathbf{w}$ with $\mathbf{u}_i$ we see that

$$\mathbf{u}_i \cdot \mathbf{w} = \mathbf{u}_i \cdot \left(\sum_{j=1}^{k} x_j \mathbf{u}_j\right) = \sum_{j=1}^{k} x_j (\mathbf{u}_i \mathbf{u}_j) = x_i$$

by another application of (4.7). Together with (4.10) this proves both (4.9) and (4.10). ■

Recall that if $\{\mathbf{v}_1, \mathbf{v}_2, \ldots, \mathbf{v}_k\}$ is a basis for S that is not orthonormal, and $\mathbf{w}$ belongs to S, we have to solve an equation, namely

$$[\mathbf{v}_1, \mathbf{v}_2, \ldots, \mathbf{v}_k] \begin{bmatrix} x_1 \\ x_2 \\ \vdots \\ x_k \end{bmatrix} = \mathbf{w} \tag{4.12}$$

to find the coordinates $\mathbf{x}_j$ of $\mathbf{w}$ in this basis. But if $\{\mathbf{v}_1, \mathbf{v}_2, \ldots, \mathbf{v}_k\}$ is an orthonormal basis, we can compute x_j by just taking a dot product: $x_j = \mathbf{v}_j \cdot \mathbf{w}$. This is much easier. Also the fact that one can compute the length of a vector directly in terms of its coordinates with respect to an orthonormal basis, as in (4.10), makes orthonormal coordinates very efficient in any problem where lengths or distances have to be computed.

Example 3 In Example 1 of Section 3, we found two bases, $\{\mathbf{v}_1, \mathbf{v}_2\}$ and $\{\mathbf{u}_1, \mathbf{u}_2\}$ for S, the plane through the origin in $I\!R^3$ given by $x_1 + x_2 + x_3 = 0$. These two bases were given by

$$\mathbf{v}_1 = \begin{bmatrix} -1 \\ 1 \\ 0 \end{bmatrix} \qquad \text{and} \qquad \mathbf{v}_2 = \begin{bmatrix} -1 \\ 0 \\ 1 \end{bmatrix}$$

and

$$\mathbf{u}_1 = \frac{1}{\sqrt{2}} \begin{bmatrix} -1 \\ 1 \\ 0 \end{bmatrix} \qquad \text{and} \qquad \mathbf{u}_2 = \frac{1}{\sqrt{6}} \begin{bmatrix} -1 \\ -1 \\ 2 \end{bmatrix} .$$

Let's find the coodinates of $\mathbf{v}_2$ with respect to the basis $\{\mathbf{u}_1, \mathbf{u}_2\}$. By Theorem 2,

$$\mathbf{v}_2 = (\mathbf{v}_2 \cdot \mathbf{u}_1)\mathbf{u}_1 + (\mathbf{v}_2 \cdot \mathbf{u}_1)\mathbf{u}_1$$

$$= \frac{1}{\sqrt{2}}\mathbf{u}_1 + \sqrt{\frac{3}{2}}\mathbf{u}_2 .$$

There is another way to look at (4.8) and (4.10): The matrix $U = [\mathbf{u}_1, \mathbf{u}_2, \ldots, \mathbf{u}_k]$ represents a length preserving transformation, since according to these equations, for any $\mathbf{x}$ in $I\!\!R^k$, $|U\mathbf{x}| = |\mathbf{x}|^2$. Now we know from Chapter One that whenever U represents a length preserving transformation, U^t is a left inverse of U. That is,

$$U^t U = I .$$

Example 4 Continuing with Example 3, the matrix $[\mathbf{u}_1, \mathbf{u}_2]$ is an isometry.

Now suppose that $k = n$ so that U is a square $n \times n$ matrix. By a key result of Chapter Two, any square matrix with a left inverse is invertible, with the inverse being the left inverse, and hence U^t is the inverse of U. This means that

$$UU^t = I$$

and so $U = (U^t)^t$ is the left inverse of U^t. This means that U^t is also an isometry, so the columns of U^t, which are the rows of U, are orthonormal. This leads to the conclusion that if U is an $n \times n$ matrix whose columns are an orthonormal basis of $I\!\!R^n$, so are the rows!

The fact that $n \times n$ matrices U can be inverted merely by taking the transpose if and only if they have orthonormal columns identifies a particularly important class of matrices:

Definition (Orthogonal Matrices) An $n \times n$ matrix U is orthogonal in case the columns of U are orthonormal.

We close this section by considering the relations between coordinates for different bases.

Change of basis

Let $\{\mathbf{v}_1, \mathbf{v}_2, \ldots, \mathbf{v}_k\}$ and $\{\mathbf{u}_1, \mathbf{u}_2, \ldots, \mathbf{u}_k\}$ be two bases for a subspace S of $I\!\!R^n$. Then any $\mathbf{w}$ in S can be expressed as a linear combination of either set of basis vectors. That is, we have both

$$\mathbf{w} = x_1\mathbf{u}_1 + x_2\mathbf{u}_2 + \cdots + x_k\mathbf{u}_k \quad \text{and} \quad \mathbf{w} = y_1\mathbf{v}_1 + y_2\mathbf{v}_2 + \cdots + y_k\mathbf{v}_k \quad (4.13)$$

for some sets of coordinates $x_1, x_2, \ldots, x_k$ and $y_1, y_2, \ldots, y_k$. *How are these two sets of coordinates related?*

To answer this question, let's organize the coordinates into coordinate vectors, putting

$$\mathbf{x} = \begin{bmatrix} x_1 \\ x_2 \\ \vdots \\ x_k \end{bmatrix} \quad \text{and} \quad \mathbf{y} = \begin{bmatrix} y_1 \\ y_2 \\ \vdots \\ y_k \end{bmatrix} .$$

If we then organize the basis vectors into the matrices

$$U = [\mathbf{u}_1, \mathbf{u}_2, \ldots, \mathbf{u}_k] \quad \text{and} \quad V = [\mathbf{v}_1, \mathbf{v}_2, \ldots, \mathbf{v}_k] ,$$

we can rewrite (4.13) in matrix form:

$$\mathbf{w} = U\mathbf{x} \quad \text{and} \quad \mathbf{w} = V\mathbf{y} \; . \tag{4.14}$$

This means that $V\mathbf{y} = U\mathbf{x}$. Since the columns of U and V are independent,

$$\text{Ker}(V) = \text{Ker}(U) = 0 \; . \tag{4.15}$$

In the special case that $k = n$ so that V and U are square matrices, this means they are invertible, and we have

$$\mathbf{y} = V^{-1}U\mathbf{x} \; . \tag{4.16}$$

If we define $R = V^{-1}U$, we can call R the *change of basis matrix* that transforms coordinates with respect to the basis $\{\mathbf{u}_1, \mathbf{u}_2, \ldots, \mathbf{u}_k\}$ into coordinates with respect to the basis $\{\mathbf{v}_1, \mathbf{v}_2, \ldots, \mathbf{v}_k\}$.

What if $k < n$, so that U and V are not square? There is still a unique $k \times k$ matrix R satisfying the equation

$$VR = U \; . \tag{4.17}$$

Indeed, $\text{Img}(V)$ is the span of the columns of V, and this is S since the columns are a basis for S. Together with the left half of (4.15), this means that $V\mathbf{y} = \mathbf{w}$ has a unique solution for each $\mathbf{w}$ in S. In particular, let $\mathbf{r}_j$ be this solution when $\mathbf{w} = \mathbf{u}_j$. That is, let $\mathbf{r}_j$ be the unique vector in $I\!\!R^k$ such that

$$V\mathbf{r}_j = \mathbf{u}_j \; .$$

Now define the $k \times k$ matrix R by

$$R = [\mathbf{r}_1, \mathbf{r}_2, \ldots, \mathbf{r}_k] \; . \tag{4.18}$$

Then

$$VR = [V\mathbf{r}_1, V\mathbf{r}_2, \ldots, V\mathbf{r}_k] = [\mathbf{u}_1, \mathbf{u}_2, \ldots, \mathbf{u}_k] = U \; ,$$

and so once again, (4.17) is satisfied. Also if $\tilde{R}$ is any $k \times k$ matrix such that $V\tilde{R} = U$, The jth column of $\tilde{R}$ must satisfy $V\mathbf{x} = \mathbf{u}_j$. By (4.15), the only solution of this is $\mathbf{r}_j$, the jth column of R. So the matrix equation $VR = U$ has a unique solution.

Combining (4.14) and (4.17),

$$\mathbf{w} = U\mathbf{x} = (VR)\mathbf{x} = V(R\mathbf{x}) \; .$$

Hence $V\mathbf{y} = \mathbf{w} = V(R\mathbf{x})$, or $V(\mathbf{y} - R\mathbf{x}) = 0$. By (4.15),

$$\mathbf{y} = R\mathbf{x} \; .$$

In summary:

• *The unique matrix matrix R satsifying $VR = U$ converts coordinates with respect to the basis $\{\mathbf{u}_1, \mathbf{u}_2, \ldots, \mathbf{u}_k\}$ to coordinates with respect to the basis $\{\mathbf{v}_1, \mathbf{v}_2, \ldots, \mathbf{v}_k\}$. Moreover, $R_{i,j}$ is the coeffcent of $\mathbf{v}_i$ when $\mathbf{u}_j$ is written as a linear combination of $\{\mathbf{v}_1, \mathbf{v}_2, \ldots, \mathbf{v}_k\}$.*

Definition Let $\{\mathbf{v}_1, \mathbf{v}_2, \ldots, \mathbf{v}_k\}$ and $\{\mathbf{u}_1, \mathbf{u}_2, \ldots, \mathbf{u}_k\}$ be two bases for a subspace S of $I\!\!R^n$. Let $V = [\mathbf{v}_1, \mathbf{v}_2, \ldots, \mathbf{v}_k]$ and let $U = [\mathbf{u}_1, \mathbf{u}_2, \ldots, \mathbf{u}_k]$. The unique $k \times k$ matrix R such that $VR = U$ is called the *change of basis matrix* from the coordinates with respect to the basis $\{\mathbf{u}_1, \mathbf{u}_2, \ldots, \mathbf{u}_k\}$ to coordinates with respect to the basis $\{\mathbf{v}_1, \mathbf{v}_2, \ldots, \mathbf{v}_k\}$.

You see the reason for the name in the definition: R converts "u coordinates" to "v coordinates. Also, the entries of the jth column of R give the coordinates in the expansion of $\mathbf{u}_j$ as a linear combination of $\{\mathbf{v}_1, \mathbf{v}_2, \ldots, \mathbf{v}_k\}$.

Example 5 Let S be the image of A, where A is the matrix given in Example 1. We found two bases for S in Example 1, namely $\{\mathbf{v}_1, \mathbf{v}_2, \mathbf{v}_3\}$ and $\{\mathbf{u}_1, \mathbf{u}_2, \mathbf{u}_3\}$. These vectors are given explicitly in (4.1) and (4.2). From (4.1) we have

$$\mathbf{v}_1 = \begin{bmatrix} 2 \\ 0 \\ 1 \\ -1 \end{bmatrix} \quad , \quad \mathbf{v}_2 = \begin{bmatrix} 0 \\ -1 \\ 0 \\ 1 \end{bmatrix} \quad \text{and} \quad \mathbf{v}_3 = \begin{bmatrix} 0 \\ 0 \\ 1 \\ 1 \end{bmatrix} ,$$

and hence

$$V = [\mathbf{v}_1, \mathbf{v}_2, \mathbf{v}_3] = \begin{bmatrix} 2 & 0 & 0 \\ 0 & -1 & 0 \\ 1 & 0 & 1 \\ -1 & 1 & 1 \end{bmatrix}$$

We could find $R = [\mathbf{r}_1, \mathbf{r}_2, \mathbf{r}_3]$ by solving the three equations

$$V\mathbf{r}_1 = \mathbf{u}_1 \qquad V\mathbf{r}_2 = \mathbf{u}_2 \qquad \text{and} \qquad V\mathbf{r}_3 = \mathbf{u}_3 ,$$

where, according to (4.2)

$$\mathbf{u}_1 = \begin{bmatrix} 1 \\ 0 \\ 0 \\ -1 \end{bmatrix} \quad , \quad \mathbf{u}_2 = \begin{bmatrix} 0 \\ 1 \\ 0 \\ -1 \end{bmatrix} \quad \text{and} \quad \mathbf{u}_3 = \begin{bmatrix} 0 \\ 0 \\ 1 \\ 1 \end{bmatrix} .$$

In concrete terms for example, solving $V\mathbf{r}_1 = \mathbf{u}_1$ maens finding numbers x, y and z so that

$$x \begin{bmatrix} 2 \\ 0 \\ 1 \\ -1 \end{bmatrix} + y \begin{bmatrix} 0 \\ -1 \\ 0 \\ 1 \end{bmatrix} + z \begin{bmatrix} 0 \\ 0 \\ 1 \\ 1 \end{bmatrix} = \begin{bmatrix} 1 \\ 0 \\ 0 \\ -1 \end{bmatrix} .$$

Without much effort, you can find the unique solution: $x = z = 1/2$, and $y = 0$. Hence $\mathbf{r}_1 = \begin{bmatrix} 1/2 \\ 0 \\ 1/2 \end{bmatrix}$. The other two, $\mathbf{r}_2$ and $\mathbf{r}_3$, can be found in this way with even less effort.

However, just as with computing inverses, we can be more efficient, and do this in "parallel". Let

$$U = [\mathbf{u}_1, \mathbf{u}_2, \mathbf{u}_3] = \begin{bmatrix} 1 & 0 & 0 \\ 0 & 1 & 0 \\ 0 & 0 & 1 \\ -1 & -1 & 1 \end{bmatrix}.$$

Then form the augmented matrix matrix $[V|U]$, and row reduce, just as in the computation of an inverse.

In the case at hand,

$$[V|U] = \begin{bmatrix} 2 & 0 & 0 & | & 1 & 0 & 0 \\ 0 & -1 & 0 & | & 0 & 1 & 0 \\ 1 & 0 & 1 & | & 0 & 0 & 1 \\ -1 & 1 & 1 & | & -1 & -1 & 1 \end{bmatrix}.$$

Row reducing all the way to echelon form, as we did to get the second basis in Example 1, we find

$$\begin{bmatrix} 1 & 0 & 0 & | & 1/2 & 0 & 0 \\ 0 & 1 & 0 & | & 0 & -1 & 0 \\ 0 & 0 & 1 & | & -1/2 & 0 & 1 \\ 0 & 0 & 0 & | & 0 & 0 & 0 \end{bmatrix}.$$

This tells us that

$$V\begin{bmatrix} 1/2 \\ 0 \\ -1/2 \end{bmatrix} = \mathbf{u}_1 \qquad V\begin{bmatrix} 0 \\ -1 \\ 0 \end{bmatrix} = \mathbf{u}_2 \qquad \text{and} \qquad V\begin{bmatrix} 0 \\ 0 \\ 1 \end{bmatrix} = \mathbf{u}_2 \ .$$

Hence the change of basis matrix R is

$$R = \begin{bmatrix} 1/2 & 0 & 0 \\ 0 & -1 & 0 \\ -1/2 & 0 & 1 \end{bmatrix},$$

which is just the upper right 3×3 matrix in our row reduction of $[V|U]$. The simplicity of R in this case reflects the close relation between the two bases; recall how we got them in Example 1.

Example 6 If $\{\mathbf{u}_1, \mathbf{u}_2, \ldots, \mathbf{u}_k\}$ is an orthonormal basis for a subspace S of $\mathbb{R}^n$, and $\{\mathbf{v}_1, \mathbf{v}_2, \ldots, \mathbf{v}_k\}$ is any other basis, it is very easy to write down the change of basis matrix from the "v" coordinates to the "u" coordinates. That is, it is very easy to write down the $k \times k$ matrix R so that

$$UR = V \ .$$

The jth column of R, $\mathbf{r}_j$ satisfies

$$U\mathbf{r}_j = \mathbf{v}_j$$

which just means that it is the coordinate vector of $\mathbf{v}_j$ in the basis $\{\mathbf{u}_1, \mathbf{u}_2, \ldots, \mathbf{u}_k\}$. Since this is an orthonormal basis, Theorem 2 tells us

$$\mathbf{v}_j = (\mathbf{v}_j \cdot \mathbf{u}_1)\mathbf{u}_1 + (\mathbf{v}_j \cdot \mathbf{u}_2)\mathbf{u}_2 + \cdots + (\mathbf{v}_j \cdot \mathbf{u}_k)\mathbf{u}_k \ .$$

This says that the solution $\mathbf{r}_j$ is

$$\mathbf{r}_j = \begin{bmatrix} \mathbf{v}_j \cdot \mathbf{u}_1 \\ \mathbf{v}_j \cdot \mathbf{u}_2 \\ \vdots \\ \mathbf{v}_j \cdot \mathbf{u}_k \end{bmatrix} \ .$$

In other words, $R_{i,j} = \mathbf{u}_i \cdot \mathbf{v}_j$, and we don't need to solve any equations to find R, we just need to compute dot products. Orthonormal bases are very convenient!

This special case is worth recording in a theorem:

Theorem 3 (Change of Basis for Orthonormal Bases) *Let* $\{\mathbf{v}_1, \mathbf{v}_2, \ldots, \mathbf{v}_k\}$ *and* $\{\mathbf{u}_1, \mathbf{u}_2, \ldots, \mathbf{u}_k\}$ *be two bases for a subspace* S *of* $\mathbb{R}^n$. *Let* $V = [\mathbf{v}_1, \mathbf{v}_2, \ldots, \mathbf{v}_k]$ *and* $U = [\mathbf{u}_1, \mathbf{u}_2, \ldots, \mathbf{u}_k]$ *as usual. The change of basis matrix* R *satisfiying* $V = UR$ *is given by*

$$R_{i,j} = \mathbf{u}_i \cdot \mathbf{v}_j \ . \tag{4.19}$$

Example 7 Let $\{\mathbf{u}_1, \mathbf{u}_2\}$ and $\{\mathbf{u}_1, \mathbf{u}_2\}$ be the two bases considered in Example 3. Since $\{\mathbf{u}_1, \mathbf{u}_2\}$ is orthonormal, the change of basis matrix R from the "u coordinates" to the "v coordinates" is

$$R = \begin{bmatrix} \mathbf{u}_1 \cdot \mathbf{v}_1 & \mathbf{u}_1 \cdot \mathbf{v}_2 \\ \mathbf{u}_2 \cdot \mathbf{v}_1 & \mathbf{u}_2 \cdot \mathbf{v}_2 \end{bmatrix} = \begin{bmatrix} \sqrt{2} & 1/\sqrt{2} \\ 0 & \sqrt{3/2} \end{bmatrix} \ .$$

Section 5: Gramm-Schmidt Orthonormalization

Suppose we have a subspace S of $I\!R^n$, and we have a set of vectors $\{v_1, v_2, \ldots, v_r\}$ for S whose span is S. It might even be a basis of S, but all we will assume is that it spans S.

For example, when S is $\text{Img}(A)$ for some matrix A, there is an obvious choice for the spanning set, namely, the columns of A.

- *There is a simple procedure for producing an orthonormal basis for S out of any spanning set for S.*

This procedure is called the *Gramm–Schmidt orthonormalization procedure*, and our main goals in this section are to explain how it works, and what it is good for.

Let's begin by stepping through the procedure in a simple example: Let S be the plane through the origin in $I\!R^3$ that is spanned by $\{v_1, v_2\}$ where

$$\mathbf{v}_1 = \begin{bmatrix} 1 \\ 1 \\ 0 \end{bmatrix} \quad \text{and} \quad \mathbf{v}_2 = \begin{bmatrix} 2 \\ 1 \\ 1 \end{bmatrix} .$$

Since these vectors are not proportional, they are linearly independent, and so $\{v_1, v_2\}$ is actualy a basis and not just a spanning set.

Here is how the Gram–Schmidt orthonormalization procedure turns this basis into an orthonormal basis $\{u_1, u_2\}$ for S.

Step One: Define the first element, $\mathbf{u}_1$, of our new basis to be the *normalization* of $\mathbf{v}_1$, the first element of our old basis. That is, we rescale $\mathbf{v}_1$ to make it a unit vector:

$$\mathbf{u}_1 = \frac{1}{|\mathbf{v}_1|}\mathbf{v}_1 . \tag{5.1}$$

Notice that $\mathbf{u}_1$ belongs to S since it is a multiple of a vector in S, namely $\mathbf{v}_1$.

Step Two: When we get to the second vector, we have to take the requirement for orthogonality into account, not just normalization. Here is what to do: Construct a vector $\mathbf{w}_2$ that is orthogonal to $\mathbf{u}_1$ using $\mathbf{v}_2$ and $\mathbf{u}_1$. This is done by defining

$$\mathbf{w}_2 = \mathbf{v}_2 - a\mathbf{u}_1 \tag{5.2}$$

and then choosing a to make $\mathbf{w}_2$ orthogonal to $\mathbf{u}_1$. Notice that for any choice of a, $\mathbf{w}_2$ will be a linear combination of $\mathbf{u}_1$ and $\mathbf{v}_2$, which are both vectors in S. Since S is a subspace, $\mathbf{w}_2$ also belongs ot S.

To get an equation for what a should be, we use the fact that two vectors are orthogonal if and only if their dot product is zero. So we must choose a to make

$$0 = \mathbf{w}_2 \cdot \mathbf{u}_1 = (\mathbf{v}_2 - a\mathbf{u}_1) \cdot \mathbf{u}_1 = \mathbf{v}_2 \cdot \mathbf{u}_1 - a|\mathbf{u}_1|^2 = \mathbf{v}_2 \cdot \mathbf{u}_1 - a .$$

Evidently, the choice $a = \mathbf{v}_2 \cdot \mathbf{u}_1$ makes $\mathbf{w}_2$ orthogonal to $\mathbf{u}_1$, and with this choice of a,

$$\mathbf{w}_2 = \mathbf{v}_2 - (\mathbf{v}_2 \cdot \mathbf{u}_1)\mathbf{u}_1 \ . \tag{5.3}$$

At this stage we have a pair of orthogonal vectors, $\{\mathbf{u}_1, \mathbf{w}_2\}$ in S.

Step Three: Divide $\mathbf{w}_2$ by its length to obtain a unit vector $\mathbf{u}_2$:

$$\mathbf{u}_2 = \frac{1}{|\mathbf{w}_2|}\mathbf{w}_2 \ . \tag{5.4}$$

This is called *normalizing* $\mathbf{w}_2$.

We now have a pair $\{\mathbf{u}_1, \mathbf{u}_2\}$ of orthonormal vectors in S. Any orthonormal set is linearly independent, and since the dimension of S is 2, $\{\mathbf{u}_1, \mathbf{u}_2\}$ must also span S. So it is indeed an orthonormal basis.

Of course we couldn't do the third step if it involved dividing by zero. But $\mathbf{w}_2 = 0$ if and only if

$$0 = \mathbf{v}_2 - (\mathbf{v}_2 \cdot \mathbf{u}_1)\mathbf{u}_1 = \mathbf{v}_2 - \frac{\mathbf{v}_2 \cdot \mathbf{v}_1}{|\mathbf{v}_1|^2}\mathbf{v}_1$$

which would mean that $\mathbf{v}_2$ is a multiple of $\mathbf{v}_1$. If $\{\mathbf{v}_1, \mathbf{v}_2\}$ is linearly independent, this cannot happen.

That's the proceedure, which clearly does what it is supposed to, and easily generalizes to higher dimensions for S.

Example 1: Now let's work through the case discussed above, and do the actual computations. First, $|\mathbf{v}_1| = \sqrt{2}$ so that by (5.1)

$$\mathbf{u}_1 = \frac{1}{\sqrt{2}}\begin{bmatrix} 1 \\ 1 \\ 0 \end{bmatrix} \ .$$

Second, we compute that $\mathbf{v}_2 \cdot \mathbf{u}_1 = 3/\sqrt{2}$ and hence by (5.3)

$$\mathbf{w}_2 = \begin{bmatrix} 2 \\ 1 \\ 1 \end{bmatrix} - \frac{3}{\sqrt{2}}\frac{1}{\sqrt{2}}\begin{bmatrix} 1 \\ 1 \\ 0 \end{bmatrix} = \frac{1}{2}\begin{bmatrix} 1 \\ -1 \\ 2 \end{bmatrix} \ .$$

Finally, by (5.4),

$$\mathbf{u}_2 = \frac{1}{|\mathbf{w}_2|}\mathbf{w}_2 = \frac{1}{\sqrt{6}}\begin{bmatrix} 1 \\ -1 \\ 1 \end{bmatrix}$$

and our orthonormal basis is $\{\mathbf{u}_1, \mathbf{u}_2\}$ where

$$\mathbf{u}_1 = \frac{1}{\sqrt{2}} \begin{bmatrix} 1 \\ 1 \\ 0 \end{bmatrix} \quad \text{and} \quad \mathbf{u}_2 = \frac{1}{\sqrt{6}} \begin{bmatrix} 1 \\ -1 \\ 2 \end{bmatrix}.$$

More than two vectors

Next, what about more than two vectors? Let $\{\mathbf{v}_1, \mathbf{v}_2, \dots, \mathbf{v}_k\}$ be a subspace S of $I\!R^n$. Let's assume for now that $\{\mathbf{v}_1, \mathbf{v}_2, \dots, \mathbf{v}_k\}$ is also linearly independent. After we've seen how this get's used, we'll see what to do without this assumption. But for now, $\{\mathbf{v}_1, \mathbf{v}_2, \dots, \mathbf{v}_k\}$ is a basis of S.

Define $\mathbf{u}_1$ and $\mathbf{u}_2$ just as above.

$$\mathbf{u}_1 = \frac{1}{|\mathbf{v}_1|} \mathbf{v}_1 \tag{5.5}$$

and

$$\mathbf{u}_2 = \frac{1}{|\mathbf{w}_2|} \mathbf{w}_2 \quad \text{where} \quad \mathbf{w}_2 = \mathbf{v}_2 - (\mathbf{v}_2 \cdot \mathbf{u}_1) \mathbf{u}_1 .$$

As before, $\mathbf{w}_2$ is a non–zero linear combination of $\mathbf{v}_1$ and $\mathbf{v}_2$ and hence is in S, and cannot be zero.

Next, define

$$\mathbf{w}_3 = \mathbf{v}_3 - (\mathbf{v}_3 \cdot \mathbf{u}_1) \mathbf{u}_1 - (\mathbf{v}_3 \cdot \mathbf{u}_2) \mathbf{u}_2 . \tag{5.6}$$

You can easily check that since $\mathbf{u}_1 \cdot \mathbf{u}_1 = \mathbf{u}_2 \cdot \mathbf{u}_2 = 1$ and $\mathbf{u}_1 \cdot \mathbf{u}_2 = 0$,

$$\mathbf{w}_3 \cdot \mathbf{u}_1 = \mathbf{w}_3 \cdot \mathbf{u}_2 = 0 .$$

Hence $\mathbf{w}_3$ is orthogonal to $\mathbf{u}_1$ and $\mathbf{u}_2$. Also, since $\mathbf{u}_1$ and $\mathbf{u}_2$ are linear combinations of $\mathbf{v}_1$ and $\mathbf{v}_2$, it is clear that $\mathbf{w}_3$ is a non–zero linear combination of $\mathbf{v}_1$, $\mathbf{v}_2$ and $\mathbf{v}_3$. Therefore it is in S, and since these vectors are linearly independent, $|\mathbf{w}_3| > 0$, and we can divide by it to define:

$$\mathbf{u}_3 = \frac{1}{|\mathbf{w}_3|} \mathbf{w}_3$$

where $\mathbf{w}_3$ is given by (5.6).

By now you see the pattern, and we make the following definition

Definition: Given a basis $\{\mathbf{v}_1, \mathbf{v}_2, \dots, \mathbf{v}_r\}$ of a subspace S of $I\!R^n$, its *Gramm–Schmidt orthonormalization* is the basis $\{\mathbf{u}_1, \mathbf{u}_2, \dots, \mathbf{u}_r\}$ where the $\mathbf{u}_j$ are defined recursively through (5.5) and then, for $j \geq 2$ by

$$\mathbf{u}_j = \frac{1}{|\mathbf{w}_j|} \mathbf{w}_j \quad \text{where} \quad \mathbf{w}_j = \mathbf{v}_j - \sum_{i=1}^{j-1} (\mathbf{v}_j \cdot \mathbf{u}_i) \mathbf{u}_i . \tag{5.7}$$

Notice that the sum on the right in (5.7) only involves $\mathbf{u}_1$ through $\mathbf{u}_{j-1}$ so once these are found, we can use (5.7) to find $\mathbf{u}_j$, and so on. Thus, this is a valid recursive definition.

Theorem 1 (Gram-Schmidt Procedure) *The Gramm-Schmidt procedure always produces an orthonormal basis $\{\mathbf{u}_1, \mathbf{u}_2, \ldots, \mathbf{u}_r\}$ for a subspace S out of an arbitrary basis $\{\mathbf{v}_1, \mathbf{v}_2, \ldots, \mathbf{v}_r\}$ for S.*

Proof: It should be clear from the discussion above that each $\mathbf{w}_j$ is a linear combination of $\{\mathbf{v}_1, \ldots, \mathbf{v}_j\}$ since we "build" the new basis out of the original basis by repeatedly taking certain linear combinations, and $\mathbf{v}_j$ doesn't enter into these until we get to the stage where we build $\mathbf{w}_j$. In particular, in the coordinate expansion of $\mathbf{w}_j$ in the original basis $\{\mathbf{v}_1, \mathbf{v}_2, \ldots, \mathbf{v}_r\}$, we have from (5.7) that the coefficient of $\mathbf{v}_j$ is 1. Hence $\mathbf{w}_j$ is a non-zero linear combination of basis elements of S. By the linear independence of the basis, it is a non–zero element of S, and we do not divide by zero in the normalization procedure. Finally, if $k < j$, we have from (5.7) that

$$\mathbf{u}_k \cdot \mathbf{w}_j = \mathbf{u}_k \cdot \left(\mathbf{v}_j - \sum_{i=1}^{j-1} (\mathbf{v}_j \cdot \mathbf{u}_i)\mathbf{u}_i \right)$$

$$= \mathbf{u}_k \cdot \mathbf{v}_j - \sum_{i=1}^{j-1} (\mathbf{v}_j \cdot \mathbf{u}_i)\mathbf{u}_k \cdot \mathbf{u}_i$$

$$= \mathbf{u}_k \cdot \mathbf{v}_j - \mathbf{u}_k \cdot \mathbf{v}_j = 0 \ .$$

(Only the $i = k$ term in the the sum is not zero by the orthogonality).

Since $\mathbf{u}_j$ is a multiple of $\mathbf{w}_j$, $\mathbf{u}_j$ is also orthogonal to $\mathbf{u}_k$ for $k = 1, 2, \ldots, j - 1$. ∎

The next theorem explains the relation between an original basis and its Gramm–Schmidt orthonormalization.

Theorem 2 (Change of Basis for Gramm–Schmidt) *Let $\{\mathbf{v}_1, \mathbf{v}_2, \ldots, \mathbf{v}_r\}$ be a basis for a subspace S of $\mathbb{R}^n$, and let $\{\mathbf{u}_1, \mathbf{u}_2, \ldots, \mathbf{u}_r\}$ be its Gramm–Schmidt orthonormalization. Let $U = [\mathbf{u}_1, \mathbf{u}_2, \ldots, \mathbf{u}_r]$ and let $V = [\mathbf{v}_1, \mathbf{v}_2, \ldots, \mathbf{v}_r]$ as usual. Then the $r \times r$ change of basis matrix R with $V = UR$ is given by*

$$R_{i,j} = \mathbf{u}_i \cdot \mathbf{v}_j \ . \tag{5.8}$$

The matrix R is upper-triangular and invertible.

Proof: The formula (5.8) for the change of basis matrix R follows from (4.19) of Theorem 3.4.3. Next, rewriting the definition of $\mathbf{w}_j$ in (5.7),

$$\mathbf{v}_j = \mathbf{w}_j + \sum_{i=1}^{j-1} (\mathbf{v}_j \cdot \mathbf{u}_i)\mathbf{u}_i = |\mathbf{w}_j|\mathbf{u}_j + \sum_{i=1}^{j-1} (\mathbf{v}_j \cdot \mathbf{u}_i)\mathbf{u}_i \ , \tag{5.9}$$

since $\mathbf{w}_j = |\mathbf{w}_j|\mathbf{u}_j$ by the left side of (5.7). This displays $\mathbf{v}_j$ as a linear combination of $\{\mathbf{u}_1, \mathbf{u}_2, \ldots, \mathbf{u}_j\}$. Since the $\{\mathbf{u}_1, \mathbf{u}_2, \ldots, \mathbf{u}_r\}$ are orthonormal, it follows from (5.9) that

$$\mathbf{u}_i \cdot \mathbf{v}_j = 0 \qquad \text{for} \qquad i > j \ . \tag{5.10}$$

This shows that R is upper triangular. Finally, change of basis matrices are always invertible. ∎

Example 2: Use the Gramm–Schmidt procedure to compute an orthonormal basis for Img(A) where A is the 4×3 matrix $A = \begin{bmatrix} 1 & 2 & 1 \\ 0 & 2 & 1 \\ 2 & -1 & 0 \\ 1 & 1 & 1 \end{bmatrix}$.

We first find a basis for Img(A): Row reducing A^t, and then cleaning out the pivotal columns with further row operations, we find the matrix

$$\begin{bmatrix} 1 & 0 & 0 & -1 \\ 0 & 1 & 0 & 2 \\ 0 & 0 & 1 & 1 \end{bmatrix} \ . \tag{5.11}$$

There are three pivots, so the rank of A^t, and hence the rank of A is 3. It follows that $\dim (\text{Img}(A)) = 3$. This means that the columns of A, which span A, are linearly independent, and hence they are a basis for Img(A).

However, having done the row reduction of A^t, we may as well use the rows of the fully row reduced form of A^t. This is the "standard procedure", explained in the previous section, for obtaining a basis for Img(A). The vectors in this basis have many zero entries, which is helpful in the computations that follow.

Hence, we have the basis:

$$\mathbf{v}_1 = \begin{bmatrix} 1 \\ 0 \\ 0 \\ -1 \end{bmatrix} \quad , \qquad \mathbf{v}_2 = \begin{bmatrix} 0 \\ 1 \\ 0 \\ 2 \end{bmatrix} \quad \text{and} \qquad \mathbf{v}_3 = \begin{bmatrix} 0 \\ 0 \\ 1 \\ 1 \end{bmatrix} \ .$$

Let $\mathbf{v}_1$, $\mathbf{v}_2$, and $\mathbf{v}_3$ be the columns of A, ordered left to right as usual. Then $|\mathbf{v}_1| = \sqrt{2}$ so that

$$\mathbf{u}_1 = \frac{1}{\sqrt{2}} \begin{bmatrix} 1 \\ 0 \\ 0 \\ -1 \end{bmatrix} \ .$$

Next, we compute that $\mathbf{v}_2 \cdot \mathbf{u}_1 = -\sqrt{2}$ and hence

$$\mathbf{w}_2 = \begin{bmatrix} 0 \\ 1 \\ 0 \\ 2 \end{bmatrix} + \begin{bmatrix} 1 \\ 0 \\ 0 \\ -1 \end{bmatrix} = \begin{bmatrix} 1 \\ 1 \\ 0 \\ 1 \end{bmatrix}$$

and so

$$\mathbf{u}_2 = \frac{1}{|\mathbf{w}_2|}\mathbf{w}_2 = \frac{1}{\sqrt{3}}\begin{bmatrix} 1 \\ 1 \\ 0 \\ 1 \end{bmatrix}$$

Next, we compute that $\mathbf{v}_3 \cdot \mathbf{u}_1 = -(1/\sqrt{2})$ and that $\mathbf{v}_3 \cdot \mathbf{u}_2 = (1/\sqrt{3})$. Therefore,

$$\mathbf{w}_3 = \begin{bmatrix} 0 \\ 0 \\ 1 \\ 1 \end{bmatrix} + \frac{1}{2}\begin{bmatrix} 1 \\ 0 \\ 0 \\ -1 \end{bmatrix} - \frac{1}{3}\begin{bmatrix} 1 \\ 1 \\ 0 \\ 1 \end{bmatrix} = \frac{1}{6}\begin{bmatrix} 1 \\ -2 \\ 6 \\ 1 \end{bmatrix}.$$

Normalizing this,

$$\mathbf{u}_3 = \frac{1}{\sqrt{42}}\begin{bmatrix} 1 \\ -2 \\ 6 \\ 1 \end{bmatrix}$$

is the third element of our orthonormal basis.

The change of basis matrix from the "v coordinates" to the "u coordinates" is given by

$$R = \begin{bmatrix} \mathbf{u}_1 \cdot \mathbf{v}_1 & \mathbf{u}_1 \cdot \mathbf{v}_2 & \mathbf{u}_1 \cdot \mathbf{v}_3 \\ 0 & \mathbf{u}_2 \cdot \mathbf{v}_2 & \mathbf{u}_2 \cdot \mathbf{v}_3 \\ 0 & 0 & \mathbf{u}_3 \cdot \mathbf{v}_3 \end{bmatrix}$$

by Theorem 2. We've already computed

$$\mathbf{u}_1 \cdot \mathbf{v}_2 = -\sqrt{2} \quad , \quad \mathbf{u}_1 \cdot \mathbf{v}_3 = -1/\sqrt{2} \quad \text{and} \quad \mathbf{u}_2 \cdot \mathbf{v}_3 = 1/\sqrt{3}$$

up above, so we only need to compute the 3 diagonal entries to find R. These are

$$\mathbf{u}_1 \cdot \mathbf{v}_1 = \sqrt{2} \quad , \quad \mathbf{u}_2 \cdot \mathbf{v}_2 = \sqrt{3} \quad \text{and} \quad \mathbf{u}_3 \cdot \mathbf{v}_3 = \sqrt{7/6}$$

hence,

$$R = \begin{bmatrix} \sqrt{2} & -\sqrt{2} & -1/\sqrt{2} \\ 0 & \sqrt{3} & 1/\sqrt{3} \\ 0 & 0 & \sqrt{7/6} \end{bmatrix}$$

How things change if $\{\mathbf{v}_1, \mathbf{v}_2, \ldots, \mathbf{v}_k\}$ only spans S

What happens if we don't bother to extract a basis from $\{\mathbf{v}_1, \mathbf{v}_2, \ldots, \mathbf{v}_r\}$? Let's try an example and see.

Example 3 Let A be the 3×3 matrix $A \begin{bmatrix} 1 & 0 & 1 \\ 0 & 1 & 1 \\ 1 & 1 & 2 \end{bmatrix}$. Writing this as $A = [\mathbf{v}_1, \mathbf{v}_2, \mathbf{v}_3]$, we

certainly have that $\{\mathbf{v}_1, \mathbf{v}_2, \mathbf{v}_3\}$ spans $\text{Img}(A)$. Let's try to transform it into an orthonormal basis using the Gramm–Schmidt procedure.

First, we normalize $\mathbf{v}_1$ and get $\mathbf{u}_1 = \frac{1}{\sqrt{2}} \begin{bmatrix} 1 \\ 0 \\ 1 \end{bmatrix}$. Then

$$\mathbf{w}_2 = \mathbf{v}_2 - (\mathbf{u}_1 \cdot \mathbf{v}_2)\mathbf{u}_1 = \begin{bmatrix} 0 \\ 1 \\ 1 \end{bmatrix} - \frac{1}{2} \begin{bmatrix} 1 \\ 0 \\ 1 \end{bmatrix} = \frac{1}{2} \begin{bmatrix} -1 \\ 2 \\ 1 \end{bmatrix} .$$

Normalizing, we have $\mathbf{u}_2 = \frac{1}{\sqrt{6}} \begin{bmatrix} -1 \\ 2 \\ 1 \end{bmatrix}$. So far, so good. Next,

$$\mathbf{w}_3 = \mathbf{v}_3 - (\mathbf{u}_1 \cdot \mathbf{v}_3)\mathbf{u}_1 - (\mathbf{u}_2 \cdot \mathbf{v}_3)\mathbf{u}_2 = \begin{bmatrix} 1 \\ 1 \\ 2 \end{bmatrix} - \frac{3}{2} \begin{bmatrix} 1 \\ 0 \\ 1 \end{bmatrix} - \frac{1}{2} \begin{bmatrix} -1 \\ 2 \\ 1 \end{bmatrix} = \begin{bmatrix} 0 \\ 0 \\ 0 \end{bmatrix} .$$

This time we can't normalize since $\mathbf{w}_3 = 0$. What should we do? Nothing! This result just tells us that

$$\mathbf{v}_3 = (\mathbf{u}_1 \cdot \mathbf{v}_3)\mathbf{u}_1 + (\mathbf{u}_2 \cdot \mathbf{v}_3)\mathbf{u}_2$$

which means that $\mathbf{v}_3$ is in the span of $\{\mathbf{u}_1, \mathbf{u}_2\}$.

Since $\mathbf{v}_1$ and $\mathbf{v}_2$ are also in the span of $\{\mathbf{u}_1, \mathbf{u}_2\}$, we see that $\{\mathbf{u}_1, \mathbf{u}_2\}$ spans $\text{Img}(A)$. Since it is linearly independent – orthonormal sets always are – it is a basis for $\text{Img}(A)$.

There is a relation between $A = [\mathbf{v}_1, \mathbf{v}_2, \mathbf{v}_3]$ and $Q = [\mathbf{u}_1, \mathbf{u}_2]$ that is very much like the situation described in Theorem 2. (In this context, Q is the traditional notation for $[\mathbf{u}_1, \mathbf{u}_2]$). We've known $A = [\mathbf{v}_1, \mathbf{v}_2, \mathbf{v}_3]$ from the start, and by the computations above, we have

$$Q = \begin{bmatrix} 1/\sqrt{2} & -1/\sqrt{6} \\ 0 & 2/\sqrt{6} \\ 1/\sqrt{2} & 1/\sqrt{6} \end{bmatrix} . \tag{5.12}$$

Since $\{\mathbf{u}_1, \mathbf{u}_2\}$ is a basis for $\text{Img}(A)$, there is a unique vector $\mathbf{r}_j$ with

$$Q\mathbf{r}_j = [\mathbf{u}_1, \mathbf{u}_2]\mathbf{r}_j = \mathbf{v}_j$$

for $j = 1, 2, 3$. The entries of $\mathbf{r}_j$ are just the coordinates of $\mathbf{v}_j$ in the basis $\{\mathbf{u}_1, \mathbf{u}_2\}$. But we know how to compute coordinates with respect to an orthonormal basis:

$$\mathbf{v}_j = (\mathbf{v}_j \cdot \mathbf{u}_1)\mathbf{u}_1 + (\mathbf{v}_j \cdot \mathbf{u}_2)\mathbf{u}_2 .$$

Hence

$$\mathbf{r}_j = \begin{bmatrix} \mathbf{u}_1 \cdot \mathbf{v}_j \\ \mathbf{u}_2 \cdot \mathbf{v}_j \end{bmatrix} .$$

Therefore, if R is the 2×3 matrix with $R_{i,j} = \mathbf{u}_i \cdot \mathbf{v}_j$, we have

$$A = [\mathbf{v}_1, \mathbf{v}_2, \mathbf{v}_3] = [Q\mathbf{r}_1, Q\mathbf{r}_2, Q\mathbf{r}_3] = QR . \qquad (5.13)$$

In the case at hand, you find from this formula that

$$R = \frac{1}{\sqrt{2}} \begin{bmatrix} 2 & 1 & 3 \\ 0 & \sqrt{3} & \sqrt{3} \end{bmatrix} . \qquad (5.14)$$

Notice that R is in row reduced form! Since Q is an isometry, we have just found a way to write A as the product of two very nice matrices:

$$A = QR$$

where Q is an isometry and where R is in row reduced form.

This "factorization" of A into these nice factors is called the *QR decomposition*. It can always be done, for any $m \times n$ matrix A, just as in the example. All you do is apply the Gramm–Schmidt procedure to the columns of $A = [\mathbf{v}_1, \mathbf{v}_2, \ldots, \mathbf{v}_n]$. As you go along, you might find that certain of the $\mathbf{w}_j$ vectors are zero. *This just means that $\mathbf{w}_j$, and hence $\mathbf{v}_j$, is already in the span of $\{\mathbf{u}_1, \mathbf{u}_2, \ldots, \mathbf{u}_{j-1}\}$, and so we can't get a new orthogonal direction out of $\mathbf{v}_j$.* Just skip it, and go on to the next vector. In the end you still get an orthonormal basis $\{\mathbf{u}_1, \mathbf{u}_2, \ldots, \mathbf{u}_r\}$ for $\mathrm{Img}(A)$, where r is the rank of A. (We know that $r = \dim(\mathrm{Img}(A))$). If you define $Q = [\mathbf{u}_1, \mathbf{u}_2, \ldots, \mathbf{u}_r]$, then Q is an isometry, and if R is the $r \times r$ matrix with $R_{i,j} = \mathbf{u}_i \cdot \mathbf{v}_j$,

$$A = QR .$$

This is very useful for solving equations. Consider the equation $A\mathbf{x} = \mathbf{b}$. Then we can rewrite this ss

$$QR\mathbf{x} = \mathbf{b}$$

and solve it in two easy steps. First, multiply on the left by Q^t. We get $Q^t(QR\mathbf{x}) = Q^t\mathbf{b}$. But since $Q^tQ = I_{r \times r}$, $Q^t(QR\mathbf{x}) = (Q^tQ)R\mathbf{x} = R\mathbf{x}$. Hence

$$QR\mathbf{x} = \mathbf{b} \Rightarrow R\mathbf{x} = Q^t\mathbf{b} . \qquad (5.15)$$

The equation on the right is easily solvable by back substitution since R is in row reduced form, and the solution always exists since the rank of R is r, and so $\mathrm{Img}(R) = I\!R^r$.

This approach is very important in applications. Therefore, routines for computing the Q and R factors of a matrix A are built into all software packages for dealing with matrices. There are very good theoretical reasons, valid in a great many circumstances, for using this method to solve matrix equations, and we will explain some of them soon.

However, notice that something curious is going on here. If $n > r$, then $\mathrm{Img}(A)$ cannot be all of $I\!R^n$, and we cannot possibly solve $A\mathbf{x} = \mathbf{b}$ for all $\mathbf{b}$ in $I\!R^n$. However, once we've reduced the equation to $R\mathbf{x} = Q^t\mathbf{b}$, there will always be a solution, and we easily find it by back substitution. What is going on here? The point is that the implication in (5.15) goes only one way, so that any solution of $A\mathbf{x} = \mathbf{b}$ must also be a solution of the easy equation $R\mathbf{x} = Q^t\mathbf{b}$, but not *vice–versa*. So what are we getting when we solve the easy equation $R\mathbf{x} = Q^t\mathbf{b}$, and no solution of $A\mathbf{x} = \mathbf{b}$ exists? It turns out that we are getting the vector $\mathbf{x}$ that makes $|A\mathbf{x} - \mathbf{b}|^2$ as small as possible – that is, the least squares solution of $A\mathbf{x} = \mathbf{b}$. We shall see why this is true in the section on least squares solutions, and what we've just said is meant to indicate that this so–called QR factorization plays an important role in that theory.

Theorem 3 (QR Factorization) *Let A be any $m \times n$ matrix with rank r. Then there is a matrix factorization*

$$A = QR$$

where Q is an $n \times r$ matrix that has orthonormal columns, and R is an $r \times r$ matrix in row reduced form that has rank r, so that $\mathrm{Img}(R) = I\!R^r$. Moreover, Q and R can be found by applying the Gramm–Schmidt procedure to the columns of A, as above.

The factorization can also be found by using a built in routine on any computer program for doing linear algebra, which is much more convenient. So in practice, finding Q and R is easy, and it is very useful if you know what to do with them!

Example 4: Let A be the matrix from Example 3, and consider the equation $A\mathbf{x} = \mathbf{b}$ where $\mathbf{b} = \begin{bmatrix} 1 \\ 1 \\ 1 \end{bmatrix}$. This equation can be written as

$$QR\mathbf{x} = \mathbf{b} \ .$$

Multiplying on the left by Q^t, and using the explicit form of Q found in (5.12),

$$R\mathbf{x} = Q^t\mathbf{b} = \begin{bmatrix} 1/\sqrt{2} & 0 & 1/\sqrt{2} \\ -1/\sqrt{6} & 2/\sqrt{6} & \sqrt{1/6} \end{bmatrix} \begin{bmatrix} 1 \\ 1 \\ 1 \end{bmatrix}$$

$$= \begin{bmatrix} \sqrt{2} \\ 2/\sqrt{6} \end{bmatrix} \ .$$

Next we solve $R\mathbf{x} = \begin{bmatrix} \sqrt{2} \\ 2/\sqrt{6} \end{bmatrix}$. Using the explicit form of R found in (5.14), this is

$$\frac{1}{\sqrt{2}} \begin{bmatrix} 2 & 1 & 3 \\ 0 & \sqrt{3} & \sqrt{3} \end{bmatrix} \begin{bmatrix} x_1 \\ x_2 \\ x_3 \end{bmatrix} = \begin{bmatrix} \sqrt{2} \\ 2/\sqrt{6} \end{bmatrix} ,$$

or

$$\begin{bmatrix} 2 & 1 & 3 \\ 0 & \sqrt{3} & \sqrt{3} \end{bmatrix} \begin{bmatrix} x_1 \\ x_2 \\ x_3 \end{bmatrix} = \begin{bmatrix} 2 \\ 2/\sqrt{3} \end{bmatrix} .$$

The variable x_3 is non-pivotal; set $x_3 = t$. Back substitution then gives us

$$x_2 = \frac{2}{3} - t \qquad \text{and} \qquad x_1 = \frac{2}{3} - t .$$

We have found the one parameter family of "solutions"

$$\begin{bmatrix} x_1 \\ x_2 \\ x_3 \end{bmatrix} = \begin{bmatrix} 2/3 - t \\ 2/3 - t \\ t \end{bmatrix} = \frac{2}{3} \begin{bmatrix} 1 \\ 1 \\ 0 \end{bmatrix} + t \begin{bmatrix} -1 \\ -1 \\ 1 \end{bmatrix} .$$

Let's check this out: Multiplying

$$A \left(\frac{2}{3} \begin{bmatrix} 1 \\ 1 \\ 0 \end{bmatrix} + t \begin{bmatrix} -1 \\ -1 \\ 1 \end{bmatrix} \right) = \frac{2}{3} A \begin{bmatrix} 1 \\ 1 \\ 0 \end{bmatrix} + t A \begin{bmatrix} -1 \\ -1 \\ 1 \end{bmatrix}$$

$$= \frac{2}{3} \begin{bmatrix} 1 \\ 1 \\ 2 \end{bmatrix}$$

(5.16)

since $A \begin{bmatrix} -1 \\ -1 \\ 1 \end{bmatrix} = 0$. Notice that

$$\frac{2}{3} \begin{bmatrix} 1 \\ 1 \\ 2 \end{bmatrix} \neq \begin{bmatrix} 1 \\ 1 \\ 1 \end{bmatrix} = \mathbf{b} .$$

So we haven't found a solution after all. Did we waste our time? Not at all. There wasn't any solution to be found, since $\mathbf{b}$ is not in $\text{Img}(A)$. To see this, row reduce $\begin{bmatrix} 1 & 0 & 1 & | & x_1 \\ 0 & 1 & 1 & | & x_2 \\ 1 & 1 & 2 & | & x_3 \end{bmatrix}$ to find $\begin{bmatrix} 1 & 0 & 1 & | & x_1 \\ 0 & 1 & 1 & | & x_2 \\ 0 & 0 & 2 & | & x_3 - x_2 - x_1 \end{bmatrix}$. Hence $\text{Img}(A)$ is the plane given by

$$x_1 + x_2 - x_3 = 0 .$$

Notice that the entries of $\mathbf{b}$ do not satisfy this equation, so $\mathbf{b}$ is not in the image of A. For this reason, $A\mathbf{x} = \mathbf{b}$ has no solution. However,

$$\mathbf{c} = \frac{2}{3} \begin{bmatrix} 1 \\ 1 \\ 2 \end{bmatrix}$$

3-52

is in the image of A. In fact, since $\text{Img}(A)$ is the plane given by $\mathbf{a} \cdot \mathbf{x} = 0$ where $\mathbf{a} = \begin{bmatrix} 1 \\ 1 \\ -1 \end{bmatrix}$,

we can apply Theorem 2.2.1 to deduce that the vector in this plane that is closest to $\mathbf{b}$ is

$$\mathbf{b} - \frac{\mathbf{a} \cdot \mathbf{b}}{|\mathbf{a}|^2}\mathbf{a} = \begin{bmatrix} 1 \\ 1 \\ 1 \end{bmatrix} - \frac{1}{3}\begin{bmatrix} 1 \\ 1 \\ -1 \end{bmatrix} = \frac{2}{3}\begin{bmatrix} 1 \\ 1 \\ 2 \end{bmatrix} = \mathbf{c} \ .$$

Comparing with (5.16), we see that the solution set we have found,

$$\frac{2}{3}\begin{bmatrix} 1 \\ 1 \\ 0 \end{bmatrix} + t\begin{bmatrix} -1 \\ -1 \\ 1 \end{bmatrix}$$

is the solution set of $A\mathbf{x} = \mathbf{c}$, which has the right hand side "corrected" to belong to $\text{Img}(A)$.

• *As we will explain in the next two sections, the QR approach to solving $A\mathbf{x} = \mathbf{b}$ leads to the usual solution set when there is one, and to the set of least squares solutions when there isn't. In the language of Section 1, it "automatically" corrects the data on the right side of the equation.*

In our demonstration of this in the present example, we used Theorem 2.2.1 for planes in $I\!R^3$. In the next section we obtain a similar result for arbitrary subspaces in $I\!R^n$.

Section 6: Orthogonal Projections

Let S be a subspace of $I\!R^n$, and let $S^\perp$ be its orthogonal complement. Suppose that $\dim(S) = k$. Then by Theorem 3.3.3, $\dim(S^\perp) = n - k$. Let $\{u_1, u_2, \ldots, u_k\}$ be an orthonormal basis for S, and let $\{u_{k+1}, u_{k+2}, \ldots, u_n\}$ be an orthonormal basis for $S^\perp$. Since every vector in S is orthogonal to every vector in $S^\perp$, the combined set $\{u_1, u_2, \ldots, u_k, u_{k+1}, \ldots, u_n\}$ is a set of n orthonormal, and hence linearly independent, vectors in $I\!R^n$. Therefore, it is a basis for $I\!R^n$. Hence for any x in $I\!R^n$, we can write $x = \sum_{\ell=1}^{n} x_\ell u_\ell$. Since the basis $\{u_1, u_2, \ldots, u_n\}$ is orthonormal, the coordinates x_ℓ are given by $x_\ell = u_\ell \cdot x$, and hence

$$x = \sum_{\ell=1}^{n} (u_\ell \cdot x)u_\ell \,. \tag{6.1}$$

Coordinates for such a basis are very useful for answering questions about S and $S^\perp$.

Example 1 Let S be the plane in $I\!R^3$ given by $x_1 + x_2 + x_3 = 0$. We found in Example 1 of Section 3 that $\{u_1, u_2\}$ is an orthonormal basis of S where

$$u_1 = \frac{1}{\sqrt{2}} \begin{bmatrix} -1 \\ 1 \\ 0 \end{bmatrix} \quad \text{and} \quad u_2 = \frac{1}{\sqrt{6}} \begin{bmatrix} -1 \\ -1 \\ 2 \end{bmatrix} \,.$$

The equation of the plane S can be written in the form $a \cdot x = 0$ where $a = \begin{bmatrix} 1 \\ 1 \\ 1 \end{bmatrix}$. Since a is, by the equation, orthogonal to every vector in S, and since $\dim\left(S^\perp\right) = (3 - \dim(S)) = 1$, $\{a\}$ is a basis for $S^\perp$. Normalizing it, we define

$$u_3 = \frac{1}{|a|}a = \frac{1}{\sqrt{3}} \begin{bmatrix} 1 \\ 1 \\ 1 \end{bmatrix} \,.$$

Then $\{u_3\}$ is our orthonormal basis for $S^\perp$. Combining these two bases, $\{u_1, u_2, u_3\}$ is an orthonormal basis for $I\!R^3$.

Why would we possibly prefer this basis to the standard basis $\{e_1, e_2, e_3\}$? Look at all those square roots! The basis $\{u_1, u_2, u_3\}$, once we have it, makes it very easy to answer geometric questions concerning S and $S^\perp$. For example, let $x = \begin{bmatrix} 1 \\ 2 \\ 3 \end{bmatrix}$. Evidently, x is not a multiple of a, so x does not belong to $S^\perp$, and $a \cdot x \neq 0$, so so x does not belong to S. But if we express x in coordinates for the basis $\{u_1, u_2, u_3\}$, we can easily find the vector in S that is closest to $\mathbf{x}$, the distance from $\mathbf{x}$ to S, and so forth. This is a problem that we solved for planes in $I\!R^3$ already in Chapter 2. Here we are introducing a different way of looking at this, using a "good" basis, that works for arbitrary subspaces.

Working out the inner products in (6.1), we have in our particular case that

$$\begin{bmatrix} 1 \\ 2 \\ 3 \end{bmatrix} = \frac{1}{\sqrt{2}}\mathbf{u}_1 + \sqrt{\frac{3}{2}}\mathbf{u}_2 + 2\sqrt{3}\mathbf{u}_3$$

This allows us to break $\mathbf{x}$ up into two orthogonal pieces:

$$\begin{bmatrix} 1 \\ 2 \\ 3 \end{bmatrix} = \left(\frac{1}{\sqrt{2}}\mathbf{u}_1 + \sqrt{\frac{3}{2}}\mathbf{u}_2 \right) + \left(2\sqrt{3}\mathbf{u}_3 \right)$$

$$= \left(\frac{1}{2}\begin{bmatrix} -1 \\ 1 \\ 0 \end{bmatrix} + \frac{1}{2}\begin{bmatrix} -1 \\ -1 \\ 2 \end{bmatrix} \right) + \left(2\begin{bmatrix} 1 \\ 1 \\ 1 \end{bmatrix} \right) \qquad (6.2)$$

$$= \begin{bmatrix} -1 \\ 0 \\ 1 \end{bmatrix} + \begin{bmatrix} 2 \\ 2 \\ 2 \end{bmatrix} .$$

The vector $\begin{bmatrix} -1 \\ 0 \\ 1 \end{bmatrix}$ belongs to S, and the vector $\begin{bmatrix} 2 \\ 2 \\ 2 \end{bmatrix}$ belongs to $S^\perp$, and these two vectors are orthogonal to each other. Using what we learned about orthogonal projections in Section 2.2, or just plain geometric intuition, we see that the first of these vectors is the orthogonal projection of $\mathbf{x}$ onto the plane S, and the second is the orthogonal projection of $\mathbf{x}$ onto the line $S^\perp$. Hence $\begin{bmatrix} -1 \\ 0 \\ 1 \end{bmatrix}$ is the vector in S that is closest to $\begin{bmatrix} 1 \\ 2 \\ 3 \end{bmatrix}$, and the distance between these vectors is the length of their difference, $2\sqrt{3}$. Notice that this is just the multiple of $\mathbf{u}_3$ in the coordinate expression for $\mathbf{x}$ in (6.2).

Notice the following: The orthogonal projection of $\mathbf{x}$ onto S was just what we got by expressing $\mathbf{x}$ as a linear combination of the basis elements $\{\mathbf{u}_1, \mathbf{u}_2, \mathbf{u}_3\}$, and deleting the coordinates that belong to $S^\perp$.

Now let's develop this procedure for subspaces S and $S^\perp$ of arbitrary dimension in $I\!R^n$. We will see that it gives us a decomposition of an arbitrary $\mathbf{x}$ in $I\!R^n$ into two orthogonal "components", one in S and one in $S^\perp$. Moreover, there is just one such decomposition, and the component of $\mathbf{x}$ in S is the vector in S that is closest to $\mathbf{x}$. In general dimension we will have to rely more on computation than pictures, but the analogy with what we did in Example 1 in $I\!R^3$ is very strong.

Orthogonal projections in $I\!R^n$

Our starting point is a k dimensional subspace S of $I\!R^n$. Let $\{\mathbf{u}_1, \mathbf{u}_2, \ldots, \mathbf{u}_k\}$ be an orthonormal basis for S, and let $\{\mathbf{u}_{k+1}, \mathbf{u}_{k+2}, \ldots, \mathbf{u}_n\}$ be an orthonormal basis for $S^\perp$. Then, as explained above, the combined set $\{\mathbf{u}_1, \mathbf{u}_2, \ldots, \mathbf{u}_k, \mathbf{u}_{k+1}, \ldots, \mathbf{u}_n\}$ is a basis for $I\!R^n$, and for any $\mathbf{x}$ in $I\!R^n$,

$$\mathbf{x} = \sum_{\ell=1}^{n} (\mathbf{u}_\ell \cdot \mathbf{x}) \mathbf{u}_\ell \ . \tag{6.3}$$

Now let's split the sum into the sum over the basis vector for S, and the sum over those for $S^\perp$:

$$\begin{aligned}
\mathbf{x} &= \sum_{\ell=1}^{n} (\mathbf{u}_\ell \cdot \mathbf{x}) \mathbf{u}_\ell \\
&= \left(\sum_{\ell=1}^{k} (\mathbf{u}_\ell \cdot \mathbf{x}) \mathbf{u}_\ell \right) + \left(\sum_{\ell=k+1}^{n} (\mathbf{u}_\ell \cdot \mathbf{x}) \mathbf{u}_\ell \right) \ .
\end{aligned} \tag{6.4}$$

Since $\{\mathbf{u}_1, \mathbf{u}_2, \ldots, \mathbf{u}_k\}$ is a basis for S, the linear combination $\left(\sum_{\ell=1}^{k} (\mathbf{u}_\ell \cdot \mathbf{x}) \mathbf{u}_\ell \right)$ belongs to S. Likewise, $\left(\sum_{\ell=k+1}^{n} (\mathbf{u}_\ell \cdot \mathbf{x}) \mathbf{u}_\ell \right)$ belongs to $S^\perp$.

We can use (6.4) to define two transformations from $I\!R^n$ to $I\!R^n$. First, define the transformation f_S by

$$f_S(\mathbf{x}) = \sum_{\ell=1}^{k} (\mathbf{u}_\ell \cdot \mathbf{x}) \mathbf{u}_\ell \ .$$

For any two vectors $\mathbf{x}$ and $\mathbf{y}$ in $I\!R^n$, and any two numbers a and b, $\mathbf{u}_\ell \cdot (a\mathbf{x} + b\mathbf{y}) = a\mathbf{u}_\ell \cdot \mathbf{x} + b\mathbf{u}_\ell \cdot \mathbf{y}$. Hence $f_S(a\mathbf{x} + b\mathbf{y}) = af_S(\mathbf{x}) + bf_S(\mathbf{y})$. It follows that f_S is a linear transformation. Let P be the corresponding matrix. Then, by definition,

$$P\mathbf{x} = \sum_{\ell=1}^{k} (\mathbf{u}_\ell \cdot \mathbf{x}) \mathbf{u}_\ell \ , \tag{6.5}$$

and so $P_{i,j}$ is the ith component of $f_S(\mathbf{e}_j)$. That is,

$$P_{i,j} = \mathbf{e}_i \cdot f_S(\mathbf{e}_j) = \mathbf{e}_i \cdot \left(\sum_{\ell=1}^{k} (\mathbf{u}_\ell \cdot \mathbf{e}_j) \mathbf{u}_\ell \right) = \sum_{\ell=1}^{k} (\mathbf{e}_i \cdot \mathbf{u}_\ell)(\mathbf{e}_j \cdot \mathbf{u}_\ell) \ . \tag{6.6}$$

In the exact same way, we define the transformation

$$f_{S^\perp}(\mathbf{x}) = \sum_{\ell=k+1}^{n} (\mathbf{u}_\ell \cdot \mathbf{x}) \mathbf{u}_\ell \ .$$

3-56

Again, and for the same reasons, this is a linear transformation, and we define the matrix $P^\perp$ by

$$P^\perp \mathbf{x} = \sum_{\ell=k+1}^{n} (\mathbf{u}_\ell \cdot \mathbf{x})\mathbf{u}_\ell \tag{6.7}$$

which gives

$$P_{i,j}^\perp = \mathbf{e}_i \cdot f_{S^\perp}(\mathbf{e}_j) = \mathbf{e}_i \cdot \left(\sum_{\ell=k+1}^{n} (\mathbf{u}_\ell \cdot \mathbf{e}_j)\mathbf{u}_\ell \right) = \sum_{\ell=k+1}^{n} (\mathbf{e}_i \cdot \mathbf{u}_\ell)(\mathbf{e}_j \cdot \mathbf{u}_\ell) . \tag{6.8}$$

Combining (6.4), (6.5) and (6.7) we have that for all $\mathbf{x}$ in $I\!\!R^n$,

$$\mathbf{x} = P\mathbf{x} + P^\perp \mathbf{x} . \tag{6.9}$$

Since $P\mathbf{x}$ is a linear combination of the basis elements for S, $P\mathbf{x}$ belongs to S. Likewise, $P^\perp \mathbf{x}$ belongs to $S^\perp$.

It might seem that P and $P^\perp$ could depend on the particular orthonormal bases $\{\mathbf{u}_1, \mathbf{u}_2, \ldots, \mathbf{u}_k\}$ and $\{\mathbf{u}_{k+1}, \mathbf{u}_{k+2}, \ldots, \mathbf{u}_n\}$ that we used for S and $S^\perp$ respectively. This is not the case. The key is the following fact:

• *For any vector $\mathbf{x}$ in $I\!\!R^n$, there is only one way to write $\mathbf{x} = \mathbf{y} + \mathbf{z}$ as the sum of a vector $\mathbf{y}$ in S and a vector $\mathbf{z}$ in $S^\perp$.*

To see why, suppose $\mathbf{x} = \mathbf{y} + \mathbf{z}$ and $\mathbf{x} = \tilde{\mathbf{y}} + \tilde{\mathbf{z}}$ where $\mathbf{x}$ and $\tilde{\mathbf{x}}$ belong to S, while $\mathbf{y}$ and $\tilde{\mathbf{y}}$ belong to $S^\perp$. Then from $\mathbf{x} + \mathbf{y} = \tilde{\mathbf{x}} + \tilde{\mathbf{y}}$ we conclude

$$\mathbf{x} - \tilde{\mathbf{x}} = \tilde{\mathbf{y}} - \mathbf{y} .$$

The left hand side belongs to S, and the right hand side to $S^\perp$. Hence both sides belong to both S and $S^\perp$. This means that $\mathbf{x} - \tilde{\mathbf{x}}$ is orthogonal to itself, so

$$|\mathbf{x} - \tilde{\mathbf{x}}|^2 = (\mathbf{x} - \tilde{\mathbf{x}}) \cdot (\mathbf{x} - \tilde{\mathbf{x}}) = 0 .$$

Hence $\mathbf{x} = \tilde{\mathbf{x}}$. In the exact same way, $\mathbf{y} = \tilde{\mathbf{y}}$.

If we use a different pair of bases $\{\tilde{\mathbf{u}}_1, \tilde{\mathbf{u}}_2, \ldots, \tilde{\mathbf{u}}_k\}$ and $\{\tilde{\mathbf{u}}_{k+1}, \tilde{\mathbf{u}}_{k+2}, \ldots, \tilde{\mathbf{u}}_n\}$ for S and $S^\perp$ respectively we could use (6.5) and (6.7) for these new bases to define matrices $\tilde{P}$ and $\tilde{P}^\perp$ respectively. Just as before we would have

$$\mathbf{x} = \tilde{P}\mathbf{x} + \tilde{P}^\perp \mathbf{x} . \tag{6.10}$$

for all $\mathbf{x}$, and $\tilde{P}\mathbf{x}$ would belong to S and $\tilde{P}^\perp \mathbf{x}$ would belong to $S^\perp$. Then

$$\left(P\mathbf{x} + P^\perp \mathbf{x} \right) = \mathbf{x} = \left(\tilde{P}\mathbf{x} + \tilde{P}^\perp \mathbf{x} \right) .$$

But since there is only one way to write $\mathbf{x}$ as a sum of a vector in S and another vector in $S^\perp$, it must be the case that

$$P\mathbf{x} = \tilde{P}\mathbf{x} \quad \text{and} \quad P^\perp\mathbf{x} = \tilde{P}^\perp\mathbf{x}$$

for all $\mathbf{x}$. This means that $P = \tilde{P}$ and $P^\perp = \tilde{P}^\perp$.

• *the conclusion is that the matrices P and $P^\perp$ defined by (6.6) and (6.8) do not depend on the choice of orthonormal bases for S and $S^\perp$ – any choise of these bases leads to the same matrices P and $P^\perp$.*

Definition (Orthogonal Projections) For any k dimensional subspace S of $I\!R^n$, the matrix P defined by (6.6) where $\{\mathbf{u}_1, \mathbf{u}_2, \ldots, \mathbf{u}_k\}$ is any orthonormal basis for S is called the *orthogonal projection onto S.* The matrix $P^\perp$ defined by (6.8) where $\{\mathbf{u}_{k+1}, \mathbf{u}_{k+2}, \ldots, \mathbf{u}_n\}$ is any basis for $S^\perp$ is called the orthogonal projection onto $S^\perp$.

Theorem 1 (Ortogonal Projections) *Let S be a subspace of $I\!R^n$ and let P be the orthogonal projection onto S. Then*

(1)

$$P = P^t \quad , \quad P^2 = P, \tag{6.11}$$

(2)

$$\text{Img}(P) = \text{Ker}(P^\perp) = S \quad \text{and} \quad \text{Ker}(P) = \text{Img}(P^\perp) = S^\perp . \tag{6.12}$$

In particular, every subspace is both an image and a kernel.

Proof: If $\mathbf{x}$ belongs to S, then $\mathbf{x}$ is a linear combination of the vectors in $\{\mathbf{u}_1, \mathbf{u}_2, \ldots, \mathbf{u}_k\}$. Since $\{\mathbf{u}_1, \mathbf{u}_2, \ldots, \mathbf{u}_k\}$ is an orthonormal basis, we know what the coordinates are:

$$\mathbf{x} = (\mathbf{x} \cdot \mathbf{u}_1)\mathbf{u}_1 + (\mathbf{x} \cdot \mathbf{u}_2)\mathbf{u}_2 + \cdots + (\mathbf{x} \cdot \mathbf{u}_k)\mathbf{u}_k .$$

But the right hand side is, by definition, $P\mathbf{x}$. Hence, $P\mathbf{x} = \mathbf{x}$ whenever $\mathbf{x}$ belongs to S.

Moreover, for any $\mathbf{x}$ in $I\!R^n$, $P\mathbf{x}$ is by definition a linear combination of basis elements of S, so $P\mathbf{x}$ belongs to S. Combining these two facts, we have $P(P\mathbf{x}) = P\mathbf{x}$ for all $\mathbf{x}$ in $I\!R^n$. This means that $P^2 = P$. It is clear from (6.6) that $P_{i,j} = P_{j,i}$, so that $P = P^t$. Hence *(1)* is proved.

Finally, we have allready observed that by definition $P\mathbf{x}$ is in S for every $\mathbf{x}$, so $\text{Img}(P) \subset S$. On the other hand, for any $\mathbf{w}$ in S, $P\mathbf{w} = \mathbf{w}$ so $\mathbf{w}$ is in $\text{Img}(P)$, which shows that $S \subset \text{Img}(P)$. Together we have $S = \text{Img}(P)$. Finally, from Theorem 3.3.5,

$$\text{Img}(P) = (\text{Ker}(P^t))^\perp .$$

Since $P^t = P$, and $S = \text{Img}(P)$, this is the same as $S = (\text{Ker}(P))^\perp$. Taking orthogonal complements of both sides, this gives us $S^\perp = \text{Ker}P$. Finally, replacing S by $S^\perp$, so that $S^\perp$ is replaced by $S = (S^\perp)^\perp$, we get $S^\perp = \text{Img}(P^\perp)$ and $S = (\text{Ker}(P^\perp))$. This proves (6.12). Notice that since S is an arbitray subspace, and since we have that $\text{Img}(P) = \text{Ker}(P^\perp) = S$, we see that every subspace is both an image and a kernel. ∎

Example 2 Let S, $S^\perp$ and $\{\mathbf{u}_1, \mathbf{u}_2, \mathbf{u}_3\}$ the same as in Example 1. Let's use (6.5) and (6.7) to compute the matrices P and $P^\perp$. We'll do this one basis vector at a time. First, notice that $(\mathbf{e}_i \cdot \mathbf{u}_1)(\mathbf{e}_j \cdot \mathbf{u}_1)$ is just the product of the ith and jth entries of $\mathbf{u}_1$. Hence, using the explicit form of $\mathbf{u}_1$ found in Example 1, the 3×3 matrix whose i,jth entry is $(\mathbf{e}_i \cdot \mathbf{u}_1)(\mathbf{e}_j \cdot \mathbf{u}_1)$ is

$$\frac{1}{2}\begin{bmatrix} 1 & -1 & 0 \\ -1 & 1 & 0 \\ 0 & 0 & 0 \end{bmatrix} .$$

Likewise, the 3×3 matrix whose i,jth entry is $(\mathbf{e}_i \cdot \mathbf{u}_2)(\mathbf{e}_j \cdot \mathbf{u}_2)$; i.e., the product of the ith and jth entries of $\mathbf{u}_2$ is

$$\frac{1}{6}\begin{bmatrix} 1 & 1 & -2 \\ 1 & 1 & -2 \\ -2 & -2 & 4 \end{bmatrix} .$$

According to (6.5) we get P, the orthogonal projection onto S, by summing these matrices:

$$P = \frac{1}{6}\begin{bmatrix} 4 & -2 & -2 \\ -2 & 4 & -2 \\ -2 & -2 & 4 \end{bmatrix} .$$

In exactly the same way, we find that $P^\perp$, the orthogonal projection onto $S^\perp$, is the matrix whose i,jth entry is $(\mathbf{e}_i \cdot \mathbf{u}_3)(\mathbf{e}_j \cdot \mathbf{u}_3)$; i.e., the product of the ith and jth entries of $\mathbf{u}_3$. Therefore,

$$P^\perp = \frac{1}{3}\begin{bmatrix} 1 & 1 & 1 \\ 1 & 1 & 1 \\ 1 & 1 & 1 \end{bmatrix} .$$

You can now easily check that $P^2 = P$ and that $\left(P^\perp\right)^2 = P^\perp$.

The way is now paved for the solution of an important geometric problem.

Theorem 2 *Let S be a subspace of $\mathbb{R}^n$, and let P be the orthogonal projection onto S. Then the vector $\mathbf{c}$ given by $\mathbf{c} = P\mathbf{b}$ is in S, and*

$$|\mathbf{c} - \mathbf{b}| < |\mathbf{w} - \mathbf{b}| \tag{6.13}$$

for any other vector $\mathbf{w}$ in S. In particular, $\mathbf{c} = P\mathbf{b}$ unique element of S at which the distance to $\mathbf{b}$ is minimized. The minimum distance $|\mathbf{c} - \mathbf{b}|$ is given by $|P^\perp\mathbf{b}|$.

Proof: Let $\mathbf{w}$ be any vector in S. Then $\mathbf{b} = P\mathbf{b} + P^\perp\mathbf{b}$ give the decomposition of $\mathbf{b}$ into the sum of its components in S and $S^\perp$. Computing the squared distance between $\mathbf{b}$ and $\mathbf{w}$, we have

$$|\mathbf{b} - \mathbf{w}|^2 = |(P\mathbf{b} + P^\perp\mathbf{b}) - \mathbf{w}|^2$$
$$= |(P\mathbf{b} - \mathbf{w}) + P^\perp\mathbf{b}|^2 . \tag{6.14}$$

Since $P\mathbf{b}$ and $\mathbf{w}$ both belong to S so does their difference. But $P^\perp\mathbf{b}$ belongs to $S^\perp$. Thus $(P\mathbf{b} - \mathbf{w})$ and $P^\perp\mathbf{b}$ are orthogonal. By the Pythagorean Theorem,

$$|(P\mathbf{b} - \mathbf{w}) + P^\perp\mathbf{b}|^2 = |P\mathbf{b} - \mathbf{w}|^2 + |P^\perp\mathbf{b}|^2 \ . \tag{6.15}$$

Together, (6.14) and (6.15) say that $|\mathbf{b} - \mathbf{w}|^2 = |P\mathbf{b} - \mathbf{w}|^2 + |P^\perp\mathbf{b}|^2$ and hence

$$|\mathbf{b} - \mathbf{w}|^2 \geq |P^\perp\mathbf{b}|^2$$

and there is equality exactly in case $\mathbf{w} = P\mathbf{b}$. Thus $\mathbf{c} = P\mathbf{b}$ is closer to $\mathbf{b}$ than any other vector in S. ∎

Example 3 Let S be the subspace of $I\!R^3$ considered in Example 1. Let's find the vector $\mathbf{c}$ in S that is closest to $\mathbf{b} = \begin{bmatrix} 1 \\ 2 \\ 3 \end{bmatrix}$, and the distance from this vector to S. We did this using geoemtric intuition about $I\!R^3$ in Example 1. Now let's do it using Theorem 2.

Since we have already computed P, the orthogonal projection onto S, it is easy to compute the closest vector $\mathbf{c}$ which is

$$\mathbf{c} = P\mathbf{b} = \frac{1}{6} \begin{bmatrix} 4 & -2 & -2 \\ -2 & 4 & -2 \\ -2 & -2 & 4 \end{bmatrix} \begin{bmatrix} 1 \\ 2 \\ 3 \end{bmatrix} = \begin{bmatrix} -1 \\ 0 \\ 1 \end{bmatrix} \ .$$

The minimum distance is given by $|P^\perp\mathbf{b}|$. By the results of Example 1,

$$P^\perp\mathbf{b} = \frac{1}{3} \begin{bmatrix} 1 & 1 & 1 \\ 1 & 1 & 1 \\ 1 & 1 & 1 \end{bmatrix} \begin{bmatrix} 1 \\ 2 \\ 3 \end{bmatrix} = \begin{bmatrix} 2 \\ 2 \\ 2 \end{bmatrix} \ .$$

This vector has length $2\sqrt{3}$, and hence the distance in question is $2\sqrt{3}$.

Example 4 Let's compute the distance from $\mathbf{b} = \begin{bmatrix} 1 \\ 2 \\ 3 \end{bmatrix}$ to the line through the origin and $\begin{bmatrix} 1 \\ 1 \\ 1 \end{bmatrix}$. This is simple to do now if we notice that the line in question is $S^\perp$, where S is the subspace in Example 1. Therefore, with P and $P^\perp$ as in Example 1, $P^\perp\mathbf{b}$ is the projection of $\mathbf{b}$ onto the line, and $P\mathbf{b}$ is the orthogonal component of $\mathbf{b}$. Hence $|P\mathbf{b}|$ is the distance from $\mathbf{b}$ to the line. We have computed $P\mathbf{b}$ in the previous example, and so all that is left to do is to compute its length, which is $\sqrt{2}$.

A better way to compute orthogonal projections

There is also another formula for P that is useful in computation. Very often, we will obtain an orthonormal basis for a subspace S by Gram–Schmidt orthonormalization. In practice, this means by computing a QR factorization. Therefore, consider the n by k matrix Q whose jth column is $\mathbf{u}_j$. That is, $Q = [\mathbf{u}_1, \mathbf{u}_2, \ldots, \mathbf{u}_k]$

For any vector $\mathbf{x}$ in $I\!\!R^n$, by Theorem 1.5.1,

$$Q^t\mathbf{x} = \begin{bmatrix} \mathbf{u}_1 \cdot \mathbf{x} \\ \mathbf{u}_2 \cdot \mathbf{x} \\ \vdots \\ \mathbf{u}_k \cdot \mathbf{x} \end{bmatrix}$$

since $\mathbf{u}_i$ is the ith row of Q^t. But then

$$Q(Q^t\mathbf{x}) = [\mathbf{u}_1, \mathbf{u}_2, \ldots, \mathbf{u}_k] \begin{bmatrix} \mathbf{u}_1 \cdot \mathbf{x} \\ \mathbf{u}_2 \cdot \mathbf{x} \\ \vdots \\ \mathbf{u}_k \cdot \mathbf{x} \end{bmatrix} = \sum_{i=1}^{k} (\mathbf{u}_i \cdot \mathbf{x})\mathbf{u}_i = P\mathbf{x} \, . \tag{6.16}$$

We record this simple but useful fact in a Theorem:

Theorem 3 Q be an $n \times k$ matrix with orthonormal columns, and let P be the orthognal projection onto $\mathrm{Img}(Q)$. Then

$$P = QQ^t \, . \tag{6.17}$$

One reason this is useful is that if you have any basis for S, or even any spanning set $[\mathbf{v}_1, \mathbf{v}_2, \ldots, \mathbf{v}_m]$, you can ask a computer linear algebra program to compute a QR decomposition of $A = [\mathbf{v}_1, \mathbf{v}_2, \ldots, \mathbf{v}_m]$. The columns of Q are then your orthonormal basis, painlessly obtained. But if you have it in Q form, you don't need to use (6.5) to work out P; just multiply out QQ^t!

Example 5 Again consider the subspace S from Example 1. Applying the Gramm-Schmidt procedure to the basis $\{\mathbf{v}_1, \mathbf{v}_2\}$ by hand (or having a computer do it), we obtain the basis $\{\mathbf{u}_1, \mathbf{u}_2\}$. With $Q = [\mathbf{u}_1, \mathbf{u}_2]$, we have

$$Q = \begin{bmatrix} 1/\sqrt{2} & 1/\sqrt{6} \\ -1/\sqrt{2} & 1/\sqrt{6} \\ 0 & -2/\sqrt{6} \end{bmatrix} = \frac{1}{\sqrt{6}} \begin{bmatrix} \sqrt{3} & 1 \\ -\sqrt{3} & 1 \\ 0 & -2 \end{bmatrix} \, .$$

Multiplying out QQ^t, we find as before, that

$$P = QQ^t = \frac{1}{6} \begin{bmatrix} 4 & -2 & -2 \\ -2 & 4 & -2 \\ -2 & -2 & 4 \end{bmatrix} \, .$$

Here one final labor saving observation on computing orthogonal projections. Notice from the fact that P and $P^\perp$ give the decomposition of any vector $\mathbf{x}$ into the sum of its components in S and $S^\perp$ respectively; i.e., $\mathbf{x} = P\mathbf{x} + P^\perp\mathbf{x}$, we have

$$P^\perp = (I - P) \quad \text{and} \quad P = (I - P^\perp) . \tag{6.18}$$

This means that if you are trying to compute P, say, but $\dim(S^\perp)$ is lower than $\dim(S)$, you are probably better off applying either (6.5) or (6.17) directly to the computation of $P^\perp$. This will involve smaller sums in (6.5), or matrices Q with fewer columns in (6.17). Then you can compute P by doing the simple subtraction $P = I - P^\perp$.

Example 6 Going back once more to Example 1, we note that $S^\perp$ is one dimensional. For this reason, it was relatively easy to compute $P^\perp$. Then using $P = I - P^\perp$, we have

$$P = \begin{bmatrix} 1 & 0 & 0 \\ 0 & 1 & 0 \\ 0 & 0 & 1 \end{bmatrix} - \frac{1}{3} \begin{bmatrix} 1 & 1 & 1 \\ 1 & 1 & 1 \\ 1 & 1 & 1 \end{bmatrix} = \frac{1}{3} \left(\begin{bmatrix} 3 & 0 & 0 \\ 0 & 3 & 0 \\ 0 & 0 & 3 \end{bmatrix} - \begin{bmatrix} 1 & 1 & 1 \\ 1 & 1 & 1 \\ 1 & 1 & 1 \end{bmatrix} \right)$$

$$= \frac{1}{3} \begin{bmatrix} 2 & -1 & -1 \\ -1 & 2 & -1 \\ -1 & -1 & 2 \end{bmatrix}$$

This is the same result as before, only written more simply.

The moral is:

- *When you need to compute an orthogonal projection P, it is always a good idea to think first about whether or not it might be much easier to directly compute $P^\perp$ than to directly compute P.*

Section 7: Least Squares Solutions

Let A be an m by n matrix. The system of equations $Ax = b$ is called *over determined* if $m > n$. Then there are more equations than unknowns, and the system will not be solvable in general. On the other hand, if $n > m$, there are more unknowns than equations, and we say that the system is *under determined*. In this case, the solutions will not be unique. What does one do with an over or under determined system of equations? The next theorem gives us the answer in terms of the orthogonal projections onto $\text{Img}(A)$ and $\text{Img}(A^t)$. Since $\text{Img}(A)$ is spanned by the columns of A, we denote the orthogonal projection onto this subspace by P_c. Likewise, since $\text{Img}(A^t)$ is spanned by the rows of A, we denote the orthogonal projection onto this subspace by P_r.

Theorem 1 (Least Squares Solutions) *Let A be an m by n matrix. Let P_c be the orthogonal projection onto $\text{Img}(A)$, and let P_r be the orthogonal projection onto $\text{Img}(A^t)$, that is, onto the span of the rows of A. Then:*

(1)

$$P_c A = A \qquad \text{and} \qquad A P_r = A . \tag{7.1}$$

(2) For every $\mathbf{b}$ in $I\!R^m$,

$$Ax = P_c \mathbf{b} \tag{7.2}$$

has at least one solution $\mathbf{x}_*$ and

$$|Ax_* - \mathbf{b}|^2 \le |Ax - \mathbf{b}|^2 \tag{7.3}$$

for all $\mathbf{x}$ in $I\!R^n$.

(3) For any $\mathbf{c}$ in $\text{Img}(A)$, if $\mathbf{x}$ is any solution of $Ax = \mathbf{c}$, then so is $\mathbf{x}_{**} = P_r\mathbf{x}$. Moreover,

$$|\mathbf{x}_{**}|^2 < |\mathbf{x}|^2 \tag{7.4}$$

for any solution $\mathbf{x}$ other than $\mathbf{x}_{**}$, so that $\mathbf{x}_{**}$ is the solution with the least length.

Proof: Since P_c is the orthogonal projection onto $\text{Img}(A)$, it follows from (6.12) that $\text{Ker}((P_c)^\perp) = \text{Img}(A)$. Therefore, $P_c^\perp A = 0$, and for any $\mathbf{x}$ in $I\!R^n$,

$$Ax = P_c(Ax) + P_c^\perp(Ax)$$
$$= (P_c A)\mathbf{x} + (P_c^\perp)Ax$$
$$= (P_c A)\mathbf{x} ,$$

which means that $P_c A = A$.

Likewise, since P_r is the orthogonal projection onto $\text{Img}(A^t)$, it follows from (6.12) and then Theorem 3.3.5,

$$\text{Img}(P_r^\perp) = \big(\text{Img}(A^t)\big)^\perp = \text{Ker}(A) .$$

Therefore, $AP_r^\perp = 0$, and for any $\mathbf{x}$ in $I\!R^n$,

$$Ax = A(P_r\mathbf{x} + P_r^\perp\mathbf{x})$$

$$= A(P_r\mathbf{x}) + (AP_r^\perp\mathbf{x})$$

$$= (AP_r)\mathbf{x} + (AP_r^\perp)\mathbf{x}$$

$$= AP_r\mathbf{x} \ .$$

Since $\mathbf{x}$ is arbitrary, this means that $AP_r = A$, and *(1)* is proved.

Part *(2)* follows directly from Theorem 3.6.2. Finally, by *(1)*, if $\mathbf{x}$ is any vector with $Ax = \mathbf{c}$, then so is $P_r\mathbf{x}$, since by (7.1),

$$A(P_r\mathbf{x}) = (AP_r)\mathbf{x} = Ax = \mathbf{c} \ .$$

But by the Pythagorean Theorem $|\mathbf{x}|^2 = |P_r\mathbf{x}|^2 + |P_r^\perp\mathbf{x}|^2$, so $|P_r\mathbf{x}| < |\mathbf{x}|$ unless $\mathbf{x} = P_r\mathbf{x}$. This proves *(3)*. ∎

Now let's do some applications. The first part of the theorem is the most important. A typical application arises in curve fitting.

Suppose that we have measured the value of a function $f(u)$ at m values of u, namely $u_1, u_2, \ldots, u_m$, yeilding the results $v_1, v_2, \ldots, v_m$ for $f(u_1), f(u_2), \ldots, f(u_m)$ respectively. Suppose that you expect a linear relation between u and $f(u)$. That is, you expect

$$f(u) = au + b \qquad (7.5)$$

for some values a and b. The problem then is to find these values of a and b.

From (7.5) we have

$$au_1 + b = v_1$$

$$au_2 + b = v_2$$

$$\vdots = \vdots$$

$$au_m + b = v_m$$

Therefore, define the matrix

$$A = \begin{bmatrix} u_1 & 1 \\ u_2 & 1 \\ \vdots & \vdots \\ u_m & 1 \end{bmatrix}$$

and the vectors

$$\mathbf{x} = \begin{bmatrix} a \\ b \end{bmatrix} \qquad \text{and} \qquad \mathbf{b} = \begin{bmatrix} v_1 \\ v_2 \\ \vdots \\ v_m \end{bmatrix} \ .$$

3-64

Then the problem of finding a and b is to find a vector $\mathbf{x}$ that solves

$$A\mathbf{x} = \mathbf{b} \ .$$

The difficulty with this problem is that in general the points

$$(u_1, v_1), (u_2, y_2), \ldots (u_m, v_m)$$

do not lie on a single line in the u, v plane due to measurement errors.

Since we can't choose a and b to make $v_i = au_i + b$ for all i, we choose them to make the sum of the squares of the differences

$$\sum_{i=1}^{m} |v_i - (au_i + b)|^2 \tag{7.6}$$

as small as possible. Theorem 1 tells us how to do this: We need to compute the solution $\mathbf{x}_*$ to

$$A\mathbf{x}_* = P_c\mathbf{b} \ . \tag{7.7}$$

Now, we know how to compute P_c, so we could compute $P_c\mathbf{b}$, and then solve (7.7) by row reduction. This would work, but it is a bit inefficient.

Recall that when we do the Gram-Schmidt orthogonalization process on the columns of an m by n matrix A, that has independent columns, we get an m by n matrix Q and an m by m matrix R so that

$$A = QR \ .$$

The columns of Q are the elements of the orthonormal basis for the column space of A, and R is a non-singular upper triangular matrix whose columns contain the information for expressing the columns of A in terms of the orthonormal basis. The next theorem tells us that once we have Q, we can proceed directly to the solution without computing $P_c = Q^tQ$. We only need to compute $Q^t\mathbf{b}$. This is good because it is n times as much work to compute Q^tQ as it is to compute $Q^t\mathbf{b}$. Better still, in the direct approach, the equation we have to solve is $R = Q^t\mathbf{b}$ which has the advantage over $A = P_c\mathbf{b}$ in that R is upper triangular. So the next Theorem gives us a real saving of effort. We are going to assume however, that A has linearly independent columns. That is the usual case in many applications like our curve fitting problem. The assumption is not essential, but for our purposes, let's say that if the columns of A are not independent, just use Theorem 1 directly.

Theorem 2 (QR and Least Squares Solutions) *Let A be an $m \times n$ matrix, and let $A = QR$ be a QR factorization of A. Let $P_c = QQ^t$ be the orthogonal projection onto $\mathrm{Img}(A)$. Then $\mathbf{x}$ solves*

$$A\mathbf{x} = P_c\mathbf{b} \tag{7.8}$$

if and only if it also solves

$$R\mathbf{x} = Q^t\mathbf{b} \ . \tag{7.9}$$

Proof: Since $P_c = QQ^t$, for any $\mathbf{x}$ in $I\!R^n$,

$$A\mathbf{x} - P_c\mathbf{b} = QR\mathbf{x} - QQ^t\mathbf{x} = Q(R\mathbf{x} - Q^t\mathbf{b}) . \tag{7.10}$$

Multiplying on the left by Q^t, and using Theorem 2 of Section 4,

$$Q^t(A\mathbf{x} - P_c\mathbf{b}) = R\mathbf{x} - Q^t\mathbf{b} . \tag{7.11}$$

Hence $A\mathbf{x} - P_c\mathbf{b} = 0 \iff R\mathbf{x} - Q^t\mathbf{b} = 0.$ ∎

Example 1 Let's first consider fitting a line to three points. Suppose the three points are

$$(1,2) \quad (2,5) \quad (3,7) .$$

$$A = \begin{bmatrix} 1 & 1 \\ 2 & 1 \\ 3 & 1 \end{bmatrix} \quad \text{and} \quad \mathbf{b} = \begin{bmatrix} 2 \\ 5 \\ 7 \end{bmatrix} .$$

Let $\mathbf{v}_1$ and $\mathbf{v}_2$ be the first and second columns of A. Doing Gramm–Schmidt, we normalize the first column of A getting

$$\mathbf{u}_1 = \frac{1}{|\mathbf{v}_1|}\mathbf{v}_1 = \frac{1}{\sqrt{14}} \begin{bmatrix} 1 \\ 2 \\ 3 \end{bmatrix} . \tag{7.12}$$

Then we compute $\mathbf{u}_1 \cdot \mathbf{v}_2 = 6/\sqrt{14}$, and form

$$\mathbf{w}_2 = \mathbf{v}_2 - (\mathbf{u}_1 \cdot \mathbf{v}_2)\mathbf{u}_1 = \begin{bmatrix} 1 \\ 1 \\ 1 \end{bmatrix} - \frac{6}{\sqrt{14}}\frac{1}{\sqrt{14}} \begin{bmatrix} 1 \\ 2 \\ 3 \end{bmatrix} = \frac{1}{7} \begin{bmatrix} 4 \\ 1 \\ -2 \end{bmatrix} . \tag{7.13}$$

Now normalize this to get

$$\mathbf{u}_2 = \frac{1}{|\mathbf{w}_2|}\mathbf{w}_2 = \frac{1}{\sqrt{21}} \begin{bmatrix} 4 \\ 1 \\ -2 \end{bmatrix} . \tag{7.14}$$

This gives us

$$Q = \frac{1}{\sqrt{42}} \begin{bmatrix} \sqrt{3} & 4\sqrt{2} \\ 2\sqrt{3} & \sqrt{2} \\ 3\sqrt{3} & -2\sqrt{2} \end{bmatrix} .$$

Finally, $\mathbf{v}_1 = \sqrt{14}\mathbf{u}_1$ so that $\mathbf{v}_1 \cdot \mathbf{u}_1 = \sqrt{14}$. Likewise, $\mathbf{u}_2 \cdot \mathbf{v}_2 = \sqrt{21}/7$, and we already computed $\mathbf{u}_1 \cdot \mathbf{v}_2 = 6/\sqrt{14}$ above, and so $R = \begin{bmatrix} \sqrt{14} & 6/\sqrt{14} \\ 0 & \sqrt{21}/7 \end{bmatrix}$.

Next, compute $Q^t\mathbf{b} = \begin{bmatrix} 33/\sqrt{14} \\ -1/\sqrt{21} \end{bmatrix}$, and then solve

$$Rx = \begin{bmatrix} 33/\sqrt{14} \\ -1/\sqrt{21} \end{bmatrix} .$$

The answer is

$$\mathbf{x} = \begin{bmatrix} 5/2 \\ -1/3 \end{bmatrix} .$$

The best fit line is therefore

$$y = \frac{5}{2}x - \frac{1}{3} .$$

Here is a picture of the best fit line plotted together with the data points:

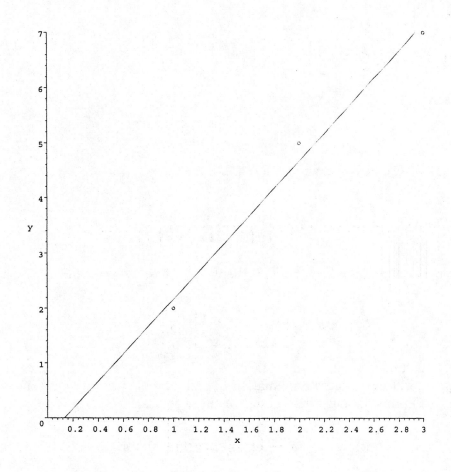

Next, consider a more realistic example, in which the data points are not all integer.

Example 2 Suppose our data points (x_i, y_i), $i = 1, 2, \ldots, 5$ are

$$(1, 1.9) \quad (2, 3.7) \quad (3, 6.2) \quad (4, 7.7) \quad (5, 10.5) \ .$$

Then

$$A = \begin{bmatrix} 1 & 1 \\ 2 & 1 \\ 3 & 1 \\ 4 & 1 \\ 5 & 1 \end{bmatrix} \quad \text{and} \quad \mathbf{b} = \begin{bmatrix} 1.9 \\ 3.7 \\ 6.2 \\ 7.7 \\ 10.5 \end{bmatrix}$$

Doing the Gram-Schmidt procedure yields $A = QR$ where

$$Q = \frac{1}{\sqrt{55}} \begin{bmatrix} 1 & 4\sqrt{2} \\ 2 & 5/\sqrt{2} \\ 3 & \sqrt{2} \\ 4 & -1/\sqrt{2} \\ 5 & -2/\sqrt{2} \end{bmatrix}$$

and

$$R = \begin{bmatrix} 1 & 3/11 \\ 0 & \sqrt{2}/11 \end{bmatrix} \ .$$

Next compute

$$Q^t \mathbf{b} = \frac{\sqrt{55}}{275} \begin{bmatrix} 556 \\ -9\sqrt{2} \end{bmatrix} \ .$$

Finally, solving $R\mathbf{x} = Q^t\mathbf{b}$ simply means solving the 2 by 2 upper triangular system

$$\begin{bmatrix} 1 & 3/11 \\ 0 & \sqrt{2}/11 \end{bmatrix} \begin{bmatrix} a \\ b \end{bmatrix} = \frac{\sqrt{55}}{275} \begin{bmatrix} 556 \\ -9\sqrt{2} \end{bmatrix} \ .$$

Since R is upper triangular, we can read off the value of b right away, and solving for a we get

$$a = \frac{53}{25} \quad \text{and} \quad b = \frac{-9}{25} \ .$$

Example 3 Let's find a good polynomial fit to $\sin(x)$ on $0 \leq x \leq \pi/2$. Since this is an odd function, we will use a linear combination of odd powers of x. If we use the first four odd powers, this means finding values of a, b, c and d so that

$$\sin(x) \approx ax + bx^3 + cx^5 + dx^7 \quad \text{for} \quad 0 \leq x \leq \frac{\pi}{2}$$

3-68

Now we know

$$\sin(\pi/6) = \frac{1}{2} \qquad \sin(\pi/4) = \frac{1}{\sqrt{2}} \qquad \sin(\pi/3) = \frac{\sqrt{3}}{2} \qquad \sin(\pi/2) = 1 .$$

From the angle addition formula we have $\sin(5\pi/12) = \sin(\pi/3 + \pi/4) = (\sqrt{3} + 1)/(2\sqrt{2})$ and $\sin(\pi/12) = \sin(\pi/3 - \pi/4) = (\sqrt{3} - 1)/(2\sqrt{2})$. This gives us our data points. Notice in this example, there are no experimental errors. The error comes in through our attempt to treat $\sin(x)$ like a polynomial of seventh degree, which it is not.

In any case, we have our data, and we define

$$x_1 = \frac{\pi}{12} \quad x_2 = \frac{\pi}{6} \quad x_3 = \frac{\pi}{4} \quad x_4 = \frac{\pi}{3} \quad x_5 = \frac{5\pi}{12} \quad x_6 = \frac{\pi}{2}$$

and

$$y_1 = \sin\frac{\pi}{12} \quad y_2 = \sin\frac{\pi}{6} \quad y_3 = \sin\frac{\pi}{4} \quad y_4 = \sin\frac{\pi}{3} \quad y_5 = \sin\frac{5\pi}{12} \quad y_6 = \sin\frac{\pi}{2}$$

The matrix A is now

$$A = \begin{bmatrix} x_1 & x_1^3 & x_1^5 & x_1^7 \\ x_2 & x_2^3 & x_2^5 & x_2^7 \\ x_3 & x_3^3 & x_3^5 & x_3^7 \\ x_4 & x_4^3 & x_4^5 & x_4^7 \\ x_5 & x_5^3 & x_5^5 & x_5^7 \\ x_6 & x_6^3 & x_6^5 & x_6^7 \end{bmatrix}$$

It would be very tedious to do this by hand, but with Maple, it is very easy to do with Maple. The answer, kept to 10 digits, is

$$\begin{bmatrix} a \\ b \\ c \\ d \end{bmatrix} = \begin{bmatrix} .9999976966 \\ -.1666522644 \\ .008309725100 \\ -.0001844086724 \end{bmatrix}$$

yielding

$$f(x) = (.9999976966)x - (.1666522644)x^3 + (.008309725100)x^5 - (.0001844086724)x^7 .$$

If we plot this from 0 to $\pi/2$, the graph is indistinguishable to the eye from that of $\sin(x)$, but if we plot from 0 to π, we can see the two graphs pull apart.

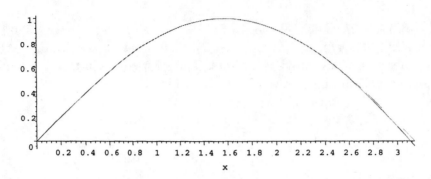

The same is true of the 7th degree Taylor polynomial approximation to $\sin(x)$. Which one is better? Let's use both to approximate

$$\int_0^{\pi/2} \sin(x)\mathrm{d}x = 1 .$$

One easily works out that

$$\int_0^{\pi/2} f(x)\mathrm{d}x = 1.000000117 .$$

which is almost accurate to one part in 10^7.

Now let $g(x)$ be the seventh degree Taylor approximation to $\sin(x)$:

$$g(x) = x - \frac{1}{6}x^3 + \frac{1}{120}x^5 - \frac{1}{5040}x^7 .$$

One easily works out that

$$\int_0^{\pi/2} g(x)\mathrm{d}x = .9999752637 .$$

This is off by more than two parts in 10^5.

In fact, the ratio of the errors, Taylor to least square, is

$$\frac{1 - .9999752637}{1.000000117 - 1} = 211.426934 .$$

The least–squares approach to curve fitting can provide extremely accurate fits that can then be used for purposes such as numerical integration.

Problems for Section 1

3.1.1 Consider the matrix

$$B = \begin{bmatrix} 1 & 2 & 3 \\ 2 & 1 & 3 \\ 1 & 1 & 2 \end{bmatrix} .$$

Each of the subspaces $\text{Img}(B)$, $\text{Ker}(B)$, $\text{Img}(B^t)$ and $\text{Ker}(B^t)$ is either a line or a plane. Which are which? Justify your answer, and find a paramteric representation of those that are lines, and the equations for those that are planes.

3.1.2 Consider the matrix

$$B = \begin{bmatrix} 1 & 2 & 1 \\ 2 & 0 & 1 \\ 3 & 2 & 2 \end{bmatrix} .$$

Each of the subspaces $\text{Img}(B)$, $\text{Ker}(B)$, $\text{Img}(B^t)$ and $\text{Ker}(B^t)$ is either a line or a plane. Which are which? Justify your answer, and find a paramteric representation of those that are lines, and the equations for those that are planes.

Problems for Section 2

3.2.1 Consider the following vectors:

$$\mathbf{v}_1 = \begin{bmatrix} 1 \\ 2 \\ 3 \\ 4 \end{bmatrix} , \quad \mathbf{v}_2 = \begin{bmatrix} 4 \\ 3 \\ 2 \\ 1 \end{bmatrix} \quad \text{and} \quad \mathbf{v}_3 = \begin{bmatrix} 2 \\ 1 \\ 4 \\ 3 \end{bmatrix}$$

Is $\{\mathbf{v}_1, \mathbf{v}_2, \mathbf{v}_3, \}$ a linearly independent set of vectors?

3.2.2 Consider the following vectors:

$$\mathbf{v}_1 = \begin{bmatrix} 1 \\ 2 \\ 3 \\ 4 \end{bmatrix} , \quad \mathbf{v}_2 = \begin{bmatrix} 4 \\ 3 \\ 2 \\ 1 \end{bmatrix} , \quad \mathbf{v}_3 = \begin{bmatrix} 2 \\ 1 \\ 4 \\ 3 \end{bmatrix} \quad \text{and} \quad \mathbf{v}_4 = \begin{bmatrix} 3 \\ 4 \\ 1 \\ 2 \end{bmatrix}$$

Is $\{\mathbf{v}_1, \mathbf{v}_2, \mathbf{v}_3, \mathbf{v}_4\}$ a linearly independent set of vectors?

3.2.3 Consider the following vectors:

$$\mathbf{v}_1 = \begin{bmatrix} 1 \\ 2 \\ 3 \\ 4 \end{bmatrix} , \quad \mathbf{v}_2 = \begin{bmatrix} 2 \\ 3 \\ 4 \\ 1 \end{bmatrix} , \quad \mathbf{v}_3 = \begin{bmatrix} 3 \\ 4 \\ 1 \\ 2 \end{bmatrix} \quad \text{and} \quad \mathbf{v}_3 = \begin{bmatrix} 4 \\ 1 \\ 2 \\ 3 \end{bmatrix}$$

Is $\{\mathbf{v}_1, \mathbf{v}_2, \mathbf{v}_3, \mathbf{v}_4\}$ a linearly independent set of vectors?

3.2.4 Let A and B be n by n matrices. Suppose that the kernel of B is the zero subspace, and that the columns of A are linearly independent. Then is AB necessarily invertible? Either explain why this is the case, or give a counterexample.

(a) For which of them, if any, are the columns independent? Justify your answer.

(b) For which of them, if any, are the rows independent? Justify your answer.

(c) For which of them, if any, do the columns span R^3? Justify your answer.

3.2.5 Consider the matrices:

$$A = \begin{bmatrix} 1 & 2 & 3 \\ 2 & 0 & 2 \\ 0 & 1 & 1 \\ 1 & 2 & 3 \end{bmatrix} \qquad B = \begin{bmatrix} 1 & 0 & 0 \\ 2 & 1 & 0 \\ 1 & 1 & 2 \end{bmatrix}$$

$$C = \begin{bmatrix} 1 & 0 & 0 & 1 \\ 0 & 2 & 1 & 1 \\ 0 & 0 & 2 & 2 \end{bmatrix} \qquad D = \begin{bmatrix} 1 & 2 & 1 \\ 1 & 0 & 1 \\ 0 & 2 & 0 \end{bmatrix}$$

(a) For which of these matrices, if any, are the columns linearly independent? Justify your answer.

(b) For which of these matrices, if any, are the rows linearly independent? Justify your answer.

(c) For which of these matrices, if any, is the kernel the zero subspace? Justify your answer.

3.2.6 Consider the following matrices:

$$A = \begin{bmatrix} 1 & 2 & 3 \\ 2 & 1 & 1 \\ 3 & 3 & 4 \end{bmatrix} \qquad B = \begin{bmatrix} 1 & 7 & -3 & -4 \\ 3 & -2 & 5 & 1 \\ 0 & 2 & 1 & 1 \end{bmatrix}$$

$$C = \begin{bmatrix} 1 & 2 & 4 \\ 1 & 3 & 9 \\ 1 & 4 & 16 \end{bmatrix} \qquad D = \begin{bmatrix} 1 & 0 & 1 & 0 \\ 0 & 1 & 0 & 1 \\ 2 & 3 & 2 & 3 \end{bmatrix}$$

Answer the questions (a), (b) and (c) from the previous problem for these matrices.
3.2.7 Let A be an m by n matrix, let B be an n by p matrix, and let $C = AB$.

(a) If the columns of A are linearly independent, and the columns of B are linearly independent, must it also be the case that the columns of C are linearly independent? Justify your answer.

(b) If the rows of A are linearly independent, and the columns of B are linearly independent, must it also be the case that the columns of C are linearly independent? Justify your answer.

3.2.8 Let A be an m by n matrix with $n > m$, and suppose that A has linearly independent rows. Consider the square matrices

$$C = A^t A \quad \text{and} \quad D = AA^t .$$

(a) Must it be the case that C is invertible? Justify your answer.

(b) Must it be the case that D is invertible? Justify your answer.

3.2.9 Consider the set of vectors $\{v_1, v_2, v_3, \}$ where

$$\mathbf{v}_1 = \begin{bmatrix} 1 \\ -1 \\ 0 \end{bmatrix} \quad, \quad \mathbf{v}_2 = \begin{bmatrix} 0 \\ 1 \\ -1 \end{bmatrix} \quad \text{and} \quad \mathbf{v}_3 = \begin{bmatrix} 1 \\ 1 \\ 1 \end{bmatrix}$$

Find numbers t_1, t_2, and t_3 so that the vector $\mathbf{b} = \begin{bmatrix} 1 \\ 2 \\ 3 \end{bmatrix}$ can be written

$$\mathbf{b} = t_1 \mathbf{v}_1 + t_2 \mathbf{v}_2 + t_3 \mathbf{v}_3 .$$

Is $\{v_1, v_2, v_3, \}$ a basis for $I\!\!R^3$?

3.2.10 Consider the set of vectors $\{v_1, v_2, v_3, \}$ where

$$\mathbf{v}_1 = \begin{bmatrix} 1 \\ 3 \\ 9 \end{bmatrix} \quad, \quad \mathbf{v}_2 = \begin{bmatrix} 1 \\ 2 \\ 4 \end{bmatrix} \quad \text{and} \quad \mathbf{v}_3 = \begin{bmatrix} 1 \\ 1 \\ 1 \end{bmatrix}$$

Find numbers t_1, t_2, and t_3 so that the vector $\mathbf{b} = \begin{bmatrix} 1 \\ 2 \\ 3 \end{bmatrix}$ can be written

$$\mathbf{b} = t_1 \mathbf{v}_1 + t_2 \mathbf{v}_2 + t_3 \mathbf{v}_3 .$$

Is $\{\mathbf{v}_1, \mathbf{v}_2, \mathbf{v}_3,\}$ a basis for $I\!R^3$?

3.2.11 Consider the set of vectors $\{\mathbf{v}_1, \mathbf{v}_2, \mathbf{v}_3,\}$ where

$$\mathbf{v}_1 = \begin{bmatrix} 1 \\ -1 \\ 0 \end{bmatrix} \quad, \quad \mathbf{v}_2 = \begin{bmatrix} 0 \\ 1 \\ -1 \end{bmatrix} \quad \text{and} \quad \mathbf{v}_3 = \begin{bmatrix} 1 \\ -2 \\ 1 \end{bmatrix}$$

Find numbers t_1, t_2, and t_3 so that the vector $\mathbf{b} = \begin{bmatrix} 1 \\ 2 \\ -3 \end{bmatrix}$ can be written

$$\mathbf{b} = t_1 \mathbf{v}_1 + t_2 \mathbf{v}_2 + t_3 \mathbf{v}_3 \; .$$

How many ways can this be done? Is $\{\mathbf{v}_1, \mathbf{v}_2, \mathbf{v}_3,\}$ a basis for $I\!R^3$?

Problems for Section 3

3.3.1 Let S be the subspace of $I\!R^4$ spanned by the vectors $\{\mathbf{v}_1, \mathbf{v}_2, \mathbf{v}_3, \mathbf{v}_4\}$ from problem 3.2.2. What is the dimension of S?

3.3.2 Let S be the subspace of $I\!R^4$ spanned by the vectors $\{\mathbf{v}_1, \mathbf{v}_2, \mathbf{v}_3, \mathbf{v}_4\}$ from problem 3.2.3. What is the dimension of S?

3.3.3 Consider the matrices:

$$A = \begin{bmatrix} 1 & 1 & 1 \\ 2 & 0 & 2 \\ 0 & 0 & 1 \\ 3 & 2 & 1 \end{bmatrix} \qquad B = \begin{bmatrix} 1 & 2 & 3 \\ 2 & 1 & 3 \\ 1 & 1 & 2 \end{bmatrix}$$

$$C = \begin{bmatrix} 1 & 2 & 0 & 1 \\ 1 & 3 & 1 & 2 \\ 1 & 2 & 2 & 2 \end{bmatrix} \qquad D = \begin{bmatrix} 1 & 2 & 4 \\ 2 & 2 & 4 \\ 0 & 2 & -1 \end{bmatrix}$$

(a) For which of these matrices, if any, are the columns linearly independent? Justify your answer.

(b) For which of these matrices, if any, are the rows linearly independent? Justify your answer.

(c) For which of these matrices, if any, is the kernel the zero subspace? Justify your answer.

3-74

(d) For which of these matrices, if any, is the dimension of the image of the transpose equal to 3? Justify your answer.

(e) For which of these matrices, if any, is the dimension of the image equal to 3? Justify your answer.

3.3.4 suppose that $\{v_1, v_2, \ldots, v_n\}$ is a basis for $I\!R^n$. For any k with $1 \le k \le n - 1$, let S be the span of $\{v_1, v_2, \ldots, v_k\}$ and let $\tilde{S}$ be the span of $\{v_{k+1}, v_{k+2}, \ldots, v_n\}$. Are A and $\tilde{A}$ always complementary subsapces of $I\!R^n$? If so, explain why. If not, give a counter example.

3.3.5 Let A be a 4×4 matrix, and suppose that $A^t x = 0$ only in case x is a multiple of
$\begin{bmatrix} 1 \\ -2 \\ 1 \\ 0 \end{bmatrix}$. Let $b = \begin{bmatrix} 1 \\ 1 \\ 1 \\ 1 \end{bmatrix}$. Does $Ax = b$ have a solution? Justify your answer.

3.3.6 Let A be a symmetric 4×4 matrix, and suppose that $Ax = 0$ only in case x is a
multiple of $\begin{bmatrix} 1 \\ -2 \\ 1 \\ 0 \end{bmatrix}$. Let $b = \begin{bmatrix} 1 \\ 1 \\ 1 \\ 1 \end{bmatrix}$. Does $Ax = b$ have a solution? Justify your answer.

Problems for Section 4

3.4.1 Find bases for $\text{Img}(A)$, $\text{Ker}(A)$, $\text{Img}(A^t)$ and $\text{Ker}(A^t)$ where

$$A = \begin{bmatrix} 1 & 2 & 4 & 1 \\ 0 & 2 & 2 & 0 \\ 2 & 3 & 7 & 1 \\ 1 & 1 & 3 & 0 \end{bmatrix}.$$

3.4.2 Find bases for $\text{Img}(A)$, $\text{Ker}(A)$, $\text{Img}(A^t)$ and $\text{Ker}(A^t)$ where

$$A = \begin{bmatrix} 1 & 2 & 3 \\ 0 & 3 & 2 \\ 2 & 0 & 1 \end{bmatrix}.$$

3.4.3 Find bases for $\text{Img}(A)$, $\text{Ker}(A)$, $\text{Img}(A^t)$ and $\text{Ker}(A^t)$ where

$$A = \begin{bmatrix} 1 & 1 & 2 \\ 1 & -1 & 0 \\ 0 & 1 & 1 \end{bmatrix}.$$

3.4.4 Find bases for $\text{Img}(A)$, $\text{Ker}(A)$, $\text{Img}(A^t)$ and $\text{Ker}(A^t)$ where

$$A = \begin{bmatrix} 1 & 0 \\ 1 & 1 \\ 0 & 1 \end{bmatrix} .$$

3.4.5 Find bases for $\text{Img}(A)$, $\text{Ker}(A)$, $\text{Img}(A^t)$ and $\text{Ker}(A^t)$ where

$$A = \begin{bmatrix} 1 & 0 \\ 0 & 1 \\ 1 & -1 \end{bmatrix} .$$

3.4.6 Let A be a 4×4 matrix, and suppose that $A^t\mathbf{x} = 0$ only in case $\mathbf{x}$ is a multiple of $\begin{bmatrix} 1 \\ -2 \\ 1 \\ 0 \end{bmatrix}$. Does $\mathbf{b} = \begin{bmatrix} 1 \\ 1 \\ 1 \\ 1 \end{bmatrix}$. Does $A\mathbf{x} = \mathbf{b}$ have a *unique* solution? Justify your answer.

3.4.7 Let A be a 4×5 matrix, and suppose that $A^t\mathbf{x} = 0$ only in case $\mathbf{x}$ is a multiple of $\begin{bmatrix} 1 \\ -2 \\ 1 \\ 0 \\ 0 \end{bmatrix}$. Does $\mathbf{b} = \begin{bmatrix} 1 \\ 1 \\ 1 \\ 1 \end{bmatrix}$. Does $A\mathbf{x} = \mathbf{b}$ have a *unique* solution? Justify your answer.

3.4.8 The sets of vectors $\{\mathbf{v}_1, \mathbf{v}_2\}$ and $\{\mathbf{u}_1, \mathbf{u}_2\}$ where

$$\mathbf{v}_1 = \begin{bmatrix} 1 \\ -1 \\ 0 \end{bmatrix} \quad \mathbf{v}_1 = \begin{bmatrix} 0 \\ 1 \\ -1 \end{bmatrix}$$

and

$$\mathbf{u}_1 = \begin{bmatrix} 2 \\ -1 \\ -1 \end{bmatrix} \quad \mathbf{u}_2 = \begin{bmatrix} 1 \\ 1 \\ -2 \end{bmatrix}$$

are both bases for S where S is the plane in $I\!R^3$ given by $x_1 + x_2 + x_3 = 0$.

(a) Find the 2×2 matrix R that is the change of basis matrix from the $\{\mathbf{u}_1, \mathbf{u}_2\}$ basis to the $\{\mathbf{v}_1, \mathbf{v}_2\}$.

(b) Express $2\mathbf{u}_1 - 3\mathbf{u}_2$ as al inear combination of $\mathbf{v}_1$ and $\mathbf{v}_2$.

3.4.9 The sets of vectors $\{\mathbf{v}_1, \mathbf{v}_2\}$ and $\{\mathbf{u}_1, \mathbf{u}_2\}$ where

$$\mathbf{v}_1 = \begin{bmatrix} 1 \\ 2 \\ 3 \end{bmatrix} \quad \mathbf{v}_1 = \begin{bmatrix} 3 \\ 2 \\ 1 \end{bmatrix}$$

and

$$\mathbf{u}_1 = \begin{bmatrix} 2 \\ 1 \\ 0 \end{bmatrix} \quad \mathbf{u}_2 = \begin{bmatrix} 0 \\ 1 \\ 2 \end{bmatrix}$$

are both bases for S where S is the plane in $I\!\!R^3$ given by $x_1 - 2x_2 + x_3 = 0$.

(a) Find the 2×2 matrix R that is the change of basis matrix from the $\{\mathbf{u}_1, \mathbf{u}_2\}$ basis to the $\{\mathbf{v}_1, \mathbf{v}_2\}$.

(b) Express $3\mathbf{u}_1 - \mathbf{u}_2$ as al inear combination of $\mathbf{v}_1$ and $\mathbf{v}_2$.

Problems for Section 5

3.5.1 Find an orthonormal basis for $\text{Img}(A)$ where

$$A = \begin{bmatrix} 1 & 2 & 4 & 1 \\ 0 & 2 & 2 & 0 \\ 2 & 3 & 7 & 1 \\ 1 & 1 & 3 & 0 \end{bmatrix}.$$

3.5.2 Find an orthonormal basis for $\text{Img}(A^t)$, where A is the matrix from problem (6.1).

3.5.3 Let A be the matrix $A = \begin{bmatrix} 1 & 2 & 3 \\ 0 & 3 & 2 \\ 2 & 0 & 1 \end{bmatrix}$. Find the QR decomposition of A. Check your result by computing the product QR.

3.5.4 Let A be the matrix

$$A = \begin{bmatrix} 1 & 1 & 2 \\ 1 & -1 & 0 \\ 0 & 1 & 1 \end{bmatrix}.$$

(a) Find an orthonormal basis for $\text{Img}A$, and find a matrix Q with orthonormal columns and another matrix R so that

$$A = QR.$$

(b) Express the second column of A as a linear combiniation of the columns of Q.

(c) Express the second column of Q as a linear combiniation of the columns of A.

(d) Give parametric descriptions of Img$A$ and Ker$A$. If either one of these is a plane, give the equation of the plane.

(e) Are the rows of of A linearly independent? Are the columns of of A linearly independent?

3.5.5 Let A be the matrix

$$A = \begin{bmatrix} 1 & 1 & 2 \\ 1 & 0 & 1 \\ 0 & 1 & 1 \end{bmatrix} .$$

(a) Find an orthonormal basis for Img(A), and find a matrix Q with orthonormal columns and another matrix R so that

$$A = QR .$$

(b) Find an orthonormal basis for Img(A^t), and find the orthogonal projection P_r onto Img(A^t).

(c) Give parametric descriptions of Img(A) and Ker(A). If either one of these is a plane, give the equation of the plane.

(d) What are the dimensions of Img(A) and Img(A^t)?

(e) Are the rows of of A linearly independent? Are the columns of of A linearly independent?

3.5.6 Let A be the matrix

$$A = \begin{bmatrix} 1 & 0 \\ 1 & 1 \\ 0 & 1 \end{bmatrix} .$$

(a) Find an orthonormal basis for Img(A), and find a matrix Q with orthonormal columns and another matrix R so that

$$A = QR .$$

(b) Img(A) is a plane. Find the equation of this plane.

(c) Let **b** be the vector

$$\mathbf{b} = \begin{bmatrix} 1 \\ 1 \\ 1 \end{bmatrix} .$$

Find the least squares solution to $A\mathbf{x} = \mathbf{b}$. What is the minimum value of $|A\mathbf{x} - \mathbf{b}|^2$ as **x** varies over R^2?

3.5.7 Let A be the matrix

$$A = \begin{bmatrix} 1 & 0 \\ 0 & 1 \\ 1 & 1 \end{bmatrix}.$$

(a) Find an orthonormal basis for $\text{Img}(A)$, and find a matrix Q with orthonormal columns and another matrix R so that

$$A = QR.$$

(b) For this matrix A, $\text{Img}(A)$ is a plane. Find the equation of the plane.

Problems for Section 6

3.6.1 Find the orthogonal projection onto S where S is the span of

$$\mathbf{v}_1 = \begin{bmatrix} 1 \\ 2 \\ 3 \end{bmatrix} \quad \text{and} \quad \mathbf{v}_1 = \begin{bmatrix} 3 \\ 2 \\ 1 \end{bmatrix}$$

Also find the orthogonal projection $P^\perp$ onto $S^\perp$, and give a one–to–one parametric representation for $S^\perp$.

3.6.2 Let A be the matrix

$$A = \begin{bmatrix} 1 & 0 & 2 \\ 2 & 2 & 1 \\ 3 & 2 & 3 \end{bmatrix}.$$

(a) Find the orthogonal projection P_c onto $\text{Img}(A)$, and $P^\perp$. (The subscript c recalls the fact that $\text{Img}(A)$ is the span of the columns of A).

(b) Find the orthogonal projection P_r onto $\text{Img}(A^t)$, and $P_r^\perp$. (The subscript r recalls the fact that $\text{Img}(A^t)$ is the span of the rows of A).

(d) Find one–to–one parametric representations of $\text{Img}(A)$, and of $\text{Ker}(A)$.

3.6.3 Let A be the matrix

$$A = \begin{bmatrix} 1 & 1 & 2 \\ 1 & -1 & 0 \\ 0 & 1 & 1 \end{bmatrix}.$$

(See problem 3.5.4).

(a) Find the orthogonal projection onto $\text{Ker}(A^t)$.

(b) Find the orthogonal projection onto $\text{Img}(A)$.

(c) How are your answers for parts (a) and (b) related?

(d) Find the orthogonal projection onto $\text{Img}(A^t)$.

3.6.4 Let A be the matrix

$$A = \begin{bmatrix} 1 & 1 & 2 \\ 1 & 0 & 1 \\ 0 & 1 & 1 \end{bmatrix}.$$

(See problem 3.5.5).

(a) Find the orthogonal projection P_c onto $\text{Img}(A)$.

(b) Find the orthogonal projection P_r onto $\text{Img}(A^t)$.

(c) Find the orthogonal projection P_c onto $\text{Ker}(A)$.

(d) Find the orthogonal projection P_r onto $\text{Ker}(A^t)$.

3.6.5 Consider the following vectors:

$$\mathbf{v}_1 = \begin{bmatrix} -1 \\ -1 \\ 0 \\ 0 \end{bmatrix} \quad \mathbf{v}_2 = \begin{bmatrix} 2 \\ 0 \\ 1 \\ 1 \end{bmatrix} \quad \mathbf{v}_3 = \begin{bmatrix} 1 \\ 0 \\ 0 \\ 1 \end{bmatrix} \quad \mathbf{v}_3 = \begin{bmatrix} 0 \\ -1 \\ 1 \\ 0 \end{bmatrix}$$

(a) Let S be the span of $\{\mathbf{v}_1, \mathbf{v}_2, \mathbf{v}_3, \mathbf{v}_4\}$. Find an orthonormal basis for S. What is the dimension of S?

(b) Find an orthonormal basis for $S^\perp$, the orthogonal complement to S. What is the dimension of $S^\perp$

(c) Find the orthogonal projections onto $S^\perp$ and onto S.

(d) Consider the vector $\mathbf{x}$ where

$$\mathbf{x} = \begin{bmatrix} 1 \\ 2 \\ 3 \\ 4 \end{bmatrix}.$$

Find vectors $\mathbf{w}$ and $\mathbf{v}$ with $\mathbf{w}$ in S and $\mathbf{v}$ in $S^\perp$ so that $\mathbf{x} = \mathbf{w} + \mathbf{v}$.

3.6.6 Consider the following vectors:

$$\mathbf{v}_1 = \begin{bmatrix} 1 \\ -1 \\ 0 \\ 0 \end{bmatrix} \quad \mathbf{v}_2 = \begin{bmatrix} 2 \\ 0 \\ -1 \\ -1 \end{bmatrix} \quad \mathbf{v}_3 = \begin{bmatrix} 1 \\ 0 \\ 0 \\ -1 \end{bmatrix} \quad \mathbf{v}_3 = \begin{bmatrix} 0 \\ -1 \\ 0 \\ 1 \end{bmatrix}$$

(a) Let S be the span of $\{v_1, v_2, v_3, v_4\}$. Find an orthonormal basis for S. What is the dimension of S?

(b) Find an orthonormal basis for $S^\perp$, the orthogonal complement to S. What is the dimension of $S^\perp$

(c) Find the orthogonal projections onto $S^\perp$ and onto S.

(d) Consider the vector $\mathbf{x}$ where

$$\mathbf{x} = \begin{bmatrix} 1 \\ 2 \\ 3 \\ 4 \end{bmatrix}.$$

Find vectors $\mathbf{w}$ and $\mathbf{v}$ with $\mathbf{w}$ in S and $\mathbf{v}$ in $S^\perp$ so that $\mathbf{x} = \mathbf{w} + \mathbf{v}$.

3.6.7 Let $C = \begin{bmatrix} 1 & 1 & 2 \\ 2 & 0 & 2 \\ 0 & 1 & 1 \end{bmatrix}$ and let $\mathbf{b} = \begin{bmatrix} 1 \\ 2 \\ 3 \end{bmatrix}$.

(a) Find vectors $\mathbf{w}$ and $\mathbf{v}$ with $\mathbf{w}$ in S and $\mathbf{v}$ in $\perp$ so that $\mathbf{b} = \mathbf{w} + \mathbf{v}$.

(b) Find the distance from $\mathbf{b}$ to $\text{Img}(C)$.

(c) Find the distance from $\mathbf{b}$ to $(\text{Img}(C))^\perp$.

3.6.8 Compute the distance from $\mathbf{b} = \begin{bmatrix} 1 \\ 0 \\ 0 \\ 1 \end{bmatrix}$ from the line in $I\!\!R^4$ through the origin in

$\begin{bmatrix} 0 \\ 1 \\ 2 \\ 0 \end{bmatrix}.$

3.6.9 Compute the distance from $\mathbf{b} = \begin{bmatrix} 1 \\ 0 \\ 0 \\ 1 \end{bmatrix}$ from the line in $I\!\!R^4$ through the $\begin{bmatrix} 0 \\ 1 \\ 1 \\ 0 \end{bmatrix}$ and

$\begin{bmatrix} 0 \\ 1 \\ 0 \\ 2 \end{bmatrix}.$

Problems for Section 7

3-81

3.7.1 Consider the matrix

$$A = \begin{bmatrix} 0 & 1 & 1 \\ 1 & 0 & 1 \\ 0 & 1 & 1 \\ 0 & 0 & 1 \end{bmatrix} .$$

(a) Give the QR factorization of A.

(b) Find the least squares solution to $A\mathbf{x} = \mathbf{b}$ where

$$\mathbf{b} = \begin{bmatrix} 1 \\ 2 \\ 3 \\ 1 \end{bmatrix} .$$

(c) Find the least squares solution to $A\mathbf{x} = \mathbf{c}$ where

$$\mathbf{c} = \begin{bmatrix} 1 \\ 0 \\ -1 \\ 0 \end{bmatrix} .$$

3.7.2 Find the values of a, b and c that give the best least squares fit of the quadratic

$$y = a + bx + cx^2$$

to the data points

$$(-1, 1) \quad (0, 1) \quad (1, 4) \quad (2, 6) .$$

3.7.3 (A computer is recommended to help with the computations here.) **(a)** Find the values of a, b and c and d that give the best least squares fit of the even polynomial

$$y = ax^6 + bx^4 + cx^2 + d$$

to the function $f(x) = e^{-x^2}$ using as data points

$$(0, f(0)) \quad (1/2, f(1/2)) \quad (1, f(1)) \quad (3/2, f(3/2)) \quad (2, f(2)) \quad (5/2, f(5/2)) .$$

(b) Use your result from part (a) to approximate the integral

$$\int_0^{5/2} e^{-x^2} \, dx .$$

3-82

3.7.4

Consider the matrix B where

$$B = \begin{bmatrix} 1 & 0 & 1 & 0 \\ 2 & -2 & 0 & 1 \\ 2 & 0 & 2 & -1 \\ -1 & 1 & 0 & 0 \\ 0 & 1 & 1 & 0 \\ 0 & -1 & -1 & 0 \end{bmatrix}$$

given that the QR factorization of B is

$$\begin{bmatrix} 1/\sqrt{10} & \sqrt{2}/6 & \sqrt{10}/15 \\ 2/\sqrt{10} & -\sqrt{2}/3 & \sqrt{10}/6 \\ 2/\sqrt{10} & \sqrt{2}/3 & -\sqrt{10}/6 \\ -1/\sqrt{10} & \sqrt{2}/6 & \sqrt{10}/15 \\ 0 & \sqrt{2}/3 & 2\sqrt{10}/15 \\ 0 & -\sqrt{2}/3 & -2\sqrt{10}/15 \end{bmatrix} \begin{bmatrix} \sqrt{10} & -1/2\sqrt{10} & 1/2\sqrt{10} & 0 \\ 0 & 3/2\sqrt{2} & 3/2\sqrt{2} & -2/3\sqrt{2} \\ 0 & 0 & 0 & 1/3\sqrt{10} \end{bmatrix}$$

and that the QR factorization of B^t is

$$\begin{bmatrix} \sqrt{2}/2 & \sqrt{7}/7 & 1/\sqrt{42} \\ 0 & -2/\sqrt{7} & -\sqrt{42}/21 \\ 1/\sqrt{2} & -1/\sqrt{7} & -1/\sqrt{42} \\ 0 & 1/\sqrt{7} & -\sqrt{42}/7 \end{bmatrix} \begin{bmatrix} \sqrt{2} & \sqrt{2} & 2\sqrt{2} & -1\sqrt{2}/2 & \sqrt{2}/2 & -\sqrt{2}/2 \\ 0 & \sqrt{7} & -1/\sqrt{7} & -3/\sqrt{7} & -3/\sqrt{7} & 3/\sqrt{7} \\ 0 & 0 & \sqrt{42}/7 & -\sqrt{42}/14 & -\sqrt{42}/14 & \sqrt{42}/14 \end{bmatrix}$$

(a) Find the orthogonal projection P_r onto the row space of A

(b) Find a solution of $B\mathbf{x} = \mathbf{b}$

$$\mathbf{b} = \begin{bmatrix} 4 \\ 2 \\ 4 \\ 1 \\ 5 \\ -5 \end{bmatrix}.$$

(c) Find the minimal length solution $\mathbf{x}_{**}$ of $B\mathbf{x} = \mathbf{b}$.

(d) Find the minimal length least squares solution of $B\mathbf{x} = \mathbf{c}$

$$\mathbf{c} = \begin{bmatrix} 0 \\ 0 \\ 0 \\ 0 \\ \sqrt{10} \\ -\sqrt{10} \end{bmatrix}.$$

3.7.5 Let A be the matrix

$$A = \begin{bmatrix} 1 & 1 & 2 \\ 1 & -1 & 0 \\ 0 & 1 & 1 \end{bmatrix}.$$

(See Problem 3.5.4)). Let $\mathbf{b}$ be the vector

$$\mathbf{b} = \begin{bmatrix} 1 \\ 2 \\ 3 \end{bmatrix}.$$

Find a least squares solution to $A\mathbf{x} = \mathbf{b}$. Which vector in $\mathrm{Img}\,A$ is closest to $\mathbf{b}$?

3.7.6 Let A be the matrix

$$A = \begin{bmatrix} 1 & 1 & 2 \\ 1 & 0 & 1 \\ 0 & 1 & 1 \end{bmatrix}.$$

(See problem 3.5.5). Let $\mathbf{b}$ be the vector

$$\mathbf{b} = \begin{bmatrix} 1 \\ 1 \\ 1 \end{bmatrix}.$$

Find a least squares solution to $A\mathbf{x} = \mathbf{b}$. Which vector in $\mathrm{Img}(A)$ is closest to $\mathbf{b}$?

3.7.7 Let A be the matrix

$$A = \begin{bmatrix} 1 & 0 \\ 1 & 1 \\ 0 & 1 \end{bmatrix}.$$

(See problem 3.5.6). Let $\mathbf{b}$ be the vector

$$\mathbf{b} = \begin{bmatrix} 1 \\ 1 \\ 1 \end{bmatrix}.$$

Find the least squares solution to $A\mathbf{x} = \mathbf{b}$. What is the minimum value of $|A\mathbf{x} - \mathbf{b}|^2$ as $\mathbf{x}$ varies over R^2?

3.7.8 Let A be the matrix

$$B = \begin{bmatrix} 1 & 0 & 2 \\ 1 & 1 & -1 \end{bmatrix}.$$

(a) Find an orthonormal basis for $\mathrm{Ker}(A)$.

(b) Find the orthogonal projection onto $\text{Img}(A^t)$

(c) Find the solution of the system

$$x + 2z = 1$$
$$x + y - z = 2$$

for which $x^2 + y^2 + z^2$ has the minimum value.

3.7.9 Let $A = \begin{bmatrix} 1 & 1 & 2 \\ 0 & 1 & 1 \\ 1 & 0 & 1 \end{bmatrix}$

(a) Find an orthonormal basis for $\text{Img}(A)$.

(b) Find P_c, the orthogonal projection onto $\text{Img}(A)$.

(c) Find a least squares solution to $A\mathbf{x} = \mathbf{b}$ where

$$\mathbf{b} = \begin{bmatrix} 1 \\ 1 \\ 1 \end{bmatrix} .$$

(d) What is the dimension of $\text{Img}(A)$? If the $\text{Img}(A)$ is a plane, give the equation of the plane. If it is a line, give a parametric representation of the line.

3.7.10 Let A be the matrix

$$A = \begin{bmatrix} 1 & 0 & 2 & 0 \\ 0 & 1 & -1 & -1 \end{bmatrix} .$$

(a) Find an orthonormal basis for the $\ker(A)$, and find the orthogonal projection onto $\ker(A)$.

(b) For this matrix A, $\text{Img}(A)$ is R^2. Explain why, and find the solution $\mathbf{x}_{**}$ of

$$A\mathbf{x} = \begin{bmatrix} 1 \\ 2 \end{bmatrix}$$

with the shortest possible length; that is, with $|\mathbf{x}_{**}| < |\mathbf{x}|$ for any other solution $\mathbf{x}$.

(c) Find two linearly independent vectors, $\mathbf{v}_1$ and $\mathbf{v}_2$, in R^4 that are orthogonal to

$$\begin{bmatrix} 1 \\ 0 \\ 2 \\ 0 \end{bmatrix} \quad \text{and} \quad \begin{bmatrix} 0 \\ 1 \\ -1 \\ -1 \end{bmatrix} .$$

Overview of Chapter 4

This chapter covers the basics of determinants, including all the "whys and wherefores", in three sections. The first section on permutations should be read through. It is useful in many subjects, and those of you that will ever take a combinatorics course will benefit from going carefully through it. It's main purpose here, that is as far as linear algebra is concerned, is to explain the definition of the determinant. This definition is given in the second section. The key results there are Theorems 2 through 5, and especially Theorem 3. For the most part, when we do computational work with determinants, we use one of these results.

One reason for computing determinants, as explained in the second section, is that computing the determinant of a square matrix A is a test for invertibility of A. But we already know how to test for invertibility: Row reduce A to upper triangular form, and see if it has a full set of pivots or not. Since, as we explain in the second section , the fast way to compute the determinant of A is through row reduction, this test is really only useful in a theoretical sense, and is somewhat superfluous.

The third section gives an explanation for why we would want to deal with determinants anyway. It explains how determinant can be used to compute areas and volumes. The three theorems in this section, as well as the ideas leading up to them, are very useful computationally, and should be understood.

Section 1: Permutations

Consider a function σ from the set $\{1, 2, \ldots, n\}$ onto itself. That is, σ "pairs" each "input" integer i in $\{1, 2, \ldots, n\}$ with an "output" integer $\sigma(i)$ in the same set. Any such function must be one–to–one, since if σ sent any two elements i and j of $\{1, 2, \ldots, n\}$ to the same integer k in $\{1, 2, \ldots, n\}$, it would have spent two of n "shots" at hitting a single target, and that precludes hitting all n. So any function σ from $\{1, 2, \ldots, n\}$ *onto* itself is also one–to–one, and hence invertible.

Definition (Permutation) A *permutation* of $\{1, 2, \ldots, n\}$ is a map σ of $\{1, 2, \ldots, n\}$ onto itself.

For example, consider the case $n = 3$. We can describe any permutation σ of $\{1, 2, 3\}$ by listing the value $\sigma(j)$ underneath j as follows:

$$\begin{array}{ccc} 1 & 2 & 3 \\ 2 & 3 & 1 \end{array}$$

This arrangement of pairs, in which the "input" integer is paired with the "output" integer below it denotes the permutation that sends 1 to 2, 2 to 3 and 3 to 1. The generalization of this way of writing permutations to higher values of n is plain, and we'll use it freely.

It is not hard to see that there are exactly $n!$ permutations of $\{1, 2, \ldots, n\}$. Indeed, consider any permutation σ of $\{1, 2, \ldots, n\}$. There are n choices for the value of $\sigma(1)$. Make this choice, and then, $\sigma(1)$ being taken, there are $n - 1$ choices remaining for value of $\sigma(2)$. Next, there are $n - 2$ choices for $\sigma(3)$, the value to be assigned to 3. Continuing in this way, there are $n(n-1)(n-2) \cdots 1 = n!$ choices to make, and each one leads to a distinct permutation.

Example 1: There are six permutations of $\{1, 2, 3\}$:

$$\sigma_a = \begin{array}{ccc} 1 & 2 & 3 \\ 1 & 2 & 3 \end{array}$$

$$\sigma_b = \begin{array}{ccc} 1 & 2 & 3 \\ 2 & 1 & 3 \end{array} \qquad \sigma_c = \begin{array}{ccc} 1 & 2 & 3 \\ 1 & 3 & 2 \end{array}$$

$$\sigma_d = \begin{array}{ccc} 1 & 2 & 3 \\ 2 & 3 & 1 \end{array} \qquad \sigma_e = \begin{array}{ccc} 1 & 2 & 3 \\ 3 & 1 & 2 \end{array} \tag{1.1}$$

$$\sigma_f = \begin{array}{ccc} 1 & 2 & 3 \\ 3 & 2 & 1 \end{array}$$

The permutation at the top of the list, σ_a is called the *identity* permutation since it just sends each element of $\{1, 2, 3\}$ to itself. All of the others "mix things up" to some extent. In fact, we have arranged these permutations in an order that reflects "how much mixing" is involved in each one, starting from no mixing at the top, to the most mixing at the bottom.

Now, what do we mean by the "amount of mixing" involved in a permutation? There are several ways in which this might be interpreted. You might for example choose to define the degree of mixing in a permutation to be the "number of integers that get assigned some place other than their own". Evidently, this is not the definition we have in mind here since then σ_f would have a lower degree of mixing than either σ_d or σ_e. The useful definition of the degree of mixing is the following:

Definition (Degree of Mixing) The *degree of mixing* of a permutation σ of $\{1, 2, \ldots, n\}$ is the number of pairs of integers (i, j) in $\{1, 2, \ldots, n\}$ with

$$ i < j \qquad \text{and} \qquad \sigma(i) > \sigma(j) \; . \tag{1.2} $$

This number is denoted $D(\sigma)$.

Evidently, the degree of mixing of a permutation is a count of the number of pairs in $\{1, 2, \ldots, n\}$ whose order gets reversed by σ. The more "reversed" pairs, the more mixing.

Example 2 Let's compute $D(\sigma)$ for each of the six permutations of $\{1, 2, 3\}$. There are just three pairs (i, j) with $i < j$, namely

$$ (1, 2) \qquad (1, 3) \qquad (2, 3) \; . $$

To compute the degree of mixing of σ, we just look at

$$ (\sigma(1), \sigma(2)) \qquad (\sigma(1), \sigma(3)) \qquad (\sigma(2), \sigma(3)) \; , $$

and count the number of times these pairs are "out of order". You can easily check that

$$ D(\sigma_a) = 0 \qquad D(\sigma_b) = 1 \qquad D(\sigma_c) = 1 \qquad D(\sigma_d) = 2 \qquad D(\sigma_e) = 2 \qquad D(\sigma_f) = 3 \; , $$

as indicated in the arrangement above.

The reason that this definition is useful has to do with how it interacts with the multiplicative properties of permutations, as we now explain. Since permutations of $\{1, 2, \ldots, n\}$ are functions from this set into itself, we can compose them: If σ_1 and σ_2 are two permutations of $\{1, 2, \ldots, n\}$, then $\sigma_2 \circ \sigma_1$ is defined by

$$ \sigma_2 \circ \sigma_1(i) = \sigma_2(\sigma_1(i)) \quad , \qquad \text{for each} \quad i = 1, \ldots, n \; . \tag{1.3} $$

The important thing to notice here is that $\sigma_2 \circ \sigma_1$ is not just any kind of function from $\{1, 2, \ldots, n\}$ to itself, but it is in fact another permutation of $\{1, 2, \ldots, n\}$. To see this we just need to check that for each j in $\{1, 2, \ldots, n\}$, there is an i such that $\sigma_2 \circ \sigma_1(i) = j$. But since σ_2 is a permutation, there is a k so that $\sigma_2(k) = j$. And since σ_1 is a permutation, there is an i so that $\sigma_1(i) = k$. Then

$$\sigma_2 \circ \sigma_1(i) = \sigma_2(\sigma_1(i)) = \sigma_2(k) = j \ .$$

Moreover, every permutation σ has an inverse, σ^{-1}, which simply sends any j in $\{1, 2, \ldots, n\}$ back to the integer i in $\{1, 2, \ldots, n\}$ from whence it came. The inverse too is a one–to–one map of the set onto itself. (It is just the original map "played in reverse".) Hence the inverse of a permutation of $\{1, 2, \ldots, n\}$ is permutation too.

Now, given two permutations σ_1 and σ_2 of $\{1, 2, \ldots, n\}$, what can we say about $D(\sigma_2 \circ \sigma_1)$? First in applying σ_1, we reverse the order of $D(\sigma_1)$ pairs. Then applying σ_2 after that, we reverse the order of $D(\sigma_2)$ pairs. So the number of pairs that are reversed by $\sigma_2 \circ \sigma_1$ is no more than $D(\sigma_1) + D(\sigma_2)$.

However, some of the pairs that σ_2 reverses may have already been put out of order by σ_1. *In this case, σ_2 puts them back in order.* An extreme case is when $\sigma_2 = (\sigma_1)^{-1}$. Then σ_2 undoes all of the mixing done by σ_1, and $D(\sigma_2 \circ \sigma_1) = 0$.

So we conclude that

$$0 \ \leq \ D(\sigma_2 \circ \sigma_1) \ \leq \ D(\sigma_1) + D(\sigma_2) \ .$$

Can we be more specific? Yes, let's take it step by step. After applying σ_1 there are $D(\sigma_1)$ out of order pairs. Now suppose that c of these are "reordered" when we apply σ_2. That leaves $D(\sigma_1) - c$ of the pairs reversed by σ_1 still reversed after σ_2 has been applied. Now by definition, σ_2 reverses the order of $D(\sigma_2)$ pairs. Exactly c of the reversals were "used up" in reordering pairs that σ_1 had put out of order. So only $D(\sigma_2) - c$ new pairs have their order reversed by σ_2.

Adding up the $D(\sigma_1) - c$ pairs put out of order by σ_1 and left out of order by σ_2 and the $D(\sigma_2) - c$ new pairs put out of order by σ_2, we have

$$D(\sigma_2 \circ \sigma_1) = (D(\sigma_1) - c) + (D(\sigma_2) - c) = D(\sigma_1) + D(\sigma_2) - 2c \ . \tag{1.4}$$

Now it might look like we haven't really learned much since we can't say much about c in general. *However, $2c$ is always an even integer,* and so

$$(-1)^{D(\sigma_2 \circ \sigma_1)} = (-1)^{D(\sigma_1)}(-1)^{D(\sigma_1)} \ . \tag{1.5}$$

Definition (Character of a Permutation) The *character* $\chi(\sigma)$ of a permutation σ is defined by

$$\chi(\sigma) = (-1)^{D(\sigma)} \ . \tag{1.6}$$

The point of the definition is that $\chi(\sigma_2 \circ \sigma_1) = \chi(\sigma_2)\chi(\sigma_1)$. That is, *the character of a product equals the product of the characters.*

It is easy to compute the characters of certain simple permutations. Permutations that simply exchange the places of a single pair of integers, leaving everything else in place, will be especially important for us.

Example 3 Consider the following permutations of $\{1, 2, 3, 4\}$:

$$
\begin{matrix}
1 & 2 & 3 & 4 \\
1 & 4 & 3 & 2
\end{matrix}
\qquad\qquad
\begin{matrix}
1 & 2 & 3 & 4 \\
4 & 1 & 3 & 2
\end{matrix}
\tag{1.7}
$$

The one on the left exchanges the places of 2 and 4, and leaves everything else in place. The one on the right moves three elements out of place.

Permutations of the first type are especially important in linear algebra. Consider a 4×4 matrix A, and let B be the 4×4 matrix obtained by exchanging rows 2 and 4 of A. Then, with σ denoting the permutation on the left in the example above, the ith row of B is the $\sigma(i)$th row of A, and so

$$
B_{i,j} = A_{\sigma(i),j}
$$

for $1 \leq i, j \leq 4$.

Definition (Pair Permutations) For each $i < j$ in $\{1, 2, \ldots, n\}$ the *pair permutation* $\sigma_{i,j}$ is defined by

$$
\sigma_{i,j}(i) = j \quad , \quad \sigma_{i,j}(j) = i \quad \text{and} \quad \sigma_{i,j}(k) = k \quad \text{for} \quad k \neq i, j . \tag{1.8}
$$

It is called an *adjacent pair permutation* in case $j = i + 1$.

Notice that each pair permutation is its own inverse – applying it twice swaps the reversed pair back into place.

Next notice that for each adjacent pair permutation $\sigma_{i,i+1}$, $D(\sigma_{i,i+1}) = 1$, and hence $\chi(\sigma_{i,i+1}) = -1$. Moreover, if $j = i + k$, one can move i to the right of j using k pair permutations. One can then move j back to the ith spot with $k - 1$ pair permutations. Only $k - 1$ are required, because the last pair permutation used to move i into the jth place already moved j back one place. In any case, this means that $\sigma_{i,j}$ can be written as the composition product of $2k - 1$ adjacent pair permutations. Since the character of a product equals the product of the characters, and the character of each adjacent pair permutation is -1, we have that $\chi(\sigma_{i,j}) = -1^{2k-1} = -1$ for all pair permutations.

We summarize the discussion that led to this definition in the following theorem:

Theorem 1 *For any two permutations σ_1 and σ_2 of $\{1, 2, \ldots, n\}$,*

$$
\chi(\sigma_2 \circ \sigma_1) = \chi(\sigma_2)\chi(\sigma_1) . \tag{1.9}
$$

Moreover, for any pair permutation $\sigma_{i,j}$,

$$
\chi(\sigma_{i,j}) = -1 . \tag{1.10}
$$

We are now ready to apply this to linear algebra.

Section 2: Determinants

Definition (Determinant) Let A be an $n \times n$ matrix. The *determinant* of A is the number $\det(A)$ defined by

$$\det(A) = \sum_{\sigma} \chi(\sigma) \left(A_{\sigma(1),1} A_{\sigma(2),2} \cdots A_{\sigma(n),n} \right) \ , \tag{2.1}$$

where the sum is taken over all of the $n!$ permutations of $\{1, 2, \ldots, n\}$.

Example 1 Consider the general 2×2 matrix $A = \begin{bmatrix} a & b \\ c & d \end{bmatrix}$. There are only two permutations of $\{1, 2\}$ to consider, namely

$$\sigma_1 = \begin{matrix} 1 & 2 \\ 1 & 2 \end{matrix} \quad \text{and} \quad \sigma_2 = \begin{matrix} 1 & 2 \\ 2 & 1 \end{matrix} \ .$$

Clearly $\chi(\sigma_1) = 1$ and $\chi(\sigma_2) = -1$. Hence

$$\det(A) = A_{1,1} A_{2,2} - A_{2,1} A_{1,2} = ad - bc \ .$$

The result,

$$\det\left(\begin{bmatrix} a & b \\ c & d \end{bmatrix} \right) = ad - bc \tag{2.2}$$

is worth bearing in mind.

Example 2 Consider a general 3×3 matrix A. We have already worked out a list of the six permutations of $\{1, 2, 3\}$ in (1.1) of the previous section, and computed the characters of each of them. In the 3×3 case then, the definition (2.1) leads to

$$\det(A) = A_{1,1} A_{2,2} A_{3,3} + A_{2,1} A_{3,2} A_{1,3} + A_{3,1} A_{1,2} A_{2,3}$$

$$- A_{2,1} A_{1,2} A_{3,3} - A_{1,1} A_{3,2} A_{2,3} - A_{3,1} A_{2,2} A_{3,1} \ .$$

It is easier to see what is going on if we introduce a better notation. Let's think of A as the matrix whose first row is $\mathbf{a}$, whose second row is $\mathbf{b}$, and whose third row is $\mathbf{c}$:

$$A = \begin{bmatrix} \mathbf{a} \\ \mathbf{b} \\ \mathbf{c} \end{bmatrix} = \begin{bmatrix} a_1 & a_2 & a_3 \\ b_1 & b_2 & b_3 \\ c_1 & c_2 & c_3 \end{bmatrix} \ .$$

In this notation, the result becomes

$$\det(A) = a_1 b_2 c_3 + b_1 c_2 a_3 + c_1 a_2 b_3 - b_1 a_2 c_3 - a_1 c_2 b_3 - c_1 b_2 a_3 \ . \tag{2.3}$$

There are several simple ways to remember this. Here is one: Form the array

$$
\begin{array}{ccc}
a_1 & a_2 & a_3 \\
b_1 & b_2 & b_3 \\
c_1 & c_2 & c_3 \\
a_1 & a_2 & a_3 \\
b_1 & b_2 & b_3
\end{array}
$$

where we have just repeated the first two rows at the bottom.

The three products entering (2.3) with a positive sign are the products of the terms on the three diagonals going down and to the right starting from the a_1, b_1 and c_1 respectively. We'll call these the "positive diagonals". And the three products entering (2.3) with a negative sign are the products of the terms on the three diagonals going down and to the left starting from the a_3, b_3 and c_3 respectively. We'll call these the "negative diagonals".

For example, consider the matrix

$$
A = \begin{bmatrix} 1 & 1 & 1 \\ 1 & 2 & 4 \\ 1 & 3 & 9 \end{bmatrix}
\tag{2.4}
$$

and let's use this device to compute its determinant. Form the array

$$
\begin{array}{ccc}
1 & 1 & 1 \\
1 & 2 & 4 \\
1 & 3 & 9 \\
1 & 1 & 1 \\
1 & 2 & 4
\end{array}
$$

and then add up the products of the terms on the "positive diagonals", and subtract off the products of the terms on the "negative diagonals". The result is

$$
\det(A) = 18 + 3 + 4 - 2 - 12 - 9 = 2
\tag{2.5}
$$

For larger values of n, it begins to look a bit complicated. For instance for $n = 4$, there are $4! = 24$ permutations, and for $n = 5$, there are $5! = 120$. We need a different approach.

So let us analyze the definition piece by piece, and see if we can find one. Consider some fixed permutation σ, and examine the product

$$
A_{\sigma(1),1} A_{\sigma(2),2} \cdots A_{\sigma(n),n} \ .
\tag{2.6}
$$

Clearly, each column index appears exactly once, and since σ is a permutation, so does each row index. *So there is exactly one element form each row in the product, and exactly one from each column.* Consider the case $n = 4$, and the permutation

$$
\sigma = \begin{array}{cccc} 1 & 2 & 3 & 4 \\ 4 & 1 & 3 & 2 \end{array} \ .
\tag{2.7}
$$

The permutation σ tells us which column indices are paired with which row indices. When σ is written in the form above, the column indices are on the top, and the row index paired with that column index is directly underneath. In this case,

$$A_{\sigma(1),1}A_{\sigma(2),2}A_{\sigma(3),3}A_{\sigma(4),4} = A_{4,1}A_{1,2}A_{3,3}A_{2,4} .$$

The factors on the right are ordered according to the column indexes. But we can reorder them according to the row indexes *without changing the value of the product*. The result is:

$$A_{1,2}A_{2,4}A_{3,3}A_{4,1}$$

Define a new permutation ρ of $\{1,2,3,4\}$ by

$$\rho = \begin{matrix} 1 & 2 & 3 & 4 \\ 2 & 4 & 3 & 1 \end{matrix} . \tag{2.8}$$

Again, ρ tells us which column indices are paired with which row indices, but this time the row indices are listed on top, and the corresponding column indices are directly underneath.

What is the relation between ρ and σ? It is easy to see that ρ is the inverse of σ. That is for each i, $\rho(\sigma(i)) = i$. This is a general fact:

Theorem 1 *For any n, any $n \times n$ matrix A, and any permutation σ of $\{1, 2, \ldots, n\}$, let $\rho = \sigma^{-1}$. Then*

$$A_{\sigma(1),1}A_{\sigma(2),2}\cdots A_{\sigma(n),n} = A_{1,\rho(1)}A_{2,\rho(2)}\cdots A_{n,\rho(n)} . \tag{2.9}$$

As a consequence,

$$\det(A) = \sum_{\sigma} \chi(\sigma)\left(A_{1,\sigma(1)}A_{2,\sigma(2)}\cdots A_{n,\sigma(n)}\right) \tag{2.10}$$

where the sum on the right is over all permutations σ of $\{1, 2, \ldots, n\}$.

Proof: Since ρ and σ "pair" the same integers as inputs and outputs, just in the reverse order, both products in (2.9) run over the same set of pairs of row and column indices. Next, since the character of the identity is 1, $\chi(\rho)\chi(\sigma) = 1$, and so $\chi(\rho) = \chi(\sigma)$.

Now, there is a one to one correspondence between permutations σ and their inverses ρ. Therefore, summing over permutations σ gives the same result as summing over their inverses – either way each permutation gets counted exactly once. Therefore, replacing $A_{\sigma(1),1}A_{\sigma(2),2}\cdots A_{\sigma(n),n}$ and $\chi(\sigma)$ by $A_{1,\rho(1)}A_{2,\rho(2)}\cdots A_{n,\rho(n)}$ and $\chi(\rho)$ in the definition of the determinant, we get the equivalent form:

$$\det(A) = \sum_{\rho} \chi(\sigma)\left(A_{1,\rho(1)}A_{2,\rho(2)}\cdots A_{n,\rho(n)}\right) . \tag{2.11}$$

Finally, whether we call the variable of summation ρ or σ doesn't matter, so we can rewrite (2.11) with σ in place of ρ, to obtain (2.10), and bring out the symmetry with the original definition (2.1). ∎

This theorem shows that it didn't matter that we put the permutation on the row indices in our definition since we would get the same thing if we used the column indices instead. This equivalence leads to the following useful fact about the determinant:

Theorem 2: *For any $n \times n$ matrix A,*

$$\det(A^t) = \det(A) . \tag{2.12}$$

Proof:

$$\det(A^t) = \sum_\sigma \chi(\sigma) \left(A^t_{1,\sigma(1)} A^t_{2,\sigma(2)} \cdots A^t_{n,\sigma(n)} \right)$$

$$= \sum_\sigma \chi(\sigma) \left(A_{\sigma(1),1} A_{\sigma(2),2} \cdots A_{\sigma(n),n} \right)$$

$$= \det(A) .$$

∎

We can get a much clearer understanding of the determinant as a function of A if write A as a column vector of its rows. Let $\mathbf{r}_i$ denote the ith row of A so that

$$A = \begin{bmatrix} \mathbf{r}_1 \\ \mathbf{r}_2 \\ \vdots \\ \mathbf{r}_n \end{bmatrix} . \tag{2.13}$$

We now list several properties of the determinant function thought of as a function of the rows of A:

First Property: The determinant of A is a linear function of each of the rows of A. That is if

$$\mathbf{r}_i = a\mathbf{v} + b\mathbf{w}$$

then

$$\det \left(\begin{bmatrix} \mathbf{r}_1 \\ \mathbf{r}_2 \\ \vdots \\ a\mathbf{v} + b\mathbf{w} \\ \vdots \\ \mathbf{r}_n \end{bmatrix} \right) = a\det \left(\begin{bmatrix} \mathbf{r}_1 \\ \mathbf{r}_2 \\ \vdots \\ \mathbf{v} \\ \vdots \\ \mathbf{r}_n \end{bmatrix} \right) + b\det \left(\begin{bmatrix} \mathbf{r}_1 \\ \mathbf{r}_2 \\ \vdots \\ \mathbf{w} \\ \vdots \\ \mathbf{r}_n \end{bmatrix} \right) . \tag{2.14}$$

This is true since each product $A_{\sigma(1),1}A_{\sigma(2),2}\cdots A_{\sigma(n),n}$ contains exactly one factor coming from the *ith* row, and hence is a linear function of the entries of the *ith* row. A linear combination of linear functions is linear, and so the determinant is a linear function of the entries of the *ith* row. Notice that the permutation characters $\chi(\sigma)$ didn't play any role in this first property.

Second Property: Suppose that B is obtained from A by interchanging the kth and ℓth rows of A. That is, with ρ denoting the pair permutation $\sigma_{k,\ell}$,

$$B_{i,j} = A_{\rho(i),j} \ . \tag{2.15}$$

Then $\det(B) = -\det(A)$. To see this, note that for any permutation σ,

$$B_{\sigma(1),1}B_{\sigma(2),2}\cdots B_{\sigma(n),n} = A_{\rho(\sigma(1)),1}A_{\rho(\sigma(2)),2}\cdots A_{\rho(\sigma(n)),n} \ .$$

Now let $\tilde{\sigma} = \rho \circ \sigma$. There is a one to one correspondence between the permutations $\tilde{\sigma}$ and the original permutations σ since in fact $\sigma = \rho^{-1} \circ \tilde{\sigma}$. Also, notice that

$$\chi(\tilde{\sigma}) = \chi(\rho)\chi(\sigma) = -\chi(\sigma) \ .$$

Therefore,

$$\det(B) = \sum_\sigma \chi(\sigma)\left(B_{\sigma(1),1}B_{\sigma(2),2}\cdots B_{\sigma(n),n}\right)$$

$$= -\sum_{\tilde{\sigma}} \chi(\tilde{\sigma})\left(A_{\tilde{\sigma}(1),1}A_{\tilde{\sigma}(2),2}\cdots A_{\tilde{\sigma}(n),n}\right)$$

$$= -\det(A) \ .$$

Third Property: If B is obtained from A by adding a multiple of row k of A to row ℓ of A, then $\det(B) = \det(A)$.

This is an easy consequence of the first two properties. By the second property, the determinant of any matrix C with two identical rows is zero. This is because interchanging the rows must change the sign of $\det(C)$. But if the two rows are the same, the matrix and hence its determinant are unchanged. The conclusion is that in this case $\det(C) = -\det(C)$, and this means that $\det(C) = 0$. Next, by the first property,

$$\det\left(\begin{bmatrix} \mathbf{r}_1 \\ \mathbf{r}_2 \\ \vdots \\ \mathbf{r}_k \\ \vdots \\ \mathbf{r}_\ell + a\mathbf{r}_k \\ \vdots \\ \mathbf{r}_n \end{bmatrix}\right) = \det\left(\begin{bmatrix} \mathbf{r}_1 \\ \mathbf{r}_2 \\ \vdots \\ \mathbf{r}_k \\ \vdots \\ \mathbf{r}_\ell \\ \vdots \\ \mathbf{r}_n \end{bmatrix}\right) + a\det\left(\begin{bmatrix} \mathbf{r}_1 \\ \mathbf{r}_2 \\ \vdots \\ \mathbf{r}_k \\ \vdots \\ \mathbf{r}_k \\ \vdots \\ \mathbf{r}_n \end{bmatrix}\right) \tag{2.16}$$

and the last determinant on the right is zero since two of its rows are the same.

Fourth Property: If A is upper triangular, then the determinant of A is the product of the diagonal entries. that is

$$\det(A) = A_{1,1}A_{2,2}\cdots A_{n,n} \tag{2.17}$$

The point here is that if A is upper triangular, there is only one way to choose one entry from each row and column without necessarily getting a factor of zero in the product.

Indeed, for the entry from the first column, we have to choose the entry in the first row since everything else is zero. Then for the entry from the second column, we must choose from the first two rows to avoid the zeros below the diagonal. But the first row is already taken, so we have no choice but to pair the second row with the second column. In the third column it is the same way. To avoid the zeros below the diagonal, we must choose from among the first three rows. But the first two are taken, so there is no choice: We must pair the third row with the third column. This pattern clearly continues, and the only permutation that does no necessarily introduce a zero factor is the identity.

These properties of the determinant lead to an efficient algorithm for computing determinants, and a reason for doing so!

Theorem 3: *Let A be an $n \times n$ matrix, and let B be an upper traingular matrix obtained from A by repeatedly adding multiples of one row to another, and interchanging of pairs of rows. Suppose that there were N interchanges of pairs of rows used. Then*

$$\det(A) = (-1)^N \left(B_{1,1}B_{2,2}\cdots B_{n,n}\right) . \tag{2.18}$$

Moreover, A is invertible if and only if $\det(A) \neq 0$.

Proof: We know that we can partially row reduce A to an upper triangular matrix B by repeatedly adding multiples of one row to another, and interchanging of pairs of rows. By the third property, the determinant is unchanged when a multiple of one row is added to another, and by the second property, the determinant changes sign each time a pair of rows is interchanged. This proves that $\det(A) = (-1)^N \det(B)$, and the fourth property tells us how to compute $\det(B)$.

Finally B is a row reduction of A, and so the number of pivots of A is the same as for B, and the latter is clearly the number of non–zero diagonal entries. Thus A has a full set of pivots if and only if B has non zero entries on the diagonal, which is the case exactly when $\det(B) \neq 0$. But A is invertible if and only if it has a full set of pivots. ∎

This result is really useful for computation. If no interchanges of rows are required, A can be reduced to upper triangular form in $n(n-1)/2$ operations: It would take $n-1$ operations to clean out the first column below the diagonal, $n-2$ for the second, and so on for a total of $(n-1) + (n-2) + \ldots + 2 + 1 = n(n-1)/2$ row operations. From here, a simple product of the diagonal entries yields the result. At worst, n row interchanges might be required, and so in any case the total number of row operations is no more than $(n+1)n/2$.

Compare this with the formula, which requires the computation of $n!$ products, and a permutation character $\chi(\sigma)$ for each one. Even if we have these in a table to look up, computing and summing the $n!$ products is a tremendous amount of work. The formula is useful for $n = 2$ or $n = 3$, especially since there is a simple graphical way to organize the computations in these cases. But even for $n = 3$, row reduction is often simpler.

Example 3 Consider the matrix A defined in (2.4). Three simple row operations (two to clean out the first column, one for the second) reduce it to upper triangular with no swapping of rows:

$$\begin{bmatrix} 1 & 1 & 1 \\ 1 & 2 & 4 \\ 1 & 3 & 9 \end{bmatrix} \rightarrow \begin{bmatrix} 1 & 1 & 1 \\ 0 & 1 & 3 \\ 0 & 2 & 8 \end{bmatrix} \rightarrow \begin{bmatrix} 1 & 1 & 1 \\ 0 & 1 & 3 \\ 0 & 0 & 2 \end{bmatrix}.$$

Form the right hand side, we set $\det(A) = 2$, as we found in (2.5).

Example 4 Consider the matrix A given by

$$A = \begin{bmatrix} a & 1 & 0 & 1 \\ 1 & a & 1 & 0 \\ 0 & 1 & a & 1 \\ 1 & 0 & 1 & a \end{bmatrix}. \tag{2.19}$$

If we swap rows 1 and 4, and then swap rows 2 and 3, the result is

$$\begin{bmatrix} 1 & 0 & 1 & a \\ 0 & 1 & a & 1 \\ 1 & a & 1 & 0 \\ a & 1 & 0 & 1 \end{bmatrix}.$$

The good thing is the first two rows are now in upper triangular form. And since we used two row swaps, we didn't change the sign of the determinant. Now simple row operations "clean out" the first two columns, resulting in the matrix

$$\begin{bmatrix} 1 & 0 & 1 & a \\ 0 & 1 & a & 1 \\ 0 & 0 & -a^2 & -2a \\ 0 & 0 & -2a & -a^2 \end{bmatrix}.$$

None of these row operations changes the value of the determinant, so the determinant of A is equal to the determinant of this simpler matrix. We can simplify further by using the linearity in each of the rows to pull out a factor of $-a$ for the two lower rows. The result is

$$\det(A) = \det\left(\begin{bmatrix} 1 & 0 & 1 & a \\ 0 & 1 & a & 1 \\ 0 & 0 & -a^2 & -2a \\ 0 & 0 & -2a & -a^2 \end{bmatrix}\right) = a^2 \det\left(\begin{bmatrix} 1 & 0 & 1 & a \\ 0 & 1 & a & 1 \\ 0 & 0 & a & 2 \\ 0 & 0 & 2 & a \end{bmatrix}\right).$$

Now one more row operation reduces the matrix on the right to upper triangular form, and we read off the result that

$$\det(A) = a^2(a^2 - 4) .$$ (2.20)

This example illustrates the computational use of properties such as "linearity in the rows".

We generally stop the row reduction of A at the upper triangular stage, because that is the place to stop if you really want to compute the determinant. But we could keep going, and there is a point to this. Suppose A is invertible. Then when we get to the upper triangular stage B, there are no zero entries on the diagonal. We can then keep going with row operations, and clean out the last column above the diagonal, and then the next to last, and so on until we have reduced to a diagonal matrix $C = \begin{bmatrix} c_1 & 0 & 0 & \dots & 0 \\ 0 & c_2 & 0 & \dots & 0 \\ \vdots & \vdots & \vdots & & 0 \\ 0 & 0 & 0 & \dots & c_n \end{bmatrix}.$

Now since the determinant is a linear function of each row,

$$\det(C) = c_1 \det\left(\begin{bmatrix} 1 & 0 & 0 & \dots & 0 \\ 0 & c_2 & 0 & \dots & 0 \\ \vdots & \vdots & \vdots & & 0 \\ 0 & 0 & 0 & \dots & c_n \end{bmatrix} \right) = c_1 c_2 \det\left(\begin{bmatrix} 1 & 0 & 0 & \dots & 0 \\ 0 & 1 & 0 & \dots & 0 \\ \vdots & \vdots & \vdots & & 0 \\ 0 & 0 & 0 & \dots & c_n \end{bmatrix} \right) .$$

Continuing in this way we see that

$$\det(C) = (c_1 c_2 \cdots c_n) \det(I) .$$

Now we saw that $\det(A) = (-1)^N \det(B)$, and then that $\det(B) = \det(C)$, and finally that $\det(C) = (c_1 c_2 \cdots c_n) \det(I)$ *just using the facts that the determinant is a linear function of the rows, and that it changes sign when rows are interchanged.* (The third property was a direct consequence of these two). Also the values $c_1, c_2, \dots, c_n$ are determined by row reduction. *So any other function $f(A)$ from the $n \times n$ matrices to the real numbers that had these properties would also satisfy $f(A) = \prod_{j=1}^{n} c_j f(I)$.* That is, we would have

$$f(A) = \det(A) f(I) .$$

We record this fact as a theorem:

Theorem 4: *Let f be any numerically valued function on the $n \times n$ matrices with the properties that $f(A)$ changes sign when any two rows of A are interchanged, and that $f(A)$ is linear in each row of A. Then*

$$f(A) = \det(A) f(I) .$$ (2.21)

As an example, fix any $n \times n$ matrix B and consider the function $f(A)$ defined by

$$f_B(A) = \det(AB) .$$

Then interchanging any two rows of A interchanges the corresponding rows of AB, and so $f_B(A)$ changes sign when any two rows of A are interchanged. Also, the rows of AB are linear combinations of the rows of A, and since the composition of linear functions is a linear function, $f_B(A) = \det(AB)$ is a linear function of the rows of A. Also, it is evident that $f_B(I) = \det(B)$. Therefore, we have the following result:

Thoerem 5: *For any two $n \times n$ matrices A and B,*

$$\det(AB) = \det(A)\det(B) . \tag{2.22}$$

Proof: $\det(AB) = f_B(A) = \det(A)f_B(I) = \det(A)\det(B)$. ■

It plainly follows from this that if A is invertible, then $\det(A^{-1}) = (\det(A))^{-1}$.

Section 3: Determinants, Area and Volume

Consider two linearly independent vectors

$$\mathbf{v}_1 = \begin{bmatrix} a \\ b \end{bmatrix} \quad \text{and} \quad \mathbf{v}_2 = \begin{bmatrix} c \\ d \end{bmatrix} \tag{3.1}$$

in the plane $I\!R^2$. It cannot be the case that $a = c = 0$ since then the two vectors would not be linearly independent. So we may assume without loss of generality, by renumbering the vectors if need be, that $a \neq 0$.

Form the matrix A whose columns are $\mathbf{v}_1$ and $\mathbf{v}_2$ respectively:

$$A = [\mathbf{v}_1, \mathbf{v}_2] = \begin{bmatrix} a & c \\ b & d \end{bmatrix} . \tag{3.2}$$

Now define S_1 by

$$S_1 = \begin{bmatrix} 1 & 0 \\ -b/a & 1 \end{bmatrix} . \tag{3.3}$$

If we multiply A on the left by S_1, the effect is to subtract b/a times the first row from the second, so that

$$S_1 A = \begin{bmatrix} a & c \\ 0 & d - bc/a \end{bmatrix} .$$

The lower right entry of this matrix is

$$\frac{ad - bc}{a} = \frac{1}{a} \det(A) \neq 0 .$$

Next define

$$S_2 = \begin{bmatrix} 1 & -ca/(ad - bc) \\ 0 & 1 \end{bmatrix} . \tag{3.4}$$

Multiplying $S_1 A$ on the left by S_2 subtracts $ca/(ad - bc)$ times the second row from the first, with the result that

$$S_2 S_1 A = \begin{bmatrix} a & 0 \\ 0 & d - bc/a \end{bmatrix} .$$

This last matrix is diagonal. Let's call it D so that

$$D = \begin{bmatrix} a & 0 \\ 0 & d - bc/a \end{bmatrix} \quad \text{and} \quad S_2 S_1 A = D . \tag{3.5}$$

Recall that S_1 and S_2 are invertible with

$$S_1^{-1} = \begin{bmatrix} 1 & 0 \\ b/a & 1 \end{bmatrix} \quad \text{and} \quad S_2^{-1} = \begin{bmatrix} 1 & ca/(ad - bc) \\ 0 & 1 \end{bmatrix} . \tag{3.6}$$

we then have the result that

$$A = S_1^{-1} S_2^{-1} D \ .$$

(3.7)

Definition (Shear Transformations in the Plane) Any linear transformation of the plane to itself corresponding to a matrix of the form

$$\begin{bmatrix} 1 & 0 \\ \alpha & 1 \end{bmatrix} \quad \text{or} \quad \begin{bmatrix} 1 & \alpha \\ 0 & 1 \end{bmatrix}$$

(3.8)

for any number α is called a *shear transformation*.

We have just seen that *any non-singular 2×2 matrix A can be written as the product of a diagonal matrix and two shear transformations, though we may need to switch the order of the columns first.*

An important property of shear transformations is that they preserve the area of regions in the plane. To make this clear, let's focus on the case

$$S = \begin{bmatrix} 1 & 0 \\ \alpha & 1 \end{bmatrix} \ .$$

(3.9)

Let R be be the unit square in the upper right quadrant of the plane. R is the set of vectors of the form

$$x\mathbf{e}_1 + y\mathbf{e}_1 \quad \text{with} \quad 0 \le x, y \le 1 \ .$$

Here is a picture of the square, together with $\mathbf{e}_1$ and $\mathbf{e}_2$:

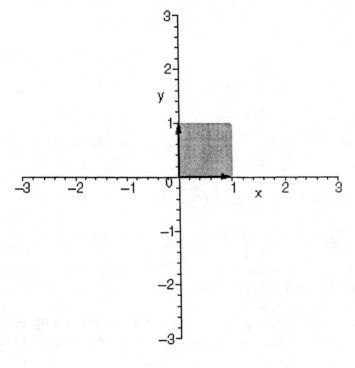

Now let $\tilde{R}$ be the set of vectors of the form $S\mathbf{v}$ for $\mathbf{v}$ in R. That is, $\tilde{R}$ is the set of vactors of the form

$$S(x\mathbf{e}_1 + y\mathbf{e}_1) = x(\mathbf{e}_1 + \alpha\mathbf{e}_2) + y\mathbf{e}_2 \qquad \text{with} \quad 0 \leq x, y \leq 1 . \qquad (3.10)$$

Here is a picture of this region, shown together with the vectors $(\mathbf{e}_1 + \alpha\mathbf{e}_2)$ and $\mathbf{e}_2$:

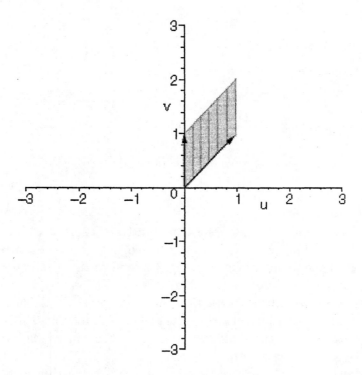

As you can see in the picture, the value of α here is close to one, but the value of α won't really enter our considerations.

The region $\tilde{R}$ is called the *image* of the region R under the transformation S, since the vectors in $\tilde{R}$ are exactly the vectors of the form $S\mathbf{v}$ for some $\mathbf{v}$ in R. Since S is invertible, the original region R is the image of $\tilde{R}$ under the inverse mapping S^{-1}.

Now, how are the areas of the two parallelograms R and $\tilde{R}$ related? *They are exactly the same.* Look at the vertical lines in the second diagram. Imagine "cutting" the parallelogram along these lines, and letting the pieces slide down so that they rest on the x axis. If we did the cutting in a fine enough way, the result would just be the unit square.

The lines are called "shear lines", and now you see why. The transformation S acts by cutting the plane into vertical lines, and sliding the points on the vertical line $x = x_0$ up by an amount αx_0. As Cavalleri observed in the sixteenth century, such a transformation has no effect on area.

This may be one thing that is even easier to understand in three dimensions than in two. Imagine holding a deck of cards with all of the cards aligned so the deck is a rectangular solid. Now slide the cards in some direction parallel to their faces, and so the amount of sliding of each card is proportional to the height of that card in the deck. This is a shear transformation in three dimensions, and the shape of the deck is now a non–rectangular parallelepiped. By sliding the cards around, we haven't in any way changed the volume that they take up, or in other words, the amount of air that they displace. The exact same conclusions apply to shear transformations of the form

$$S = \begin{bmatrix} 1 & \alpha \\ 0 & 1 \end{bmatrix} .$$

Next, consider any diagonal matrix

$$D = \begin{bmatrix} \gamma & 0 \\ 0 & \delta \end{bmatrix} . \tag{3.11}$$

Clearly the image of R under D is the set of vectors of the form

$$x\gamma \mathbf{e}_1 + y\delta \mathbf{e}_2 \qquad \text{with} \quad 0 \leq x, y \leq 1 .$$

This is a rectangle of area $|\gamma\delta|$. So under the diagonal transformation D, the area of the unit square is multiplied by $|\gamma\delta|$. In fact, it is easy to see that this transformation changes the area of any region in the plane by the same multiple: Imagine the region divided into many little rectangular cells – "tiles" if you like. The transformation D multiplies the horizontal dimension of each of these tiles by $|\gamma|$, and the vertical dimension by $|\delta|$. The result is that the area of the tile, just like the unit square, is multiplied by $|\gamma\delta|$. Since the area of every "tile" changes by a the same factor, so does the sum of the area of the tiles, and hence so does the area of the region.

It takes a little effort to make this precise, and you will carry this through when you study area and volume in multivariable calculus. But the idea is clear enough, and Cavalleri used it, before the invention of calculus to compute the area of ellipses by employing such a "stretching" transformation to transform ellipses into circles.

Now let's go back to our general invertible matrix A as in (3.2). As we have seen in (3.7),

$$A = S_1^{-1} S_2^{-1} D .$$

Now D multiples the area of a region R by absolute value of the product of its diagonal elements. These are given in (3.7), and we see that the absolute value of their product is $|ad - bc|$. Then the two shear transformations leave the volume unchanged. The net effect is that A itself changes the volume of a region in the plane by a factor of $|ad - bc|$, which is the absolute value of the determinant of A.

If the matrix A were singular, then Img(A) is one dimensional (or zero if A is the zero matrix). In this case, the image of the *any* region in the plane is contained in a line, namely Img(A), and its area is zero. So in this case also, the transformation A changes the area of a region by multiplying it by $|\det(A)|$.

We summarize this in the following theorem:

Theorem 1: *Let A be any 2×2 matrix. Let R be a region in the plane with a well–defined area. Let $\tilde{R}$ be the set of vectors $\mathbf{w}$ of the form $\mathbf{w} = A\mathbf{v}$ for some $\mathbf{v}$ in R. That is, $\tilde{R}$ is the image of R under A. Then $\tilde{R}$ has a well defined area, and the area of $\tilde{R}$ is $|\det(A)|$ times the area of R.*

This analysis generalizes easily to higher dimensions. Let A be any 3×3 matrix with linearly independent columns. After a possible reordering of the columns, we can row reduce A to a diagonal matrix D by applying a sequence of shear transformations. None of the shear transformations has any effect on volume, and the diagonal matrix D multiplies volumes by the absolute value of the product of its diagonal entries, which is its determinant. Since every shear transformation is either upper or lower triangular, with each diagonal entry equal to 1, the determinant of each shear transformation is 1. And since the determinant of a product of matrices is the product of their determinants, $|\det(A)| = |\det(D)|$.

We summarize this in the following theorem:

Theorem 2: *Let A be any 3×3 matrix. Let R be a region in $I\!R^3$ with a well–defined area. Let $\tilde{R}$ be the set of vectors $\mathbf{w}$ of the form $\mathbf{w} = A\mathbf{v}$ for some $\mathbf{v}$ in R. That is, $\tilde{R}$ is the image of R under A. Then $\tilde{R}$ has a well defined volume, and the area of $\tilde{R}$ is $|\det(A)|$ times the area of R.*

Definition (Parallelepiped) Given two vectors $\mathbf{v}_1$ and $\mathbf{v}_2$ in $I\!R^2$, the *parallelepiped spanned by $\mathbf{v}_1$ and $\mathbf{v}_2$* is the set of vectors of the from

$$x\mathbf{v}_1 + y\mathbf{v}_2 \qquad \text{with} \qquad 0 \leq x, y \leq 1 \;.$$

Given three vectors $\mathbf{v}_1$, $\mathbf{v}_2$ and $\mathbf{v}_3$ in $I\!R^3$, the *parallelepiped spanned by $\mathbf{v}_1$, $\mathbf{v}_2$ and $\mathbf{v}_3$* is the set of vectors of the from

$$x\mathbf{v}_1 + y\mathbf{v}_2 + z\mathbf{v}_3 \qquad \text{with} \qquad 0 \leq x, y, z \leq 1 \;.$$

It is clear from the definition that if $A = [\mathbf{v}_1, \mathbf{v}_2, \mathbf{v}_3]$, the 3×3 matrix whose jth column is $\mathbf{v}_j$, then

$$x\mathbf{v}_1 + y\mathbf{v}_2 + z\mathbf{v}_3 = A\begin{bmatrix} x & y & z \end{bmatrix}$$

and so the parallelepiped spanned by $\mathbf{v}_1$, $\mathbf{v}_2$ and $\mathbf{v}_3$ is the image under A of the unit cube. Since the unit cube has unit volume, it follows immediately from the theorem that the volume of the parallelepiped spanned by $\mathbf{v}_1$, $\mathbf{v}_2$ and $\mathbf{v}_3$ is $|\det([\mathbf{v}_1, \mathbf{v}_2, \mathbf{v}_3])|$. The analogous result holds in two dimensions, and we record both results in the following theorem:

Theorem 3: (Volumes of Parallelepipeds) *The area of the parallelepiped panned by $\mathbf{v}_1$ and $\mathbf{v}_2$ is $|\det([\mathbf{v}_1, \mathbf{v}_2])|$. The volume of the parallelepiped panned by $\mathbf{v}_1$, $\mathbf{v}_2$ and $\mathbf{v}_3$ is $|\det([\mathbf{v}_1, \mathbf{v}, \mathbf{v}_3])|$.*

Section 4: Determinants and the Cross Product

Let $\mathbf{a}$ and $\mathbf{b}$ be any two vectors in $\mathbb{R}^3$. Then given any other vector $\mathbf{v}$ in $\mathbb{R}^3$, form the matrix

$$A = \begin{bmatrix} \mathbf{a} \\ \mathbf{b} \\ \mathbf{v} \end{bmatrix} .$$

We know from Section 2 that that $\det(A)$ is a linear function of each of its rows, so in particular, with $\mathbf{a}$ and $\mathbf{b}$ fixed and $\mathbf{v}$ considered as a variable,

$$f(\mathbf{v}) = \det\left(\begin{bmatrix} \mathbf{a} \\ \mathbf{b} \\ \mathbf{v} \end{bmatrix} \right)$$

is a linear form on $\mathbb{R}^3$. That is, for any numbers s and t, and any two vectors $\mathbf{v}_1$ and $\mathbf{v}_2$ in $\mathbb{R}^3$,

$$f(s\mathbf{v}_1 + t\mathbf{v}_2) = sf(\mathbf{v}_1) + tf(\mathbf{v}_2) .$$

Writing

$$\mathbf{v} = x\mathbf{e}_1 + y\mathbf{e}_2 + z\mathbf{e}_3 ,$$

we have

$$\begin{aligned} f(\mathbf{v}) &= f(x\mathbf{e}_1 + y\mathbf{e}_2 + z\mathbf{e}_3) \\ &= xf(\mathbf{e}_1) + yf(\mathbf{e}_2) + zf(\mathbf{e}_3) \\ &= \begin{bmatrix} x \\ y \\ z \end{bmatrix} \cdot \begin{bmatrix} f(\mathbf{e}_1) \\ f(\mathbf{e}_2) \\ f(\mathbf{e}_3) \end{bmatrix} . \end{aligned}$$

Hence there is a vector $\mathbf{c}$ so that

$$f(\mathbf{v}) = \mathbf{c} \cdot \mathbf{v} .$$

This is just what the very first theorem in Chapter Two tells us: Every linear form can be written as a dot product. We have rederived this here because the derivation is so simple, and because we want a formula for the vector $\mathbf{c}$. The derivation will provide the formula.

The vector $\mathbf{c}$ depends on $\mathbf{a}$ and $\mathbf{b}$, and we will see that it gives us useful geometric information about $\mathbf{a}$ and $\mathbf{b}$. That this vector $\mathbf{c}$ will convey geometric information about $\mathbf{a}$ and $\mathbf{b}$ should not be surprising: It is constructed out of $\mathbf{a}$ and $\mathbf{b}$ using the determinant and the dot product, both of which have a geometric character.

To get a formula for $\mathbf{c}$, all we have to do is to explicitly compute $\begin{bmatrix} f(\mathbf{e}_1) \\ f(\mathbf{e}_2) \\ f(\mathbf{e}_3) \end{bmatrix}$. For this

purpose, let's introduce the notation

$$\mathbf{a} = \begin{bmatrix} a_1 \\ a_2 \\ a_3 \end{bmatrix} \qquad \text{and} \qquad \mathbf{b} = \begin{bmatrix} b_1 \\ b_2 \\ b_3 \end{bmatrix} .$$

Then,

$$f(\mathbf{e}_1) = \det\left(\begin{bmatrix} a_1 & a_2 & a_3 \\ b_1 & b_2 & b_3 \\ 1 & 0 & 0 \end{bmatrix}\right) = a_2b_3 - a_3b_2$$

by the formula (2.3). In the same way we find

$$f(\mathbf{e}_2) = \det\left(\begin{bmatrix} a_1 & a_2 & a_3 \\ b_1 & b_2 & b_3 \\ 0 & 1 & 0 \end{bmatrix}\right) = a_1b_3 - a_3b_1$$

and

$$f(\mathbf{e}_3) = \det\left(\begin{bmatrix} a_1 & a_2 & a_3 \\ b_1 & b_2 & b_3 \\ 0 & 0 & 1 \end{bmatrix}\right) = a_1b_2 - a_2b_1 \ .$$

Definition (Cross Product) The *cross product* of two vectors

$$\mathbf{a} = \begin{bmatrix} a_1 \\ a_2 \\ a_3 \end{bmatrix} \qquad \text{and} \qquad \mathbf{b} = \begin{bmatrix} b_1 \\ b_2 \\ b_3 \end{bmatrix}$$

is the vector $\mathbf{a} \times \mathbf{b}$ defined by

$$\mathbf{a} \times \mathbf{b} = \begin{bmatrix} a_2b_3 - a_3b_2 \\ a_3b_1 - a_1b_3 \\ a_1b_2 - a_2b_1 \end{bmatrix} \ . \tag{4.1}$$

The definition of $\mathbf{a} \times \mathbf{b}$ has been made so that for any vector $\mathbf{v}$,

$$(\mathbf{a} \times \mathbf{b}) \cdot \mathbf{v} = \det\left(\begin{bmatrix} \mathbf{a} \\ \mathbf{b} \\ \mathbf{v} \end{bmatrix}\right) \ . \tag{4.2}$$

Moreover, since $\det(A^t) = \det(A)$ for any sqaure matrix A, it is also the case that

$$(\mathbf{a} \times \mathbf{b}) \cdot \mathbf{v} = \det\left([\mathbf{a}, \mathbf{b}, \mathbf{v}]\right) \ . \tag{4.3}$$

The pattern in (4.1) is not hard to remember, but you may find it easier to remeber the formula in the form

$$\mathbf{a} \times \mathbf{b} = \det\left(\begin{bmatrix} a_1 & a_2 & a_3 \\ b_1 & b_2 & b_3 \\ \mathbf{e}_1 & \mathbf{e}_2 & \mathbf{e}_3 \end{bmatrix}\right) \tag{4.4}$$

$$= (a_2b_3 - a_3b_2)\mathbf{e}_1 + (a_3b_1 - a_1b_3)\mathbf{e}_2 + (a_1b_2 - a_2b_1)\mathbf{e}_3 \ .$$

This may look like a strange sort of determinant since the entries of the bottom row are vectors and not numbers. But since we can multiply vectors and numbers, and since each of the products in the definition of the determinant contains only one entry from each row, and hence only one vector, the formula makes sense, and the result of (4.4) is the same as (4.1).

Example 1 Let $\mathbf{a} = \begin{bmatrix} 1 \\ 2 \\ 3 \end{bmatrix}$ and $\mathbf{b} = \begin{bmatrix} 3 \\ 2 \\ 1 \end{bmatrix}$. Then from (4.1),

$$\mathbf{a} \times \mathbf{b} = \begin{bmatrix} 2-6 \\ 9-1 \\ 2-6 \end{bmatrix} = \begin{bmatrix} -4 \\ 8 \\ -4 \end{bmatrix} .$$

Example 2 Let $\mathbf{a} = \mathbf{e}_1$ and $\mathbf{b} = \mathbf{e}_2$. Then from (4.1) you find $\mathbf{a} \times \mathbf{b} = \mathbf{e}_3$. In fact in the same way, you find

$$\mathbf{e}_1 \times \mathbf{e}_2 = \mathbf{e}_3 \qquad \mathbf{e}_2 \times \mathbf{e}_3 = \mathbf{e}_1 \qquad \text{and} \qquad \mathbf{e}_3 \times \mathbf{e}_1 = \mathbf{e}_2 . \tag{4.5}$$

The next theorem records some important properties of the cross product.

Theorem 1 (Properties of the Cross Product) *The cross product of two vectors* $\mathbf{a}$ *and* $\mathbf{b}$ *in* $\mathbb{R}^3$ *has the following properties:*

(i) It is anticommutatitve; that is,

$$\mathbf{a} \times \mathbf{b} = -\mathbf{b} \times \mathbf{a} . \tag{4.6}$$

(ii) It is linear in both $\mathbf{a}$ *and* $\mathbf{b}$. *In particular, for any numbers* s *and* t, *and any vectors* $\mathbf{a}_1$, $\mathbf{a}_2$ *and* $\mathbf{b}$,

$$(s\mathbf{a}_1 + t\mathbf{a}_2) \times \mathbf{b} = s(\mathbf{a}_1 \times \mathbf{b}) + t(\mathbf{a}_2 \times \mathbf{b}) . \tag{4.7}$$

(iii) The cross product $\mathbf{a} \times \mathbf{b}$ *is orthogonal to both* $\mathbf{a}$ *and* $\mathbf{b}$. *That is,*

$$(\mathbf{a} \times \mathbf{b}) \cdot \mathbf{a} = 0 \qquad \text{and} \qquad (\mathbf{a} \times \mathbf{b}) \cdot \mathbf{b} = 0 . \tag{4.8}$$

(iv) The length of $\mathbf{a} \times \mathbf{b}$ *is the area of the parallelogram with edges along* $\mathbf{a}$ *and* $\mathbf{b}$. *That is,*

$$|\mathbf{a} \times \mathbf{b}| = |\mathbf{a}||\mathbf{b}||\sin(\theta)| \tag{4.9}$$

where θ *is the angle between* $\mathbf{a}$ *and* $\mathbf{b}$. *In particular,* $\mathbf{a} \times \mathbf{b} = 0$ *if* $\mathbf{b}$ *is a multiple of* $\mathbf{a}$.

Proof: The anticommutivity (4.6) follows directly from (4.4) and the fact that the determinant changes sign when two rows are swapped. The linearity (4.7) follows directly from (4.4) and the fact that the determinant is a linear function of each of its rows. The fact that (4.8) is true follows from (4.2) and the fact that a determinant with two identical rows is zero.

Finally, let $\mathbf{n}$ be given by

$$\mathbf{n} = \frac{1}{|\mathbf{a} \times \mathbf{b}|}\mathbf{a} \times \mathbf{b}.$$

This is a unit vector pointing in the direction of $\mathbf{a} \times \mathbf{b}$, and so

$$|\mathbf{a} \times \mathbf{b}| = \mathbf{n} \cdot \mathbf{a} \times \mathbf{b}.$$

By (4.2),

$$|\mathbf{a} \times \mathbf{b}| = \det\left(\begin{bmatrix} \mathbf{a} \\ \mathbf{b} \\ \mathbf{n} \end{bmatrix}\right).$$

From the previous section, we know that this determinant is the volume of the parallelipiped spanned by the vectors $\mathbf{a}$, $\mathbf{b}$ and $\mathbf{n}$. This in thurn is the area of the parallelogram spanned by $\mathbf{a}$ and $\mathbf{b}$ times the height if the paralellepiped in the direction riging out of the plane spanned by $\mathbf{a}$ and $\mathbf{b}$. Since $\mathbf{n}$ is a unit vector in this direction, the height is one, and so $|\mathbf{a} \times \mathbf{b}|$ equals the area of the parallelogram spanned by $\mathbf{a}$ and $\mathbf{b}$. By elementary planar geometry, this is the product of the lengths of the sides and the sine of the angle θ between them, which is to say, $|\mathbf{a}||\mathbf{b}|\sin(\theta)$. ■

Here is a picture representing the orthogonality property of the cross product:

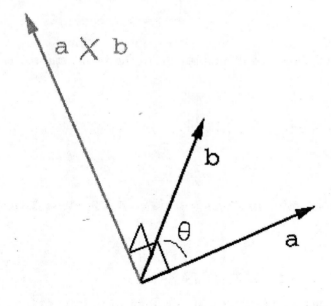

The fact that $|\mathbf{a} \times \mathbf{b}| = |\mathbf{a}||\mathbf{b}|| \sin(\theta)|$ is at least as important as the fact that $\mathbf{a} \times \mathbf{b}$ is orthogonal to both $\mathbf{a}$ and $\mathbf{b}$, as we shall see. We have derived this in the proof just above by considering volume, but it can also be seen from (4.1) by direct computation. From (4.1) we have

$$
\begin{aligned}
|\mathbf{a} \times \mathbf{b}|^2 &= (a_2b_3 - a_3b_2)^2 + (a_3b_1 - a_1b_3)^2 + (a_2b_1 - a_1b_2)^2 \\
&= (a_2b_3)^2 + (a_3b_2)^2 + (a_3b_1)^2 + (a_1b_3)^2 + (a_2b_1)^2 + (a_1b_2)^2 \\
&\quad - 2a_2b_3a_3b_2 - 2a_3b_1a_1b_3 - 2a_2b_1a_1b_2 \\
&= |\mathbf{a}|^2|\mathbf{b}|^2 - (\mathbf{a} \cdot \mathbf{b})^2 \\
&= |\mathbf{a}|^2|\mathbf{b}|^2 - |\mathbf{a}|^2|\mathbf{b}|^2 \cos^2(\theta) \\
&= |\mathbf{a}|^2|\mathbf{b}|^2 \sin^2(\theta)
\end{aligned}
$$

where in the next ot the last line we have used the fact that $\mathbf{a} \cdot \mathbf{b} = |\mathbf{a}||\mathbf{b}| \cos(\theta)$.

The cross product gives us another way to compute the equation of the plane in $I\!R^3$ spanned by two linearly independent vectors.

Example 3 Let's find the equation for the plane spanned by $\mathbf{a} = \begin{bmatrix} 1 \\ 2 \\ 3 \end{bmatrix}$ and $\mathbf{b} = \begin{bmatrix} 3 \\ 2 \\ 1 \end{bmatrix}$. We have already computed in Example 1 that

$$
\mathbf{a} \times \mathbf{b} = \begin{bmatrix} -4 \\ 8 \\ -4 \end{bmatrix} .
$$

Since $\{\mathbf{a}, \mathbf{b}\}$ is a basis for the plane, and since $\mathbf{a} \times \mathbf{b}$ is orthogonal to each of these vectors, $\{\mathbf{a} \times \mathbf{b}\}$ is a basis for the orthogonal complement of the plane, and so the equation of the plane is

$$
\begin{bmatrix} x \\ y \\ z \end{bmatrix} \cdot \begin{bmatrix} -4 \\ 8 \\ -4 \end{bmatrix} = 0 ,
$$

or

$$
x - 2y + z = 0 .
$$

We have two other ways to do this already; the cross product gives us a third. Let's review the other two:

Form the matrix

$$
\begin{bmatrix} \mathbf{a} \\ \mathbf{b} \end{bmatrix} = \begin{bmatrix} 1 & 2 & 3 \\ 3 & 2 & 1 \end{bmatrix} .
$$

The vectors in the kernel of this matrix are orthogonal to each of the rows, namely bfa and $\mathbf{b}$. so finding a non-zero vector in the kernel of this matrix gives us the orthogonal vector we seek.

To find it, row reduce the matrix. In one step,

$$\begin{bmatrix} 1 & 2 & 3 \\ 3 & 2 & 1 \end{bmatrix} \rightarrow \begin{bmatrix} 1 & 2 & 3 \\ 0 & -4 & -8 \end{bmatrix} .$$

The variable z is non–pivotal, and so we set $z = 1$, and solve for x and y, finding $y = -2$ and then $x = 1$. Hence $\{\mathbf{v}\}$ where

$$\mathbf{v} = \begin{bmatrix} 1 \\ -2 \\ 1 \end{bmatrix}$$

is the vector we seek. Notice that it is $-1/4$ times $\mathbf{a} \times \mathbf{b}$. Hence we find the equation

$$\begin{bmatrix} x \\ y \\ z \end{bmatrix} \cdot \begin{bmatrix} 1 \\ -2 \\ 1 \end{bmatrix} = 0$$

or

$$x - 2y + z = 0$$

as before.

In both of these two approaches we found the equation by finding a non–zero vector that was orthogonal to both $\mathbf{a}$ and $\mathbf{b}$: First by computing $\mathbf{a} \times \mathbf{b}$, and then by using row reduction to find a basis for the kernel of $\begin{bmatrix} \mathbf{a} \\ \mathbf{b} \end{bmatrix}$. There is yet another way that we already know: Row reduce $\begin{bmatrix} 1 & 3 & x \\ 2 & 2 & y \\ 3 & 1 & z \end{bmatrix}$. In two steps you find

$$\begin{bmatrix} 1 & 3 & x \\ 2 & 2 & y \\ 3 & 1 & z \end{bmatrix} \rightarrow \begin{bmatrix} 1 & 3 & x \\ 0 & -4 & -2x+y \\ 0 & -8 & -3x+z \end{bmatrix} \rightarrow \begin{bmatrix} 1 & 3 & x \\ 0 & -4 & -2x+y \\ 0 & 0 & x-2y+z \end{bmatrix} .$$

Evidently, $\begin{bmatrix} x \\ y \\ z \end{bmatrix}$ is in the image of $[\mathbf{a}, \mathbf{b}]$ exactly when

$$x - 2y + z = 0 .$$

As far as computing equations of planes goes, the cross product just gives us another way to do something we already knew how to do. In fact, it isn't even faster in most case; usually finding the kernel, as in our second approach, involves the least amount of computation.

The real value of the cross product $\mathbf{a} \times \mathbf{b}$ is not that it gives us *some* vector orthogonal to both $\mathbf{a}$ and $\mathbf{b}$: The real value lies in the the fact that $\mathbf{a} \times \mathbf{b}$ *and* the legth of $\mathbf{a} \times \mathbf{b}$ is the area of the parallelogram spanned by $\mathbf{a}$ and $\mathbf{b}$. This is very useful in multivariable calculus when one has to compute the area of a *curved* two dimesnional surface in $I\!\!R^3$: You break the surface up into a large number of parallelograms and then use the cross product to compute each of their areas. Then you add these areas up, and taking an appropriate limit, the sum becomes an integral.

Pursuing this line of inquiry would take us outside the field of linear algebra, and we won't go there. However, we will illustrate the use of the cross product to compute the areas of parallelograms in $I\!\!R^3$.

Example 4 Let $\mathbf{a} = \begin{bmatrix} 1 \\ -1 \\ 1 \end{bmatrix}$ and $\mathbf{b} = \begin{bmatrix} 3 \\ 2 \\ 1 \end{bmatrix}$. What is the area of the parallelogram spanned by $\mathbf{a}$ and $\mathbf{b}$? From (4.1),

$$\mathbf{a} \times \mathbf{b} = \begin{bmatrix} -1 - 2 \\ 3 - 1 \\ 2 + 3 \end{bmatrix} = \begin{bmatrix} -3 \\ 2 \\ 5 \end{bmatrix} ,$$

and so

$$|\mathbf{a} \times \mathbf{b}| = \sqrt{38} .$$

This is the area of the parallelogram. (Though it is not required to do so to answer the question, it is a good idea to check that $\mathbf{a} \times \mathbf{b} \cdot \mathbf{a}$ and $\mathbf{a} \times \mathbf{b} \cdot \mathbf{b}$ are both zero, as the must be, and will be, if $\mathbf{a} \times \mathbf{b} \cdot \mathbf{a}$ has been computed correctly.

Example 5 Let's compute the total surface area of the parallelepiped spanned by the three vectors

$$\mathbf{a} = \begin{bmatrix} 1 \\ -1 \\ 1 \end{bmatrix} \qquad \mathbf{b} = \begin{bmatrix} 3 \\ 2 \\ 1 \end{bmatrix} \qquad \text{and} \qquad \mathbf{c} = \begin{bmatrix} 1 \\ 2 \\ 3 \end{bmatrix} .$$

Any such parallelepiped in $I\!\!R^3$ has six sides, like the cube, which come in three parallel pairs. There is the parallelogram spanned by $\mathbf{a}$ and $\mathbf{b}$, and another parallelogram of the same area on the opposite side, and so on, for each of the three pairs of vectors. Hence the total area is

$$2|\mathbf{a} \times \mathbf{b}| + 2|\mathbf{a} \times \mathbf{c}| + 2|\mathbf{b} \times \mathbf{c}| .$$

In Examples 1 and 4 we've already computed $\mathbf{a} \times \mathbf{b}$ and $\mathbf{b} \times \mathbf{c}$ (using different names for the same vectors). In example 4 we found $\mathbf{a} \times \mathbf{b} = \begin{bmatrix} -3 \\ 2 \\ 5 \end{bmatrix}$ and $|\mathbf{a} \times \mathbf{b}| = \sqrt{38}$. In Example

1 we found $\mathbf{b} \times \mathbf{c} = \begin{bmatrix} -4 \\ 8 \\ -4 \end{bmatrix}$ and so $\mathbf{b} \times \mathbf{c} = \sqrt{86}$. Finally,

$$\mathbf{a} \times \mathbf{c} = \begin{bmatrix} -3 - 2 \\ 1 - 3 \\ 2 + 1 \end{bmatrix} = \begin{bmatrix} -5 \\ -2 \\ 3 \end{bmatrix},$$

and so

$$|\mathbf{a} \times \mathbf{b}| = \sqrt{38}.$$

Hence the total surface area of the parallepiped is $4\sqrt{38} + 2\sqrt{86}$.

We close with a final remark on computing cross products. Let $\mathbf{v}$ and $\mathbf{w}$ be any two vectors in $I\!R^3$, and write them in terms of the standard basis vectors:

$$\mathbf{v} = \sum_{i=1}^{3} v_i \mathbf{e}_i \qquad \text{and} \qquad \mathbf{v} = \sum_{j=1}^{3} w_j \mathbf{e}_j.$$

Then since the cross product is linear in each of its factors,

$$\mathbf{v} \times \mathbf{w} = \sum_{i,j=1}^{3} v_i b_j \mathbf{e}_i \times \mathbf{e}_j.$$

Now by the antisymmetry of the cross product,

$$\mathbf{e}_i \times \mathbf{e}_j = -\mathbf{e}_j \times \mathbf{e}_i$$

for each i and j. In particular, $\mathbf{e}_i \times \mathbf{e}_i = 0$ for each i, and so of the nine terms $\mathbf{e}_i \times \mathbf{e}_j$, the three with $i = j$ are al zero, and the other three come in pairs differing only in sign. Hence

$$\sum_{i,j=1}^{3} v_i w_j \mathbf{e}_i \times \mathbf{e}_j = (v_1 w_2 - v_2 w_1)\mathbf{e}_1 \times \mathbf{e}_2 + (v_2 w_3 - v_3 w_2)\mathbf{e}_2 \times \mathbf{e}_3 + (v_3 w_1 - v_1 w_3)\mathbf{e}_3 \times \mathbf{e}_1.$$

Hence, because of the bilinearity and the antisymmetry, if you know how to compute the three cross products

$$\mathbf{e}_1 \times \mathbf{e}_2 \qquad \mathbf{e}_2 \times \mathbf{e}_3 \qquad \text{and} \qquad \mathbf{e}_3 \times \mathbf{e}_1,$$

you can compute the cross products of *any* pair of vectors. So if you remember (4.5), which has a memorable cyclic pattern to it, you can compute the cross product of any pair of vectors.

Problems for Sections 1, 2 and 3

4.3.1 Compute the determinant of

$$A = \begin{bmatrix} 1 & 2 & 4 & 8 \\ 1 & -2 & 4 & -8 \\ 1 & 3 & 9 & 27 \\ 1 & -3 & 9 & -27 \end{bmatrix} .$$

4.3.2 Compute the determinant of

$$A = \begin{bmatrix} a & 2 & 0 & 0 \\ 2 & a & 2 & 0 \\ 0 & 2 & a & 2 \\ 0 & 0 & 2 & a \end{bmatrix} .$$

For which values of a is A invertible?

4.3.3 Compute the determinant of

$$A = \begin{bmatrix} 0 & 0 & 2 & a \\ 0 & 2 & a & 2 \\ 2 & a & 2 & 0 \\ a & 2 & 0 & 0 \end{bmatrix} .$$

For which values of a is A invertible?

4.3.4 Let $\mathbf{v}_1$ and $\mathbf{v}_2$ be the vectors

$$\mathbf{v}_1 = \begin{bmatrix} 1 \\ -2 \end{bmatrix} \quad \text{and} \quad \mathbf{v}_2 = \begin{bmatrix} 3 \\ 1 \end{bmatrix} .$$

Find the area of the parallelepiped spanned by $\mathbf{v}_1$ and $\mathbf{v}_2$.

4.3.5 Consider the three points $(2,3)$, $(3,5)$ and $(-1,-1)$. what is the area of the triangle with these vertices?

4.3.6 Let $\mathbf{v}_1$, $\mathbf{v}_2$ and $\mathbf{v}_3$ be the vectors

$$\mathbf{v}_1 = \begin{bmatrix} 1 \\ -2 \\ 1 \end{bmatrix} , \quad \mathbf{v}_2 = \begin{bmatrix} 0 \\ -1 \\ 2 \end{bmatrix} \quad \text{and} \quad \mathbf{v}_3 = \begin{bmatrix} 3 \\ -1 \\ 0 \end{bmatrix} .$$

Find the volume of the parallelepiped spanned by $\mathbf{v}_1$, $\mathbf{v}_2$ and $\mathbf{v}_2$.

4.3.7 Consider the three points $(0,0,0,)$, $(2,3,1)$, $(3,0,1)$ and $(-1,-1,0)$. what is the area of the simplex with these vertices? (Note the simplex determined by four points is the set of all convex combinations of these points, or in other words, all weighted averages of these points.)

Problems for Section 4

(4.4.1) Let **u**, **v** and **w** the three vectors

$$\mathbf{u} = \begin{bmatrix} 2 \\ 0 \\ 1 \end{bmatrix} \qquad \mathbf{v} = \begin{bmatrix} -1 \\ 1 \\ 1 \end{bmatrix} \quad \text{and} \quad \mathbf{w} = \begin{bmatrix} 2 \\ 0 \\ -2 \end{bmatrix} .$$

(a) Compute the cross products $\mathbf{u} \times \mathbf{v}$, $\mathbf{u} \times \mathbf{w}$ and $\mathbf{v} \times \mathbf{w}$.

(b) Compute the total surface area of the parallelepiped spanned by **u**, **v** and **w**.

(4.4.2) Let **u**, **v** and **w** the three vectors

$$\mathbf{u} = \begin{bmatrix} 1 \\ 1 \\ 1 \end{bmatrix} \qquad \mathbf{v} = \begin{bmatrix} 1 \\ -1 \\ 1 \end{bmatrix} \quad \text{and} \quad \mathbf{w} = \begin{bmatrix} 1 \\ 2 \\ 4 \end{bmatrix} .$$

(a) Compute the cross products $\mathbf{u} \times \mathbf{v}$, $\mathbf{u} \times \mathbf{w}$ and $\mathbf{v} \times \mathbf{w}$.

(b) Compute the total surface area of the parallelepiped spanned by **u**, **v** and **w**.

(4.4.3) Let **a** and **b** the two vectors

$$\mathbf{a} = \begin{bmatrix} 1 \\ 3 \\ 2 \end{bmatrix} \quad \text{and} \quad \mathbf{b} = \begin{bmatrix} 1 \\ 2 \\ -1 \end{bmatrix} .$$

(a) Compute the cross product $\mathbf{a} \times \mathbf{b}$, and use this computation to find the equation of the plane spanned by **a** and **b**.

(b) Use row reduction to find a non–zero vector in $\text{Ker}\left(\begin{bmatrix} \mathbf{a} \\ \mathbf{b} \end{bmatrix} \right)$, and use this computation to find the equation of the plane spanned by **a** and **b**.

(c) what is the area of the parallelogram spanned by **a** and **b**?

(4.4.4) Let **a** and **b** the two vectors

$$\mathbf{a} = \begin{bmatrix} 1 \\ 2 \\ 4 \end{bmatrix} \quad \text{and} \quad \mathbf{b} = \begin{bmatrix} 1 \\ -2 \\ 4 \end{bmatrix} .$$

(a) Compute the cross product $\mathbf{a} \times \mathbf{b}$, and use this computation to find the equation of the plane spanned by **a** and **b**.

(b) Use row reduction to find a non–zero vector in Ker $\left(\begin{bmatrix} \mathbf{a} \\ \mathbf{b} \end{bmatrix} \right)$, and use this computation to find the equation of the plane spanned by $\mathbf{a}$ and $\mathbf{b}$.

(c) what is the area of the parallelogram spanned by $\mathbf{a}$ and $\mathbf{b}$?

(4.4.5) Let B be a three by three matrix. What conditions on B guarantee that

$$(B\mathbf{u} \times B\mathbf{v}) \cdot B\mathbf{w} = (\mathbf{u} \times \mathbf{v}) \cdot \mathbf{w}$$

for all vectors $\mathbf{u}$, $\mathbf{v}$ and $\mathbf{w}$ in $I\!\!R^3$? (Don't just give the trivial ones like $B = I$ or $B = 0$; there are other more interesting cases).

(4.4.6) Let B be a three by three matrix. What conditions on B guarantee that

$$(B\mathbf{u} \times B\mathbf{v}) = B(\mathbf{u} \times \mathbf{v})$$

for all vectors $\mathbf{u}$ and $\mathbf{v}$ in $I\!\!R^3$? (Don't just give the trivial ones like $B = I$ or $B = 0$; there are other more interesting cases).

(4.4.7) Given any three vectors $\mathbf{v}_1$, $\mathbf{v}_2$ and $\mathbf{v}_3$ in $I\!\!R^3$, define a *modified cross product* $\mathbf{a} \ast\!\!\ast \mathbf{b}$ by requiring that

$$\mathbf{a} \ast\!\!\ast \mathbf{b} = -\mathbf{b} \ast\!\!\ast \mathbf{a}$$

for all $\mathbf{a}$ and $\mathbf{b}$, and that $\mathbf{a} \ast\!\!\ast \mathbf{b}$ be linear in both $\mathbf{a}$ and $\mathbf{b}$. Finally, we specify that

$$\mathbf{e}_1 \ast\!\!\ast \mathbf{e}_2 = \mathbf{v}_3 \qquad \mathbf{e}_2 \ast\!\!\ast \mathbf{e}_3 = \mathbf{v}_1 \qquad \text{and} \qquad \mathbf{e}_3 \ast\!\!\ast \mathbf{e}_1 = \mathbf{v}_2 \; .$$

(a) Explain why this is enough information to compute $\mathbf{a} \ast\!\!\ast \mathbf{b}$ for any $\mathbf{a}$ and $\mathbf{b}$ in $I\!\!R^3$. (If you aren't sure how to explain this, consider part (b) first).

(b) A bit more concretely, let

$$\mathbf{v}_1 = \begin{bmatrix} 1 \\ 0 \\ 0 \end{bmatrix} \qquad \mathbf{v}_2 = \begin{bmatrix} 1 \\ 1 \\ 0 \end{bmatrix} \qquad \text{and} \qquad \mathbf{v}_3 = \begin{bmatrix} 1 \\ 1 \\ 1 \end{bmatrix},$$

and

$$\mathbf{a} = \begin{bmatrix} 1 \\ 2 \\ 4 \end{bmatrix} \qquad \text{and} \qquad \mathbf{b} = \begin{bmatrix} 1 \\ -2 \\ 4 \end{bmatrix} \; .$$

Compute $\mathbf{a} \ast\!\!\ast \mathbf{b}$.

(c) Let B be the matrix $B = [\mathbf{v}_1, \mathbf{v}_2, \mathbf{v}_3]$. How are $\mathbf{a} \ast\!\!\ast \mathbf{b}$ and $B(\mathbf{a} \times \mathbf{b})$ related? (Answer this in general, not just for the specific case conssidered in part (b)).

(4.4.8) Define a generalized "tripple cross product" of *three* vectors in $I\!\!R^4$, and explain why you consider your definition to be a natural generalization.

Overview of Chapter 5

A main theme of this course has been finding ways to "take matrices apart" into products of simpler matrices. Diagonal matrices are particularly simple. All of the questions that we generally ask about matrices can be answered at a glance for diagonal matrices.

Unfortunately, matrices that come up in problems we'd like to solve tend not to be diagonal. However, it turns out that *every* symmetric matrix A can be written as a product

$$A = UDU^{-1}$$

where D is a diagonal matrix, and U is an orthogonal matrix. The assertion we are making then is that any symmetric matrix A can be "taken apart" into a product of an orthogonal matrix U, a diagonal matrix D, and the transpose of U. (Since U is orthogonal, $U^{-1} = U^t$). As we shall see, this factorization is the key to a number of matrix computations. It is very useful since very often the matrices arising in problems we'd like to solve are symmetric. Better yet, even when A is not symmetric, but is at least square, it is often possible to find an invertible matrix V and a diagonal matrix D so that $A = VDV^{-1}$. As we shall see, this factorization renders many questions concerning A transparent.

The key to producing these factorizations is to find numbers μ and non–zero vectors $\mathbf{v}$ so that

$$A\mathbf{v} = \mu\mathbf{v} \ .$$

Such numbers are called "eigenvalues", and such vectors are called "eigenvectors". (The fully translated versions of these hybrid German–English terms would be "characteristic values" and "characteristic vectors", which are occasionally used.)

The problem of finding all numbers μ and all vectors $\mathbf{v}$ so that $A\mathbf{v} = \mu\mathbf{v}$ is called *the eigenvalue problem*. Our aim in this part of the course is to explain how to solve the eigenvalue problem, and then to apply the "diagonalizations" we so obtain. One new sort of problem to which these methods may be profitably applied concerns *systems of linear differential equations*. An example would be

$$x'(t) = 3x(t) - 2y(t) \qquad \text{and} \qquad y'(t) = -2x(t) + y(t) \ .$$

The object here is to find functions $x(t)$ and $y(t)$ so that when one computes $x'(t)$ and $y'(t)$, one finds that these equations are satisfied. As we shall see, the key to this is to diagonalize the matrix

$$A = \begin{bmatrix} 3 & -2 \\ -2 & 1 \end{bmatrix} ,$$

which entails finding all of its eigenvectors and eigenvalues.

Section 1: The Eigenvalue Problem and Diagonalization

Definition (Eigenvalues and Eigenvectors) Let A be an $n \times n$ matrix. A number μ is an *eigenvalue* of A if there is a non–zero vector $\mathbf{v}$ such that

$$A\mathbf{v} = \mu\mathbf{v} . \tag{1.1}$$

Any such vector $\mathbf{v}$ is called an *eigenevector of A with eigenvalue μ.*

It is crucial that the vector $\mathbf{v}$ in the definition is a non-zero vector. Since $A0 = \mu 0$ for *any* number μ, the definition of eigenvalue would be vacuous if we allowed $\mathbf{v}$ to be the zero vector.

Given an $n \times n$ matrix A, the *eigenvalue problem* for A is to find all of the eigenvalues μ for A, and all of the non-zero vectors $\mathbf{v}$ so that (1.1) holds.

Compare this problem with the problem of solving $A\mathbf{x} = \mathbf{b}$, which has been our major concern until now. The eigenvalue problem is quite different. While only $\mathbf{x}$ is unknown when we are trying to solve $A\mathbf{x} = \mathbf{b}$, there are two unknowns – μ and $\mathbf{v}$ – to be solved for in the eigenvalue problem.

The following theorem relates the eigenvalue problem to something we've already studied: determining the kernel of a matrix.

Theorem 1: (Kernels, Eigenvalues and Eigenvectors) *Let A be an $n \times n$ matrix. Then μ is an eigenvalue of A if and only if $\mathrm{Ker}(A - \mu I) \neq 0$. Moreover, suppose that μ is an eigenvalue of A. Then a non-zero vector $\mathbf{v}$ is an eigenvector of A with eigenvalue μ if and only if $\mathbf{v}$ belongs to $\mathrm{Ker}(A - \mu I)$.*

Proof: $A\mathbf{v} = \mu\mathbf{v} \quad \Longleftrightarrow \quad (A - \mu I)\mathbf{v} = 0 \quad \Longleftrightarrow \quad \mathbf{v}$ belongs to $\mathrm{Ker}(A - \mu I)$. ∎

So finding the eigenvalues of A amounts to finding the numbers μ so that $\mathrm{Ker}(A - \mu I) \neq 0$, and we know how to do this. And once we know the eigenvalues μ of A, finding the eigenvectors is just the problem of determining the kernel of $(A - \mu I)$. We know how to do this too.

The best way to do these things depends on the size of the matrix A, among other things. But when the size is small, computing the determinant of $A - \mu I$ is a convenient way to check whether or not $\mathrm{Ker}(A - \mu I) \neq 0$. We know that

$$\mathrm{Ker}(A - \mu I) \neq 0 \quad \Longleftrightarrow \quad \det(A - \mu I) = 0 . \tag{1.2}$$

Example 1 Let's find all of the eigenvalues and eigenvectors of $A = \begin{bmatrix} 1 & 2 \\ 2 & 1 \end{bmatrix}$. In this case,

we have $A - \mu I = \begin{bmatrix} 1 - \mu & 2 \\ 2 & 1 - \mu \end{bmatrix}$. Then $\det(A - \mu I) = (1 - \mu)^2 - 4$, and so

$$\det(A - \mu I) = 0 \quad \Longleftrightarrow \quad (1 - \mu)^2 - 4 = 0 .$$

The only solutions of the equation on the right are

$$\mu = 1 \pm 2 \ .$$

Hence the eigenvalues of A are 3 and -1. This takes care of the eigenvalues.

Now for the eigenvectors. The eigenvectors of A with eigenvalue 3 are the non-zero vectors in the kernel of $A - 3I = \begin{bmatrix} -2 & 2 \\ 2 & -2 \end{bmatrix}$. This matrix row–reduces to $\begin{bmatrix} 1 & -1 \\ 0 & 0 \end{bmatrix}$, and hence the kernel of $A - 3I$ consists of all multiples of $\mathbf{v}_1 = \begin{bmatrix} 1 \\ 1 \end{bmatrix}$. The eigenvectors of A with eigenvalue 3 are exactly the non–zero multiples of $\mathbf{v}_1$.

Next consider $\mu = -1$. The corresponding eigenvectors are the non-zero vectors in the kernel of $A + I = \begin{bmatrix} 2 & 2 \\ 2 & 2 \end{bmatrix}$. You can easily see that the kernel of this matrix is all multiples of $\mathbf{v}_2 = \begin{bmatrix} -1 \\ 1 \end{bmatrix}$. Hence the eigenvalues of A with eigenvalue -1 are exactly the non–zero multiples of $\mathbf{v}_2$.

If we examine the results obtained in Example 1, we learn some interesting things. First notice that $\det(A - \mu I) = \mu^2 - 2\mu - 3$ is a quadratic polynomial in μ. This is no accident. In general for any $n \times n$ matrix A, we have from the definition of the determinant that

$$\det(A - \mu I) = \sum_\sigma \chi(\sigma)(A - \mu I)_{\sigma(1),1}(A - \mu I)_{\sigma(2),2} \cdots (A - \mu I)_{\sigma(n),n} \ .$$

Now, each factor $(A - \mu I)_{\sigma(j),j}$ in the product

$$(A - \mu I)_{\sigma(1),1}(A - \mu I)_{\sigma(2),2} \cdots (A - \mu I)_{\sigma(n),n}$$

is either simply $A_{\sigma(j),j}$, which is the case if $\sigma(j) \neq j$, or is $(A_{j,j} - \mu)$ in case $\sigma(j) = j$. Therefore, for each σ, the product is a polynomial in μ whose degree is at most n, which occurs only when $\sigma(j) = j$ for all j; i.e., when σ is the identity permutation. In this case, the term is

$$(A_{1,1} - \mu)(A_{2,2} - \mu) \cdots (A_{n,n} - \mu) = (-1)^n \mu^n + \text{ lower order in } \mu \ .$$

Every other term in the sum is a polynomial of degree at most $n - 1$ in μ, so $\det(A - \mu I)$ is a polynomial of degree n in μ, and the coefficient of μ^n is $(-1)^n$.

Definition (Characteristic Polynomial) For any $n \times n$ matrix A, the nth degree polynomial $p_A(\mu)$ defined by

$$p_A(\mu) = \det(A - \mu I)$$

is called the *characteristic polynomial* of A.

Theorem 1 tells us that the eigenvalues of A are exactly the roots of the characteristic polynomial of A. The Fundamental Theorem of Algebra says that every nth degree polynomial has n roots, $\{z_1, z_2, \ldots, z_n\}$, in the complex plane. Of course, they don't have to be distinct. But for each different root, we can find non–zero eigenvectors. These vectors may have complex entries themselves but that won't mean they are useless, *even in a problem where we expect the final answer will only involve real numbers.* We will see examples of this in the fourth section.

In any case, as long as we allow complex eigenvalues and complex eigenvectors, every matrix A has at least one eigenvalue.

Example 2 Let $A = \begin{bmatrix} 0 & 1 \\ -1 & 0 \end{bmatrix}$. The characteristic polynomial $p_A(\mu)$ is

$$p_A(\mu) = \det\left(\begin{bmatrix} -\mu & 1 \\ -1 & -\mu \end{bmatrix}\right) = \mu^2 + 1 \ .$$

Evidently, there are no real solutions to $p_A(\mu) = 0$. However, we do have the two complex solutions, $\mu_1 = i$ and $\mu_2 = -i$, where i denotes $\sqrt{-1}$. These are the eigenvalues of A, and to find the eigenvectors with eigenvalue i, we form

$$A - iI = \begin{bmatrix} -i & 1 \\ -1 & -i \end{bmatrix} \ .$$

The kernel of this matrix is spanned by $\mathbf{v}_1 = \begin{bmatrix} -i \\ 1 \end{bmatrix}$, and hence this complex vector satisfies $A\mathbf{v}_1 = \mu_1 \mathbf{v}_1$. Likewise,

$$A + iI = \begin{bmatrix} i & 1 \\ -1 & i \end{bmatrix} \ ,$$

and so the kernel of this matrix is spanned by $\mathbf{v}_2 = \begin{bmatrix} i \\ 1 \end{bmatrix}$, and hence this complex vector satisfies $A\mathbf{v}_2 = \mu_2 \mathbf{v}_2$. So as long as we are willing to admit complex eigenvalues and eigenvectors with complex entries, then every $n \times n$ matrix has at least one eigenvalue.

Now let's go back to Example 1, and look more closely at the eigenvectors we found. These were

$$\mathbf{v}_1 = \begin{bmatrix} 1 \\ 1 \end{bmatrix} \qquad \text{and} \qquad \mathbf{v}_2 = \begin{bmatrix} 1 \\ -1 \end{bmatrix} \ .$$

Notice that $\mathbf{v}_1 \cdot \mathbf{v}_2 = 0$. We know that non-zero orthogonal vectors are linearly independent, hence $\{\mathbf{v}_1, \mathbf{v}_2\}$ is a basis for R^2. Better yet, if we define

$$\mathbf{u}_1 = \frac{1}{\sqrt{2}} \mathbf{v}_1 \qquad \text{and} \qquad \mathbf{u}_2 = \frac{1}{\sqrt{2}} \mathbf{v}_2$$

we have in $\{\mathbf{u}_1, \mathbf{u}_2\}$ an orthonormal basis of R^2 consisting of eigenvectors of A.

Now let U be the matrix whose first column is $\mathbf{u}_1$ and whose second column is $\mathbf{u}_2$; that is, $U = [\mathbf{u}_1, \mathbf{u}_2]$. Now we know that $A\mathbf{u}_1 = 3\mathbf{u}_1$ and $A\mathbf{u}_2 = -1\mathbf{u}_2$, and that

$$A[\mathbf{u}_1, \mathbf{u}_2] = [A\mathbf{u}_1, A\mathbf{u}_2] = [3\mathbf{u}_1, -\mathbf{u}_2] = [\mathbf{u}_1, \mathbf{u}_2] \begin{bmatrix} 3 & 0 \\ 0 & -1 \end{bmatrix} .$$

If we define $D = \begin{bmatrix} 3 & 0 \\ 0 & -1 \end{bmatrix}$ we can rewrite this as

$$AU = UD \quad \text{or} \quad A = UDU^{-1} .$$

Notice that not only is U invertible, but because its columns are orthonormal, U is an orthogonal matrix, and hence $U^{-1} = U^t$. *Thus, our solution of the eigenvalue problem for A has led to a way to write A as the product of an orthogonal matrix U, a diagonal matrix D, and U^{-1}.*

Diagonal matrices are "nice", and orthogonal matrices are "nice". Anytime we have "taken a matrix apart into nice pieces" we have accomplished something of considerable computational value. Was it something very special about this example that made this possible, or is there something more general going on here? Let's investigate this.

The next theorem says that eigenvectors corresponding to different eigenvalues are guaranteed to be linearly independent. That is, as soon as we had determined that we had two different eigenvalues, namely 3 and -1, we could be sure that the corresponding eigenvectors were going to be linearly independent.

Theorem 2: (Distinct Eigenvalues and Independence) *Let A be an $n \times n$ matrix, and let $\{\mathbf{v}_1, \mathbf{v}_2, \ldots, \mathbf{v}_m\}$ be m eigenvectors of A with distinct eigenvalues $\{\mu_1, \mu_2, \ldots, \mu_m\}$. Then $\{\mathbf{v}_1, \mathbf{v}_2, \ldots, \mathbf{v}_m\}$ is linearly independent.*

Proof: Clearly the theorem is true for $m = 1$. Now make the inductive hypothesis that it is true for $m - 1$ eigenvectors with distinct eigenvalues.

Suppose that $a_1\mathbf{v}_1 + a_2\mathbf{v}_2 + \ldots + a_m\mathbf{v}_m = 0$. Then

$$\begin{aligned} 0 &= A(a_1\mathbf{v}_1 + a_2\mathbf{v}_2 + \ldots + a_m\mathbf{v}_m) \\ &= a_1 A\mathbf{v}_1 + a_2 A\mathbf{v}_2 + \ldots + a_m A\mathbf{v}_m \\ &= a_1\mu_1\mathbf{v}_1 + a_2\mu_2\mathbf{v}_2 + \ldots + a_m\mu_m\mathbf{v}_m . \end{aligned} \quad (1.3)$$

Now multiplying $a_1\mathbf{v}_1 + a_2\mathbf{v}_2 + \ldots + a_m\mathbf{v}_m = 0$ through by μ_n, we get $a_1\mu_m\mathbf{v}_1 + a_2\mu_m\mathbf{v}_2 + \ldots + a_m\mu_m\mathbf{v}_m = 0$, and subtracting this from (1.3) we get

$$a_1(\mu_1 - \mu_m)\mathbf{v}_1 + a_2(\mu_2 - \mu_m)\mathbf{v}_2 + \ldots + a_{m-1}(\mu_{m-1} - \mu_m)\mathbf{v}_{m-1} = 0 .$$

The inductive hypothesis implies that $a_j(\mu_j - \mu_m) = 0$ for all $j = 1, 2, \ldots m - 1$. Since the eigenvalues are distinct, $(\mu_j - \mu_m) \neq 0$, and hence $a_j = 0$ for each $j = 1, 2, \ldots, m - 1$.

We then have $a_m \mathbf{v}_m = 0$, and so $a_m = 0$ too. Thus we have seen that $a_j = 0$ for each $j = 1, 2, \ldots, m$, which is what it means for $\{\mathbf{v}_1, \mathbf{v}_2, \ldots, \mathbf{v}_m\}$ to be linearly independent.

∎

Now if A is an $n \times n$ matrix with n distinct eigenvalues $\{\mu_1, \mu_2, \ldots, \mu_n\}$, then there is a basis $\{\mathbf{v}_1, \mathbf{v}_2, \ldots, \mathbf{v}_n\}$ of R^n consisting of eignevectors of A. All we have to do to produce it is to find, for each j, a non-zero solution $\mathbf{v}_j$ of $(A - \mu_j)\mathbf{v} = 0$. Let $\{\mathbf{v}_1, \mathbf{v}_2, \ldots, \mathbf{v}_n\}$ be the set of vectors found this way. By the Theorem 2, this is a set of n linearly independent vectors in R^n, and that means it is a basis.

Not every matrix has enough linearly independent eigenvectors to make a basis, even if complex numbers are admitted:

Example 3 Consider the matrix $A = \begin{bmatrix} 0 & 1 \\ 0 & 0 \end{bmatrix}$. Notice that $A^2 = \begin{bmatrix} 0 & 0 \\ 0 & 0 \end{bmatrix}$. This is something that doesn't happen when we multiply numbers: You never get zero when you square a non-zero number. But, evidently, this can happen with matrices.

Computing

$$\det(A - \mu I) = \det\left(\begin{bmatrix} -\mu & 1 \\ 0 & -\mu \end{bmatrix} \right) = \mu^2 ,$$

we see that the only eigenvalue μ of A is $\mu = 0$. This means that $\mathbf{v}$ is an eigenvector of A if and only if $\mathbf{v}$ is a non-zero vector in the kernel of A. But the kernel of A consists of all non-zero multiples of $\mathbf{v} = \begin{bmatrix} 1 \\ 0 \end{bmatrix}$. Hence the kernel of A is a one dimensional subspace of R^n, and there cannot be a basis of R^n consisting of eigenvectors of A. However, by Theorem 2, every $n \times n$ matrix does have at least as many linearly independent eigenvectors, possibly with complex entries, as its characteristic polynomial has distinct roots.

So for an $n \times n$ matrix A, there may or may not be a basis $\{\mathbf{v}_1, \mathbf{v}_2, \ldots, \mathbf{v}_n\}$ of R^n consisting of eignevectors of A. But when there is, we can form the matrix

$$V = [\mathbf{v}_1, \mathbf{v}_2, \ldots, \mathbf{v}_n]$$

whose jth column is $\mathbf{v}_j$. Then, if $A\mathbf{v}_j = \mu_j \mathbf{v}_j$, we have

$$A[\mathbf{v}_1, \mathbf{v}_2, \ldots, \mathbf{v}_n] = [A\mathbf{v}_1, A\mathbf{v}_2, \ldots, A\mathbf{v}_n] = [\mu_1 \mathbf{v}_1, \mu_2 \mathbf{v}_2, \ldots, \mu_n \mathbf{v}_n] .$$

Now we know that

$$[\mu_1 \mathbf{v}_1, \mu_2 \mathbf{v}_2, \ldots, \mu_n \mathbf{v}_n] = [\mathbf{v}_1, \mathbf{v}_2, \ldots, \mathbf{v}_n] \begin{bmatrix} \mu_1 & 0 & 0 & \ldots & 0 \\ 0 & \mu_2 & 0 & \ldots & 0 \\ \vdots & \vdots & \vdots & \ddots & \vdots \\ 0 & 0 & 0 & \ldots & \mu_n \end{bmatrix} .$$

Therefore, if we define

$$D = \begin{bmatrix} \mu_1 & 0 & 0 & \ldots & 0 \\ 0 & \mu_2 & 0 & \ldots & 0 \\ \vdots & \vdots & \vdots & \ddots & \vdots \\ 0 & 0 & 0 & \ldots & \mu_n \end{bmatrix}$$

we have that

$$AV = VD \ .$$

Next, since the columns of V are independent, V is invertible, and multiplying on the right by V^{-1}, we get

$$A = VDV^{-1} \ .$$

Definition (Diagonalizable): An $n \times n$ matrix A is *diagonalizable* in case there is an invertible $n \times n$ matrix V and a diagonal $n \times n$ matrix D so that $A = VDV^{-1}$.

The next theorem tells us when we can diagonalize a matrix, how to do it, and what this is good for. The "good for" part is the last part. The point is that it is easy to compute powers of diagonal matrices – you just have to apply the power to each of the diagonal elements.

Theorem 3: (Diagonalization and Eigenvectors) *If A is an $n \times n$ matrix for which there is a basis $\{\mathbf{v}_1, \mathbf{v}_2, \ldots, \mathbf{v}_n\}$ of R^n consisting of eignevectors of A, then A is diagonalizable and $A = VDV^{-1}$ where the jth column of V is $\mathbf{v}_j$ and where the jth diagonal entry of D is μ_j, the egienvalue corresponding to $\mathbf{v}_j$.*

Conversely, if $A = VDV^{-1}$ where D is diagonal, then the jth column of V is an eigenvector of A, and the corresponding eigenvalue is the jth diagonal entry of D, and the columns of V form a basis of R^n.

Finally, in case case $A = VDV^{-1}$, where the jth diagonal entry of D is μ_j, one has that for every positive integer k,

$$A^k = VD^kV^{-1} = V \begin{bmatrix} \mu_1^k & 0 & 0 & \ldots & 0 \\ 0 & \mu_2^k & 0 & \ldots & 0 \\ \vdots & \vdots & \vdots & \ddots & \vdots \\ 0 & 0 & 0 & \ldots & \mu_n^k \end{bmatrix} V^{-1} \ .$$

Proof: We have seen above that if there is such a basis, then A is diagonalizable in the manner described in the theorem.

For the converse part, suppose that A is diagonalizable, and $A = VDV^{-1}$. Then $AV = VD$. If $\mathbf{v}_j$ is the jth column of V, and μ_j is the jth diagonal entry of D, then

$$AV = [A\mathbf{v}_1, A\mathbf{v}_2, \ldots, A\mathbf{v}_n] \quad \text{and} \quad VD = [\mu_1\mathbf{v}_1, \mu_2\mathbf{v}_2, \ldots, \mu_n\mathbf{v}_n] \ .$$

Hence $A\mathbf{v}_j = \mu_j\mathbf{v}_j$, so that $\mathbf{v}_j$ is an eigenvector of A. Moreover, since V is invertible, the columns of V are linearly independent, and hence, since there are n of them, they constitute a basis of R^n.

For the final part, notice that

$$A^2 = (VDV^{-1})(VDV^{-1}) = VD(V^{-1}V)DV^{-1} = VD^2V^{-1} \ .$$

Now make the inductive assumption that $A^{k-1} = VD^{k-1}V^{-1}$. Then

$$A^k = AA^{k-1} = (VDV^{-1})(VD^{k-1}V^{-1})$$
$$= VD(V^{-1}V)D^{k-1}V^{-1} = VDD^{k-1}V^{-1}$$
$$= VD^kV^{-1} \ .$$

We can use this theorem to compute functions of diagonalizable matrices.

Example 5: Let $A = \begin{bmatrix} 4 & 3 \\ 2 & 3 \end{bmatrix}$ Consider the problem of finding a matrix B with

$$B^2 = A \ .$$

Such a matrix is a "square root" of A. How can we go about solving this equation for B? Diagonalization of course! We compute

$$\det(A - \mu I) = (4 - \mu)(3 - \mu) - 6$$
$$= \mu^2 - 7\mu + 6$$
$$= (\mu - 6)(\mu - 1) \ .$$

The eigenvalues are $\mu_1 = 6$ and $\mu_2 = 1$. Since there are two distinct eigenvalues, there will be two linearly independent eigenvectors, and so A will be invertible.

Now let's find the eigenvectors. $A - 6I = \begin{bmatrix} -2 & 3 \\ 2 & -3 \end{bmatrix}$. Evidently

$$\mathbf{v}_1 = \begin{bmatrix} 3 \\ 2 \end{bmatrix}$$

is orthogonal to the rows, and is in the kernel. So that is one eigenvector.

Next, $A - I = \begin{bmatrix} 3 & 3 \\ 2 & 2 \end{bmatrix}$. Evidently

$$\mathbf{v}_2 = \begin{bmatrix} -1 \\ 1 \end{bmatrix}$$

is orthogonal to the rows, and is in the kernel. So that is the other eigenvector. Hence we have

$$V = \begin{bmatrix} 3 & -1 \\ 2 & 1 \end{bmatrix} \quad \text{and} \quad D = \begin{bmatrix} 6 & 0 \\ 0 & 1 \end{bmatrix} \ .$$

We easily compute that

$$V^{-1} = \frac{1}{5} \begin{bmatrix} 1 & 1 \\ -2 & 3 \end{bmatrix} .$$

Now it is easy to find a square root of D:

$$\begin{bmatrix} \sqrt{6} & 0 \\ 0 & 1 \end{bmatrix} \begin{bmatrix} \sqrt{6} & 0 \\ 0 & 1 \end{bmatrix} = \begin{bmatrix} 6 & 0 \\ 0 & 1 \end{bmatrix} ,$$

and hence

$$V \begin{bmatrix} \sqrt{6} & 0 \\ 0 & 1 \end{bmatrix} V^{-1} V \begin{bmatrix} \sqrt{6} & 0 \\ 0 & 1 \end{bmatrix} V^{-1} = V \begin{bmatrix} 6 & 0 \\ 0 & 1 \end{bmatrix} V^{-1} = A ,$$

Therefore, we define

$$B = V \begin{bmatrix} \sqrt{6} & 0 \\ 0 & 1 \end{bmatrix} V^{-1} = \frac{1}{5} \begin{bmatrix} 3 & -1 \\ 2 & 1 \end{bmatrix} \begin{bmatrix} \sqrt{6} & 0 \\ 0 & 1 \end{bmatrix} \begin{bmatrix} 1 & 1 \\ -2 & 3 \end{bmatrix}$$

$$= \frac{1}{5} \begin{bmatrix} 3\sqrt{6} + 2 & 3\sqrt{6} - 3 \\ 2\sqrt{6} - 2 & 2\sqrt{6} + 3 \end{bmatrix} .$$

You can multiply B by itself, and verify that indeed $B^2 = A$. This example is fundamental. Make sure you understand every detail.

So now we know how to take the square roots of diagonalizable matrices. There are other functions, besides the square root function, that it is even more useful to consider – the exponential function in particular. As we will see, if you can diagonalize a matrix, you can exponentiate it, and this is very important for solving systems of linear differential equations.

Section 2: Vector Valued Functions and their Derivatives

Consider a vector valued function $\mathbf{v}(t)$ of the real variable t with values in R^n. For example, let's consider $n = 3$, and

$$\mathbf{v}(t) = \begin{bmatrix} \cos(t) \\ \sin(t) \\ 1/t \end{bmatrix} . \tag{2.1}$$

Here is a three dimensional plot of the curve traced out by $\mathbf{v}(t)$ as t varies from $t = 1$ to $t = 20$:

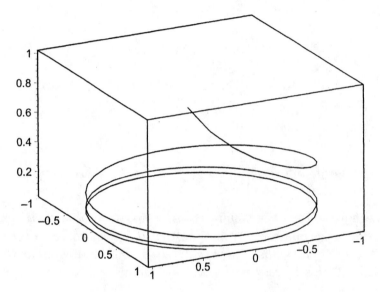

Such vector valued functions arise whenever we need to describe the position of a particle as a function of time. But more generally, we might have any sort of system that is described by n parameters. These could be, for example, the voltages across n points in an electric circuit. We can arrange this data into a vector, and if the data is varying with time, as is often the case in applications, we then have a time dependent vector $\mathbf{v}(t)$ in R^n.

When quantities are varying in time, it is often useful to consider their rates of change; i.e., derivatives.

Definition (Derivatives of Vector Valued Functions) Let $\mathbf{v}(t)$ be a vector valued function of the variable t. We say that $\mathbf{v}(t)$ is *differentiable at* $t = t_0$ with derivative $\mathbf{v}'(t_0)$ in case

$$\lim_{h \to 0} \frac{1}{h} \left(\mathbf{v}(t_0 + h) - \mathbf{v}(t_0) \right) = \mathbf{v}'(t_0)$$

in the sense that this limit exists for each of the n entries separately. A vector valued function is *differentiable* in some interval (a, b) if it is differentiable for each t_0 in (a, b).

There is nothing new really going on here. To compute the derivative of $\mathbf{v}(t)$, you just differentiate it entry by entry in the usual way.

Example 1 Let $\mathbf{v}(t)$ be given by (2.1). Then for any $t \neq 0$,

$$\mathbf{v}'(t) = \begin{bmatrix} -\sin(t) \\ \cos(t) \\ -1/t^2 \end{bmatrix} . \tag{2.2}$$

We will need to know just a few things about derivatives of vector valued functions in this chapter, and these are summarized in the next two theorems.

Theorem 1 *Suppose that $\mathbf{v}(t)$ is a differentiable vector valued function for t in (a, b) with values in R^n. Then for any $m \times n$ matrix A, $A\mathbf{v}(t)$ is a differentiable vector valued function for t in (a, b) with values in R^m, and moreover,*

$$(A\mathbf{v}(t))' = A\mathbf{v}'(t) \tag{2.3}$$

for each t in (a, b).

In other words, multiplying by A first and then differentiating yields the same result as differentiating first and then multiplying by A.

Proof: For each i with $i = 1, 2, \ldots, m$,

$$\lim_{h \to 0} \frac{1}{h} \left(A\mathbf{v}(t+h) - A\mathbf{v}(t) \right)_i = \lim_{h \to 0} \frac{1}{h} \left(\sum_{j=1}^{n} A_{i,j} v_j(t+h) - \sum_{j=1}^{n} A_{i,j} v_j(t) \right)$$

$$= \lim_{h \to 0} \frac{1}{h} \sum_{j=1}^{n} A_{i,j} (v_j(t+h) - v_j(t))$$

$$= \sum_{j=1}^{n} A_{i,j} \left(\lim_{h \to 0} (v_j(t+h) - v_j(t)) \right)$$

$$= \sum_{j=1}^{n} A_{i,j} v_j'(t)$$

$$= (A\mathbf{v}'(t))_i .$$

■

The second result we will need is a "product rule" for the dot product.

Theorem 2: *Suppose that $\mathbf{v}(t)$ and $\mathbf{w}(t)$ are differentiable vector valued functions for t in (a, b) with values in R^n. Then $\mathbf{v}(t) \cdot \mathbf{w}(t)$ is differentiable for t in (a, b), and*

$$\frac{\mathrm{d}}{\mathrm{d}t} \mathbf{v}(t) \cdot \mathbf{w}(t) = \mathbf{v}'(t) \cdot \mathbf{w}(t) + \mathbf{v}(t) \cdot \mathbf{w}'(t) . \tag{2.4}$$

Proof: For each i, we have by the usual product rule

$$\frac{d}{dt}v_i(t)w_i(t) = v_i'(t)w_i(t) + v_i(t)w_i'(t) \ .$$

Summing on i now gives us (2.4). ∎

Systems of differential equations and the superposition principle

Consider the system of differential equations

$$x'(t) = 4x(t) + 3y(t)$$
$$y'(t) = 2x(t) + 3y(t) \tag{2.5}$$

with the initial conditions $x(0) = 1$ and $y(0) = 4$. We can use what we know about eigenvectors to find a pair of functions satisfying all these conditions, which is what it means to *solve this system of differential equations for the given initial data*. Here is how: Introduce the vector $\mathbf{x}(t) = \begin{bmatrix} x(t) \\ y(y) \end{bmatrix}$ and the matrix $A = \begin{bmatrix} 4 & 3 \\ 2 & 3 \end{bmatrix}$. Then we can rewrite (2.5) as

$$\mathbf{x}'(t) = A\mathbf{x}(t) \ . \tag{2.6}$$

We saw in Example 5 of the previous section that

$$A\mathbf{v}_1 = 6\mathbf{v}_1 \qquad \text{and} \qquad A\mathbf{v}_2 = \mathbf{v}_2$$

where

$$\mathbf{v}_1 = \begin{bmatrix} 3 \\ 2 \end{bmatrix} \qquad \text{and} \qquad \mathbf{v}_2 = \begin{bmatrix} -1 \\ 1 \end{bmatrix} \ .$$

Now define

$$\mathbf{v}_1(t) = e^{6t}\mathbf{v}_1 = e^{6t} \begin{bmatrix} 3 \\ 2 \end{bmatrix} \ ,$$

and

$$\mathbf{v}_2(t) = e^t\mathbf{v}_2 = e^t \begin{bmatrix} -1 \\ 1 \end{bmatrix} \ .$$

Then,

$$\mathbf{v}_1'(t) = e^{6t}6\mathbf{v}_1 = e^{6t}A\mathbf{v}_1 = A(e^{6t}\mathbf{v}_1) = A\mathbf{v}_1(t) \ .$$

Thus, $\mathbf{v}_1(t)$ is a solution of our system of differential equations (2.6). In the same way, we see that $\mathbf{v}_2(t)$ is also a solution.

Now let's see if we can take a linear combination of these solutions to get a solution that satisfies the initial data $x(0) = 2$ and $y(0) = 3$. Since $\mathbf{v}_1(0) = \mathbf{v}_1$ and $\mathbf{v}_2(0) = \mathbf{v}_2$, all we have to do is to choose a and b so that

$$a\mathbf{v}_1 + b\mathbf{v}_2 = \begin{bmatrix} x(0) \\ y(0) \end{bmatrix} = \begin{bmatrix} 1 \\ 4 \end{bmatrix}. \tag{2.7}$$

In the case at hand, you see that $a = 1$ and $b = 2$ works. Since $\mathbf{v}_1$ and $\mathbf{v}_2$ we would have been able to solve for a and b no matter what the right hand side was.

Now form the *superposition*

$$\mathbf{x}(t) = a\mathbf{v}_1(t) + b\mathbf{v}_2(t) = \mathbf{v}_1(t) + 2\mathbf{v}_2(t) = e^{6t} \begin{bmatrix} 3 \\ 2 \end{bmatrix} + 2e^t \begin{bmatrix} -1 \\ 1 \end{bmatrix}. \tag{2.8}$$

Since,

$$\begin{aligned} \mathbf{x}'(t) &= (a\mathbf{v}_1(t) + \mathbf{v}_2(t))' \\ &= a\mathbf{v}_1'(t) + b\mathbf{v}_2'(t) \\ &= aA\mathbf{v}_1(t) + bA\mathbf{v}_2(t) \\ &= A(a\mathbf{v}_1(t) + b\mathbf{v}_2(t)) \\ &= A\mathbf{x}(t) , \end{aligned}$$

$\mathbf{x}(t)$, as given in (2.8) satisfies both the equation (2.6), and the initial condition $\begin{bmatrix} x(0) \\ y(0) \end{bmatrix} = \begin{bmatrix} 1 \\ 4 \end{bmatrix}$.

Writing $\mathbf{x}(t)$ out explicitly, we have

$$\mathbf{x}(t) = \begin{bmatrix} 3e^{6t} - 2e^t \\ 2e^{6t} + 2e^t \end{bmatrix}$$

Hence the pair of functions

$$x(t) = 3e^{6t} - 2e^t$$

and

$$y(t) = 2e^{6t} + 2e^t$$

solves (2.5) and the initial conditions $x(0) = 1$ and $y(0) = 4$. We will soon see that this is the only solution, but for now we want to focus on this method of finding solutions of systems of differential equations. The idea behind this example is called the *superposition principle*.

It can be applied in great generality. Suppose A is an $n \times n$ matrix with n linearly independent eigenvectors $\mathbf{v}_1, \mathbf{v}_2, \ldots, \mathbf{v}_n$. Let $\mu_1, \mu_2, \ldots, \mu_n$ be the corresponding eigenvalues. Then for each j,

$$v_j(t) = e^{\mu_j t}\mathbf{v}_j$$

solves the vector differential equation

$$\mathbf{x}'(t) = A\mathbf{x}(t) \tag{2.9}$$

just as in the example above. Since $\{\mathbf{v}_1, \mathbf{v}_2, \ldots, \mathbf{v}_n\}$ is a basis of $I\!R^n$, we can choose an appropriate linear combination of these solutions to match any given initial condition. Just as in the example, this linear combination is still a solution of (2.9), and so we have a solution that has the given initial data. You will get practice with this in the exercises, and see how effective it is!

Section 3: Diagonalization of Symmetric Matrices

There is a particularly important case in which we can be sure that there is a basis of eigenvectors – the case in which $A = A^t$. Moreover, in this case, all of the eigenvalues and eigenvectors are real; there is no need to bring complex numbers into the game to diagonalize symmetric matrices. Explaining this is the main goal of this section.

A key to this is provided by Theorem 1.5.3, relating the dot product and the transpose: Recall that if A is an $n \times n$ matrix, and $\mathbf{v}$ and $\mathbf{w}$ are any two vectors in R^n, then

$$\mathbf{v} \cdot (A\mathbf{w}) = (A^t\mathbf{v}) \cdot \mathbf{w} . \tag{3.1}$$

Here is our first use of this:

Theorem 1: (Distinct Eigenvalues and Orthogonality) *Let A be a symmetric $n \times n$ matrix. If $\mathbf{v}$ and $\mathbf{w}$ are eigenvectors of A with distinct eigenvalies, then $\mathbf{v}$ and $\mathbf{w}$ are orthogonal.*

Proof: Consider two eigenvectors $\mathbf{v}$ and $\mathbf{w}$ of A corresponding to distinct eigenvalues μ and ν respectively. Then

$$\nu\mathbf{v} \cdot \mathbf{w} = \mathbf{v} \cdot (A\mathbf{w}) = (A^t\mathbf{v}) \cdot \mathbf{w} = (A\mathbf{v}) \cdot \mathbf{w} = \mu\mathbf{v} \cdot \mathbf{w} ,$$

and hence

$$(\nu - \mu)\mathbf{v} \cdot \mathbf{w} = 0 .$$

By hypothesis, $\nu - \mu \neq 0$, and hence $\mathbf{v} \cdot \mathbf{w} = 0$. ■

With these simple results in hand, we are ready for the main point. *For symmetric matrices, the eigenvalue problem is a maximization–minimization problem.* This point deserves a careful explanation, and it is the basis of several methods for finding eigenvectors and eigenvalues that work in real applications. (Computing the characteristic polynomial, and then trying to factor it is generally not an effective strategy for larger matrices.) We begin with a careful analysis of the planar case, $n = 2$, where everything can be clearly visualized.

Consider a 2×2 symmetric matrix

$$A = \begin{bmatrix} a & b \\ b & d \end{bmatrix} . \tag{3.2}$$

Consider the time dependent vector

$$\mathbf{v}(t) = \begin{bmatrix} \cos(t) \\ \sin(t) \end{bmatrix} . \tag{3.3}$$

Notice that for each t, $|\mathbf{v}(t)| = 1$. As t varies, $\mathbf{v}(t)$ traces out the unit circle.

Now put the two together and consider the function $f(t)$ given by

$$f(t) = \mathbf{v}(t) \cdot A\mathbf{v}(t) \ . \tag{3.4}$$

We ask the following question: *For what values of t does $f(t)$ take on its maximum and minimum values?* The question is properly formulated since computing $f(t)$, we find

$$f(t) = a\cos^2(t) + c\sin^2(t) + 2b\sin(t)\cos(t)$$

$$= \left(\frac{a+d}{2}\right) + \left(\frac{a-d}{2}\right)(\cos^2(t) - c\sin^2(t)) + 2b\sin(t)\cos(t)$$

$$= \left(\frac{a+d}{2}\right) + \left(\frac{a-d}{2}\right)\cos(2t) + b\sin(2t) \ .$$

Evidently, $f(t)$ is a continuous function of t that is periodic with period π, which means that that is $f(t+\pi) = f(t)$ for all t. Under these circumstances, $f(t)$ takes on both its maximum and minimum somewhere in the interval $[0, \pi)$.

To find out *where*, we differentiate:

$$f'(t) = -(a-d)\sin(2t) + 2b\cos(2t)$$

and so for $b \neq 0$

$$f'(t_0) = 0 \quad \Longleftrightarrow \quad \tan(2t_0) = \frac{a-d}{2b} \ . \tag{3.5}$$

(If $b = 0$, A is diagonal.) If we graph the function $y = \tan(2t)$ for $0 \le t < \pi$, we get

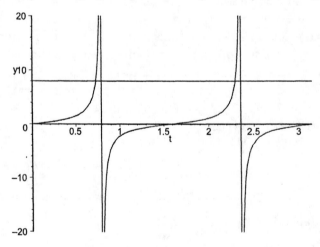

The horizontal line, shown here at $y = 8$, represents the value $(a - d)/2b$. No matter what this value is, you see the horizontal line crosses the graph of $y = \tan(2t)$ exactly twice for t in $[0, \pi)$, once in $[0, \pi/2)$, and once again in $[\pi/2, \pi)$. (The two vertical lines are asymptotes and not part of the graph, and of course, we have to exclude the case $b = 0$, in which case A is already diagonal.)

By (3.5), the two values of t at which the graph of $y = \tan(2t)$ crosses the line $y = (a - d)/2b$ are the two solutions to $f'(t) = 0$. One gives the minimum of $f(t)$, and the other gives the maximum. Let t_0 be the first solution; that is, the one in $[0, \pi/2)$. Then, since $\tan(2t)$ is periodic with period $\pi/2$, the second solution occurs at $t_0 + \pi/2$. We have

$$f'(t_0) = f'(t_0 + \pi/2) = 0 . \tag{3.6}$$

Now by (2.3) and (2.4),

$$f'(t_0) = \mathbf{v}'(t_0) \cdot A\mathbf{v}(t_0) + \mathbf{v}(t_0) \cdot A\mathbf{v}'(t_0) . \tag{3.7}$$

By (3.1) and the symmetry of A, $\mathbf{v}(t_0) \cdot A\mathbf{v}'(t_0) = A\mathbf{v}(t_0) \cdot \mathbf{v}'(t_0)$, and so by (3.6) and (3.7),

$$\mathbf{v}'(t_0) \cdot A\mathbf{v}(t_0) = 0 , \tag{3.8}$$

so that $\mathbf{v}'(t_0)$ and $A\mathbf{v}(t_0)$ are orthogonal.

Next notice that

$$\mathbf{v}'(t) \cdot \mathbf{v}(t) = 0 ,$$

and that $\mathbf{v}'(t)$ is also a unit vector for each t. Indeed, by direct computation,

$$\mathbf{v}'(t) = \begin{bmatrix} -\sin(t) \\ \cos(t) \end{bmatrix} = \begin{bmatrix} \cos(t + \pi/2) \\ \sin(t + \pi/2) \end{bmatrix} = \mathbf{v}(t_0 + \pi/2) . \tag{3.9}$$

It follows that if we define the pair of vectors $\{\mathbf{u}_1, \mathbf{u}_2\}$ by

$$\mathbf{u}_1 = \mathbf{v}(t_0) \qquad \text{and} \qquad \mathbf{u}_2 = \mathbf{v}'(t_0) ,$$

we get an orthonormal basis for R^2. Moreover, can rewrite (3.8) as $\mathbf{u}_2 \cdot A\mathbf{u}_1 = 0$, and by (3.1) again, this means $A\mathbf{u}_2 \cdot \mathbf{u}_1 = 0$. Hence:

$$\mathbf{u}_2 \cdot A\mathbf{u}_1 = 0 \qquad \text{and} \qquad \mathbf{u}_1 \cdot A\mathbf{u}_2 = 0 . \tag{3.10}$$

We know how to work out coordinates with respect to an orthonormal basis, and so let's work out the coordinates for $A\mathbf{u}_1$:

$$A\mathbf{u}_1 = (\mathbf{u}_1 \cdot A\mathbf{u}_1)\mathbf{u}_1 + (\mathbf{u}_2 \cdot A\mathbf{u}_1)\mathbf{u}_2 . \tag{3.11}$$

Combining this with (3.10), we see that

$$A\mathbf{u}_1 = (\mathbf{u}_1 \cdot A\mathbf{u}_1)\mathbf{u}_1 . \tag{3.12}$$

This says that $\mathbf{u}_1$ is an eigenvector of A with eigenvalue

$$\mu_1 = \mathbf{u}_1 \cdot A\mathbf{u}_1 = \mathbf{v}(t_0) \cdot A\mathbf{v}(t_0) = f(t_0) . \tag{3.13}$$

In the same way,

$$A\mathbf{u}_2 = (\mathbf{u}_1 \cdot A\mathbf{u}_2)\mathbf{u}_1 + (\mathbf{u}_2 \cdot A\mathbf{u}_2)\mathbf{u}_2 , \tag{3.14}$$

and by (3.10) this simplifies to

$$A\mathbf{u}_2 = (\mathbf{u}_2 \cdot A\mathbf{u}_2)\mathbf{u}_2 , \tag{3.15}$$

which says that $\mathbf{u}_2$ is an eigenvector of A with eigenvalue

$$\mu_2 = \mathbf{u}_2 \cdot A\mathbf{u}_2 = \mathbf{v}(t_0 + \pi/2) \cdot A\mathbf{v}(t_0 + \pi/2) = f(t_0 + \pi/2) . \tag{3.16}$$

We have found an orthonormal basis of eigenvectors of A by seeking the maximum and minimum of the function $f(t)$ defined in (3.3). *Moreover, the eigenvalues are exactly the maximum and minimum values of $f(t)$, namely $f(t_0)$ and $f(t_0+\pi/2)$.* The same approach works in n dimensions, with a few small twists.

Theorem 2: (Eigenvalues and Maximization) *Let A be a symmetric $n \times n$ matrix. There is a unit vector $\mathbf{v}$ in R^n so that*

$$\mathbf{w} \cdot A\mathbf{w} \leq \mathbf{v} \cdot A\mathbf{v} . \tag{3.17}$$

Moreover, $\mu = \mathbf{v} \cdot A\mathbf{v}$ is an eigenvalue of A, and $\mathbf{v}$ is an eigenvector with $A\mathbf{v} = \mu\mathbf{v}$.

This theorem says that by considering the problem of maximizing $\mathbf{w} \cdot A\mathbf{w}$ over all possible unit vectors $\mathbf{w}$, we find at *least one* eigenvector and eigenvalue, both of which are real, and not complex.

Proof: Suppose that a maximizing vector $\mathbf{v}$, as in (3.17) does exist. We'll postpone this part of the proof to the appendix, and concentrate on the understanding why (3.17) says that $\mathbf{v}$ is an eigenvector of A with eigenvalue $\mu = \mathbf{v} \cdot A\mathbf{v}$.

Let $\mathbf{u}$ be any unit vector orthogonal to $\mathbf{v}$. Then for any t, $|\mathbf{v}+t\mathbf{u}|^2 = (\mathbf{v}+t\mathbf{u}) \cdot (\mathbf{v}+t\mathbf{u}) = 1 + t^2$, since $\mathbf{v}$ and $\mathbf{u}$ are orthonormal. Therefore, if we define $c(t) = (1 + t^2)^{-1/2}$ and $\mathbf{w}(t) = c(t)(\mathbf{v} + t\mathbf{u})$, we have that $\mathbf{w}(t)$ is a unit vector for each t, and that $\mathbf{w}(0) = \mathbf{v}$. Then from (3.17), we have

$$\mathbf{w}(t) \cdot A\mathbf{w}(t) \leq \mathbf{w}(0) \cdot A\mathbf{w}(0) \qquad \text{for all} \quad t . \tag{3.18}$$

This means that the function $f(t) = \mathbf{w}(t) \cdot A\mathbf{w}(t)$ has a maximum at $t = 0$. Differentiating,

$$0 = f'(0) = \mathbf{w}'(0) \cdot A\mathbf{w}(0) + \mathbf{w}(0) \cdot A\mathbf{w}'(0) = 2\mathbf{w}'(0) \cdot A\mathbf{w}(0) . \tag{3.19}$$

We already know that $\mathbf{w}(0) = \mathbf{v}$. And by the definition of $\mathbf{w}(t)$, $\mathbf{w}'(t) = c'(t)(\mathbf{v}+t\mathbf{u})+c(t)\mathbf{u}$. Since $c'(0) = 0$, we get $\mathbf{w}'(0) = u$. Hence (3.19) tells us that

$$\mathbf{u} \cdot A\mathbf{v} = 0 . \tag{3.20}$$

The point is that (3.20) holds for *any* **u** that is orthogonal to **v**.

Now consider an orthonormal basis $\{\mathbf{u}_1, \mathbf{u}_2, \ldots, \mathbf{u}_n\}$ of R^n such that $\mathbf{u}_1 = \mathbf{v}$. Then by (3.20),

$$\mathbf{u}_j \cdot A\mathbf{v} = 0 \qquad \text{for} \quad j = 2, 3, \ldots, n . \tag{3.21}$$

Computing the coordinates of $A\mathbf{v}$ in the basis $\{\mathbf{u}_1, \mathbf{u}_2, \ldots, \mathbf{u}_n\}$ in the usual way , and using (3.21) we have

$$A\mathbf{v} = \sum_{j=1}^{n} (\mathbf{u}_j \cdot A\mathbf{v})\mathbf{u}_j = (\mathbf{u}_1 \cdot A\mathbf{v})\mathbf{u}_1 .$$

Since $\mathbf{u}_1 = \mathbf{v}$ by definition of the basis, $A\mathbf{v} = (\mathbf{v} \cdot A\mathbf{v})\mathbf{v}$, which is what we had to show. ∎

We now come to the main result of this section.

Theorem 3: (Diagonalizability of Symmetric Matrices) *Let A be a symmetric* $n \times n$ *matrix. Then there is an orthonormal basis* $\{\mathbf{u}_1, \mathbf{u}_2, \ldots, \mathbf{u}_n\}$ *of* R^n *consisting of eigenvectors of A*

Proof: Let ℓ be any integer with $1 \leq \ell < n$, and suppose that we have already found ℓ orthonormal $\{\mathbf{u}_1, \mathbf{u}_2, \ldots, \mathbf{u}_\ell\}$ eigenvectors of A. The previous theorem tell us how to find the first one, so we can always get to this point. Let S be the span of $\{\mathbf{u}_1, \mathbf{u}_2, \ldots, \mathbf{u}_\ell\}$, which is evidently an ℓ dimensional subspace of R^n. Therefore, $S^\perp$ is an $n - \ell$ dimensional subspace of R^n. Let $\{\mathbf{v}_1, \mathbf{v}_2, \ldots, \mathbf{v}_{n-\ell}\}$ be an orthonormal basis for $S^\perp$. (We are not assuming anything about these vectors being eigenvectors – just that they are some orthonormal basis. Any such basis will do.)

Let **w** be any vector in $S^\perp$. Suppose that $(x_1, x_2, \ldots, x_{n-\ell})$ are its coordinates with respect to our basis, so that

$$\mathbf{w} = \sum_{j=1}^{n-\ell} x_j \mathbf{v}_j . \tag{3.22}$$

By eqv(4SD1), $A\mathbf{w}$ is orthogonal to each $\mathbf{u}_j$ for $j \leq \ell$. Indeed,

$$(A\mathbf{w}) \cdot \mathbf{u}_j = \mathbf{w} \cdot (A\mathbf{u}_j) = \mu_j \mathbf{w} \cdot \mathbf{u}_j = 0$$

since **w** belongs to $S^\perp$. This shows that $A\mathbf{w}$ also belongs to $S^\perp$, and so it has a coordinate expansion

$$A\mathbf{w} = \sum_{i=1}^{n-\ell} y_i \mathbf{v}_i . \tag{3.23}$$

Let $\mathbf{x} = \begin{bmatrix} x_1 \\ x_2 \\ \vdots \\ x_{n-\ell} \end{bmatrix}$ and $\mathbf{y} = \begin{bmatrix} y_1 \\ y_2 \\ \vdots \\ y_{n-\ell} \end{bmatrix}$. We can find the relation between **x** and **y** by applying A to (3.22):

$$A\mathbf{w} = \sum_{j=1}^{n-\ell} x_j (A\mathbf{v}_j) .$$ (3.24)

Now for each $i = 1, 2 \ldots, n - \ell$,

$$A\mathbf{v}_j = \sum_{i=1}^{n-\ell} (\mathbf{v}_i \cdot A\mathbf{v}_j)\mathbf{v}_i .$$

Substituting this into (3.24),

$$A\mathbf{w} = \sum_{j=1}^{n-\ell} x_j \left(\sum_{i=1}^{n-\ell} (\mathbf{v}_i \cdot A\mathbf{u}_j)\mathbf{v}_i \right) = \sum_{i=1}^{n-\ell} \left(\sum_{j=1}^{n-\ell} B_{i,j} x_j \right) \mathbf{v}_i ,$$ (3.25)

where

$$B_{i,j} = \mathbf{v}_i \cdot A\mathbf{u}_j .$$ (3.26)

Comparing (3.25) and (3.23), we see that $\mathbf{y} = B\mathbf{x}$ where B is the $(n - \ell) \times (n - \ell)$ matrix with entries given by (3.26). Now, (3.1) says that B is a symmetric matrix. Theorem 4 says that we can find at least one eigenvalue μ of B and one unit vector $\mathbf{x}$ such that $B\mathbf{x} = \mu\mathbf{x}$. Now let $\mathbf{w}$ be vector in R^n given in terms of $\mathbf{x}$ by (3.22). Then with $\mathbf{y} = B\mathbf{x} = \mu\mathbf{x}$,

$$A\mathbf{w} = \sum_{i=1}^{n-\ell} y_i \mathbf{v}_i = \mu \left(\sum_{i=1}^{n-\ell} x_i \mathbf{v}_i \right) = \mu\mathbf{w} .$$

Therefore, μ is an eigenvalue of A, and $\mathbf{w}$ satisfies $A\mathbf{w} = \mu\mathbf{w}$. Also, since $\{\mathbf{v}_1, \mathbf{v}_2, \ldots, \mathbf{v}_{n-\ell}\}$ is an orthonormal basis of $S^\perp$,

$$|\mathbf{w}|^2 = \sum_{j=1}^{n-\ell} x_j^2 = 1$$

so $\mathbf{w}$ is a unit vector. Finally, since it is in $S^\perp$, it is orthogonal to each vector in $\{\mathbf{u}_1, \mathbf{u}_2, \ldots, \mathbf{u}_\ell\}$.

Therefore, we can define $\mathbf{u}_{\ell+1} = \mathbf{w}$, and we have produced a set $\{\mathbf{u}_1, \mathbf{u}_2, \ldots, \mathbf{u}_\ell, \mathbf{u}_{\ell+1}\}$ of $\ell + 1$ orthonormal eigenvectors of A. Repeating as necessary, we arrive at a full basis $\{\mathbf{u}_1, \mathbf{u}_2, \ldots, \mathbf{u}_n\}$. ■

The ideas of this section have an important application to non–symmetric matrices, and even non–square matrices. Let A be any $m \times n$ matrix. The maximum value of $|A\mathbf{v}|$ as $\mathbf{v}$ ranges over all of the unit vectors on R^n is a measure of how much "streching" A does. It is the maximum ratio of the "output length" to the "input length", and has a name:

Definition (Norm of a Matrix) For any $m \times n$ matrix A, the norm of A, denoted $\|A\|$, is defined by

$$\|A\| = \text{least upper bound of } \{|A\mathbf{v}| \; : \; \mathbf{v} \text{ in } R^n \, , \quad |\mathbf{v}| = 1\} \; .$$

Theorem 4 (Norms and Eigenvalues) $m \times n$ *matrix* A, $\|A\| = \sqrt{\mu_1}$ *where μ_1 is the largest eigenvalue of $A^t A$.*

Proof: For any unit vectro $\mathbf{v}$ in R^n, note that

$$|A\mathbf{v}|^2 = A\mathbf{v} \cdot A\mathbf{v} = \mathbf{v} \cdot (A^t A)\mathbf{v} \; .$$

The result now follows directly from Theorem 4. ∎

Example 1 *Compute the norm of* $A = \begin{bmatrix} 1 & 2 \\ 3 & 4 \end{bmatrix}$. We compute $A^t A = \begin{bmatrix} 10 & 14 \\ 14 & 20 \end{bmatrix}$, which has the characteristic polynomial $\mu^2 - 3 - \mu + 4$. The two eigenvalues are then $15 \pm \sqrt{221}$, and hence

$$\|A\| = \sqrt{15 + \sqrt{221}} \; .$$

Section 4: The Exponential of a Square Matrix

Consider the differential equation

$$x'(t) = ax(t) \quad , \quad x(0) = x_0,$$

$$(4.1)$$

where a is a number. The function

$$x(t) = \exp(at)x_0$$

$$(4.2)$$

satisfies (4.1) That is, if we differentiate, we find

$$\frac{d}{dt} \exp(at)x_0 = a \exp(at)x_0 ,$$

and $x(0) = x_0$, which is what it means for $x(t) = \exp(at)x_0$ to satisfy (4.1).

Very often in applications, we must consider systems of linear differential equations. The simplest case concern two unknown functions of t, $x(t)$ and $y(t)$, and the system of equations has the form

$$x'(t) = ax(t) + by(t) \quad , \quad x(0) = x_0$$
$$y'(t) = cx(t) + dy(t) \quad , \quad y(0) = y_0 .$$

$$(4.3)$$

Just as we can think of systems of linear algebraic equations as a single algebraic equation for a vector valued variable $\mathbf{x}$, we can think of this system of linear differential equations as a single equation for a vector valued variable $\mathbf{x}(t)$. Let's introduce

$$\mathbf{x}(t) = \begin{bmatrix} x(t) \\ y(t) \end{bmatrix} \quad \text{and} \quad \mathbf{x}_0 = \begin{bmatrix} x_0 \\ y_0 \end{bmatrix} .$$

Then (4.3) can be rewritten as

$$\mathbf{x}'(t) = A\mathbf{x}(t) \quad , \quad \mathbf{x}(0) = \mathbf{x}_0 .$$

$$(4.4)$$

This form brings out the similarity to (4.1). Does the solution look anything like (4.2)? Yes it does. It turns out that it is possible to makes sense of the exponential of a square matrix, and that (4.4) is solved by

$$\mathbf{x}(t) = e^{tA}\mathbf{x}_0 .$$

Recall that for numbers a,

$$e^{ta} = \sum_{k=0}^{\infty} \frac{1}{k!}(ta)^k .$$

This suggests that for an $n \times n$ matrix A, we should define

$$e^{tA} = \sum_{k=0}^{\infty} \frac{1}{k!}(t)^k A^k \ .$$
(4.5)

We interpret the infinite sum on the right hand side as

$$\lim_{N \to \infty} \sum_{k=0}^{N} \frac{1}{k!}(t)^k A^k \ .$$
(4.6)

For each fixed N, the matrix on the right hand side of (4.6) is a well–defined $n \times n$ matrix. We say that a sequence of matrices converges to a limit if for *each* i and j, the corresponding numerical sequence of the entries converges. It is just the familiar issue of convergence for numerical sequences, with the matrix considered one entry at a time.

Now, for which matrices A does the limit in (4.6) exist, so that e^{tA} can be defined by (4.5)? The answer is that the limit in (4.6) exists for *every* $n \times n$ matrix A. However, we will only be able to compute a closed form expression for it in special cases, the most important being when A is diagonalizable. So what we will actually show is that the limit in (4.6) converges whenever A is diagonalizable, and so e^{tA} is defined in this case. Moreover, we will see that the matrix exponential behaves like an exponential should, namely:

$$e^{sA}e^{tA} = e^{(s+t)A} \ ,$$
(4.7)

$$\lim_{t \to 0} e^{tA} = I \ ,$$
(4.8)

and

$$\frac{\mathrm{d}}{\mathrm{d}t}e^{tA} = Ae^{tA} \ .$$
(4.9)

As soon as we have seen that (4.9) holds, we can define

$$\mathbf{x}(t) = e^{tA}\mathbf{x}_0 \ ,$$
(4.10)

and then

$$\mathbf{x}'(t) = \left(\frac{\mathrm{d}}{\mathrm{d}t}e^{tA}\right)\mathbf{x}_0 = Ae^{tA}\mathbf{x}_0 = A\mathbf{x}(t)$$
(4.11)

and moreover, once we have (4.8), we have

$$\lim_{t \to 0} \mathbf{x}(t) = \left(\lim_{t \to 0} e^{tA}\right)\mathbf{x}_0 = \mathbf{x}_0 \ .$$
(4.12)

Therefore, according to (4.12), $\mathbf{x}(t)$ as defined in (4.10) satsifies the differential equation, and has the right initial condition.

If A is diagonalizable, it is easy not only to verify (4.7), (4.8) and (4.9), but to compute a closed form expression for e^{tA}. Here is how this works.

Recall that when A is diagonalizaible, there is an invertible matrix V whose columns are eigenvectors of A, and a diagonal matrix D, whose diagonal entries are the corresponding eigenvalues of A, so that $A = VDV^{-1}$. We have also seen that in this case,

$$A^k = VD^kV^{-1} \ .$$

But then for any N,

$$\sum_{k=0}^{N} \frac{1}{k!}(t)^k A^k = \sum_{k=0}^{N} \frac{1}{k!}(t)^k VD^kV^{-1} = V\left(\sum_{k=0}^{N} \frac{1}{k!}(t)^k D^k\right) V^{-1} \ .$$

Now, if $D = \begin{bmatrix} \mu_1 & 0 & 0 & \dots & 0 \\ 0 & \mu_2 & 0 & \dots & 0 \\ \vdots & \vdots & \vdots & \ddots & \vdots \\ 0 & 0 & 0 & \dots & \mu_n \end{bmatrix}$, then $D^k = \begin{bmatrix} \mu_1^k & 0 & 0 & \dots & 0 \\ 0 & \mu_2^k & 0 & \dots & 0 \\ \vdots & \vdots & \vdots & \ddots & \vdots \\ 0 & 0 & 0 & \dots & \mu_n^k \end{bmatrix}$. Therefore,

$$\sum_{k=0}^{N} \frac{1}{k!}(t)^k D^k = \begin{bmatrix} \sum_{k=0}^{N}(t^k\mu_1^k/k!) & 0 & 0 & \dots & 0 \\ 0 & \sum_{k=0}^{N}(t^k\mu_2^k/k!) & 0 & \dots & 0 \\ \vdots & \vdots & \vdots & \ddots & \vdots \\ 0 & 0 & 0 & \dots & \sum_{k=0}^{N}(t^k\mu_n^k/k!) \end{bmatrix} \ .$$

It is easy to take the limit entry by entry since, for example,

$$\lim_{N\to\infty} \sum_{k=0}^{N}(t^k\mu_1^k/k!) = e^{t\mu_1} \ .$$

The result is that

$$\sum_{k=0}^{\infty} \frac{1}{k!}(t)^k D^k = \lim_{N\to\infty} \sum_{k=0}^{N} \frac{1}{k!}(t)^k D^k = \begin{bmatrix} e^{t\mu_1} & 0 & 0 & \dots & 0 \\ 0 & e^{t\mu_2} & 0 & \dots & 0 \\ \vdots & \vdots & \vdots & \ddots & \vdots \\ 0 & 0 & 0 & \dots & e^{t\mu_n} \end{bmatrix} \ .$$

Finally, we have that

$$e^{tA} = V \begin{bmatrix} e^{t\mu_1} & 0 & 0 & \dots & 0 \\ 0 & e^{t\mu_2} & 0 & \dots & 0 \\ \vdots & \vdots & \vdots & \ddots & \vdots \\ 0 & 0 & 0 & \dots & e^{t\mu_n} \end{bmatrix} V^{-1} \ .$$

With this in hand it is clear that

$$
e^{sA}e^{tA} = V \begin{bmatrix} e^{s\mu_1} & 0 & 0 & \dots & 0 \\ 0 & e^{s\mu_2} & 0 & \dots & 0 \\ \vdots & \vdots & \vdots & \ddots & \vdots \\ 0 & 0 & 0 & \dots & e^{s\mu_n} \end{bmatrix} V^{-1} V \begin{bmatrix} e^{t\mu_1} & 0 & 0 & \dots & 0 \\ 0 & e^{t\mu_2} & 0 & \dots & 0 \\ \vdots & \vdots & \vdots & \ddots & \vdots \\ 0 & 0 & 0 & \dots & e^{t\mu_n} \end{bmatrix} V^{-1}
$$

$$
= V \begin{bmatrix} e^{s\mu_1} & 0 & 0 & \dots & 0 \\ 0 & e^{s\mu_2} & 0 & \dots & 0 \\ \vdots & \vdots & \vdots & \ddots & \vdots \\ 0 & 0 & 0 & \dots & e^{s\mu_n} \end{bmatrix} \begin{bmatrix} e^{t\mu_1} & 0 & 0 & \dots & 0 \\ 0 & e^{t\mu_2} & 0 & \dots & 0 \\ \vdots & \vdots & \vdots & \ddots & \vdots \\ 0 & 0 & 0 & \dots & e^{t\mu_n} \end{bmatrix} V^{-1}
$$

$$
= V \begin{bmatrix} e^{(s+t)\mu_1} & 0 & 0 & \dots & 0 \\ 0 & e^{(s+t)\mu_2} & 0 & \dots & 0 \\ \vdots & & \vdots & \ddots & \vdots \\ 0 & 0 & 0 & \dots & e^{(s+t)\mu_n} \end{bmatrix} V^{-1}
$$

$$
= e^{(s+t)A} .
$$

Next,

$$
\frac{\mathrm{d}}{\mathrm{d}t} e^{tA} = V \left(\frac{\mathrm{d}}{\mathrm{d}t} \begin{bmatrix} e^{t\mu_1} & 0 & 0 & \dots & 0 \\ 0 & e^{t\mu_2} & 0 & \dots & 0 \\ \vdots & \vdots & \vdots & \ddots & \vdots \\ 0 & 0 & 0 & \dots & e^{t\mu_n} \end{bmatrix} \right) V^{-1}
$$

$$
= V \begin{bmatrix} \mu_1 e^{t\mu_1} & 0 & 0 & \dots & 0 \\ 0 & \mu_2 e^{t\mu_2} & 0 & \dots & 0 \\ \vdots & \vdots & \vdots & \ddots & \vdots \\ 0 & 0 & 0 & \dots & \mu_n e^{t\mu_n} \end{bmatrix} V^{-1}
$$

$$
= V \begin{bmatrix} \mu_1 & 0 & 0 & \dots & 0 \\ 0 & \mu_2 & 0 & \dots & 0 \\ \vdots & \vdots & \vdots & \ddots & \vdots \\ 0 & 0 & 0 & \dots & \mu_n \end{bmatrix} \begin{bmatrix} e^{t\mu_1} & 0 & 0 & \dots & 0 \\ 0 & e^{t\mu_2} & 0 & \dots & 0 \\ \vdots & \vdots & \vdots & \ddots & \vdots \\ 0 & 0 & 0 & \dots & e^{t\mu_n} \end{bmatrix} V^{-1}
$$

$$
= V \begin{bmatrix} \mu_1 & 0 & 0 & \dots & 0 \\ 0 & \mu_2 & 0 & \dots & 0 \\ \vdots & \vdots & \vdots & \ddots & \vdots \\ 0 & 0 & 0 & \dots & \mu_n \end{bmatrix} V V^{-1} \begin{bmatrix} e^{t\mu_1} & 0 & 0 & \dots & 0 \\ 0 & e^{t\mu_2} & 0 & \dots & 0 \\ \vdots & \vdots & \vdots & \ddots & \vdots \\ 0 & 0 & 0 & \dots & e^{t\mu_n} \end{bmatrix} V^{-1}
$$

$$
= A e^{tA} .
$$

It is an even simpler matter to verify (4.8):

$$\lim_{t \to 0} e^{tA} = V \left(\lim_{t \to 0} \begin{bmatrix} e^{t\mu_1} & 0 & 0 & \cdots & 0 \\ 0 & e^{t\mu_2} & 0 & \cdots & 0 \\ \vdots & \vdots & \vdots & \ddots & \vdots \\ 0 & 0 & 0 & \cdots & e^{t\mu_n} \end{bmatrix} \right) V^{-1} = VIV^{-1} = I .$$

We have proved the following result:

Theorem 1: (Matrix Exponentials) *Let A be an $n \times n$ diagonalizable matrix. Then for any t, e^{tA}, defined as in (4.5) exists and satisfies (4.7), (4.8) and (4.9). Moreover if $A = VDV^{-1}$ where*

$$D = \begin{bmatrix} \mu_1 & 0 & 0 & \cdots & 0 \\ 0 & \mu_2 & 0 & \cdots & 0 \\ \vdots & \vdots & \vdots & \ddots & \vdots \\ 0 & 0 & 0 & \cdots & \mu_n \end{bmatrix} ,$$

we have

$$e^{tA} = V \begin{bmatrix} e^{t\mu_1} & 0 & 0 & \cdots & 0 \\ 0 & e^{t\mu_2} & 0 & \cdots & 0 \\ \vdots & \vdots & \vdots & \ddots & \vdots \\ 0 & 0 & 0 & \cdots & e^{t\mu_n} \end{bmatrix} V^{-1} .$$

Finally, if $\mathbf{x}_0$ is any vector in R^n, and we define $\mathbf{x}(t) = e^{At}\mathbf{x}_0$, then $\mathbf{x}(t)$ solves the initial value problem

$$\mathbf{x}'(t) = A\mathbf{x}(t) \quad , \quad \mathbf{x}(0) = \mathbf{x}_0 .$$

Example 1 Consider the matrix $A = \begin{bmatrix} 1 & 2 \\ 2 & 1 \end{bmatrix}$ from Example 1 of Section 1. We found in Section 1 that $A = VDV^{-1}$ where

$$V = \frac{1}{\sqrt{2}} \begin{bmatrix} 1 & -1 \\ 1 & 1 \end{bmatrix} \quad \text{and} \quad D = \begin{bmatrix} 3 & 0 \\ 0 & -1 \end{bmatrix} .$$

Since V is orthogonal, V^{-1} is the transpose of V, and we have

$$\begin{aligned} e^t A &= \frac{1}{2} \begin{bmatrix} 1 & -1 \\ 1 & 1 \end{bmatrix} \begin{bmatrix} e^{3t} & 0 \\ 0 & e^{-t} \end{bmatrix} \begin{bmatrix} 1 & 1 \\ -1 & 1 \end{bmatrix} \\ &= \frac{1}{2} \begin{bmatrix} 1 & -1 \\ 1 & 1 \end{bmatrix} \begin{bmatrix} e^{3t} & e^{3t} \\ -e^{-t} & e^{-t} \end{bmatrix} \\ &= \frac{1}{2} \begin{bmatrix} e^{3t} + e^{-t} & e^{3t} - e^{-t} \\ e^{3t} - e^{-t} & e^{3t} + e^{-t} \end{bmatrix} . \end{aligned}$$

Now let $\mathbf{x}_0 = \begin{bmatrix} 2 \\ 1 \end{bmatrix}$. Then

$$e^{tA}\mathbf{x}_0 = \frac{1}{2} \begin{bmatrix} e^{3t} + e^{-t} & e^{3t} - e^{-t} \\ e^{3t} - e^{-t} & e^{3t} + e^{-t} \end{bmatrix} \begin{bmatrix} 2 \\ 1 \end{bmatrix} = \frac{1}{2} \begin{bmatrix} 3e^{3t} + e^{-t} \\ 3e^{3t} - e^{-t} \end{bmatrix}$$

Now you can easily check that with

$$x(t) = \frac{1}{2}(3e^{3t} + e^{-t}) \qquad \text{and} \qquad y(t) = \frac{1}{2}(3e^{3t} - e^{-t}) \,,$$

$$x'(t) = x(t) + 2y(t) \qquad , \qquad x(0) = 2$$
$$y'(t) = 2x(t) + y(t) \qquad , \qquad y(0) = 1$$

and so we have solved this system of linear differential equations. As t varies, $\mathbf{x}(t) = e^{tA}\mathbf{x}_0$ traces out a curce in the x, y plane. Here is a graph of that curve for $-2 \le t \le 1/4$.

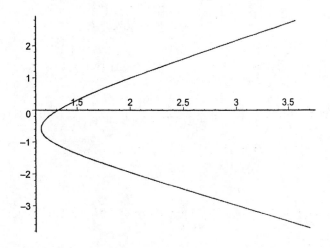

As you see, the curve passes through the point $(2, 1)$. Also, when t is positive and large, e^{-t} is negligible compares to e^{3t}, and so for such t $x(t) \approx y(t) \approx e^{3t}$. Therefore, we expect the curve to be asymptotic to the line $y = x$ for large t. Indeed, we see this already near $t = 1/4$! (Notice the different scales on the x and y axes.) Also, when t is negative and large, e^{3t} is negligible compares to e^{-t}, and so for such t $x(t) \approx -y(t) \approx e^{-t}$. Therefore, we expect the curve to be asymptotic to the line $y = -x$ for such t. Again this is what we see.

Example 2 Let's find the solution to

$$x'(t) = x(t) + 3y(t) \qquad , \qquad x(0) = -4$$
$$y'(t) = 2x(t) \qquad\qquad , \qquad y(0) = 5$$

In matrix form the differential equation is $\mathbf{x}'(t) = A\mathbf{x}(t)$ where

$$A = \begin{bmatrix} 1 & 3 \\ 2 & 0 \end{bmatrix} .$$

This matrix differs form the matrix in Example 4 of Section 1 by a multiple of the identity, and hence it has the same eigenvectors. From the results of that example, we have $A = VDV^{-1}$ where

$$V = \begin{bmatrix} 3 & -1 \\ 2 & 1 \end{bmatrix} \qquad D = \begin{bmatrix} 3 & 0 \\ 0 & -2 \end{bmatrix} \qquad \text{and} \qquad V^{-1} = \frac{1}{5}\begin{bmatrix} 1 & 1 \\ -2 & 3 \end{bmatrix} .$$

(Notice that this time, since A was not symmetric, the inverse was not simply the transpose.) Therefore

$$e^{tA} = \frac{1}{5}\begin{bmatrix} 3 & -1 \\ 2 & 1 \end{bmatrix}\begin{bmatrix} e^{3t} & 0 \\ 0 & e^{-2t} \end{bmatrix}\begin{bmatrix} 1 & 1 \\ -2 & 3 \end{bmatrix}$$

$$= \frac{1}{5}\begin{bmatrix} 3 & -1 \\ 2 & 1 \end{bmatrix}\begin{bmatrix} e^{3t} & e^{3t} \\ -2e^{-2t} & 3e^{-2t} \end{bmatrix}$$

$$= \frac{1}{5}\begin{bmatrix} 3e^{3t} + 2e^{-2t} & 3e^{3t} - 3e^{-2t} \\ 2e^{3t} - 2e^{-2t} & 2e^{3t} + 3e^{-2t} \end{bmatrix} .$$

The data for $x(0)$ and $y(0)$ tells us that $\mathbf{x}_0 = \begin{bmatrix} -4 \\ 5 \end{bmatrix}$, and our solution is

$$\mathbf{x}(t) = e^{tA}\mathbf{x}_0 = \frac{1}{5}\begin{bmatrix} 3e^{3t} + 2e^{-2t} & 3e^{3t} - 3e^{-2t} \\ 2e^{3t} - 2e^{-2t} & 2e^{3t} + 3e^{-2t} \end{bmatrix}\begin{bmatrix} -4 \\ 5 \end{bmatrix}$$

$$= \frac{1}{5}\begin{bmatrix} 3e^{3t} - 23e^{-2t} \\ 2e^{3t} + 23e^{-2t} \end{bmatrix} .$$

The solution therefore is

$$x(t) = \frac{1}{5}(3e^{3t} - 23e^{-2t}) \qquad \text{and} \qquad y(t) = \frac{1}{5}(2e^{3t} + 23e^{-2t}) .$$

For $-1/2 \le t \le 1$, this traces out the following curve:

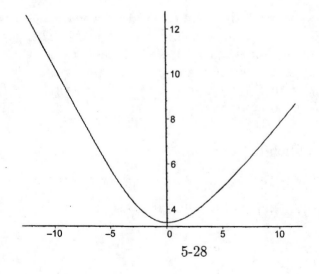

Again we see the negative eigenvalue dominating for negative times, and the positive eigenvalue dominating for positive times, with a transition between the two asymptotic behaviors near $t = 0$.

We can use these ideas to solve an even more important class of systems of linear differential equations: Let A be an $n \times n$ matrix, and consider the vector differential equation

$$\mathbf{x}'(t) = A\mathbf{x}(t) + \mathbf{r}(t) , \tag{4.13}$$

where $\mathbf{r}(t)$ is a given time dependent vector. Let's suppose that there is a solution $\mathbf{x}(t)$ that satisfies the initial condition $\mathbf{x}(t0) = \mathbf{x}_0$, and try to find a formula for it. Introduce a new variable $\mathbf{z}(t)$ by

$$\mathbf{z}(t) = e^{-tA}\mathbf{x}(t) . \tag{4.14}$$

Then we have

$$z'(t) = -Ae^{-tA}\mathbf{x}(t) + e^{-tA}\mathbf{x}'(t) . \tag{4.15}$$

Using (4.13) to eliminate $\mathbf{x}'(t)$ from (4.15), we obtain

$$\begin{aligned}
z'(t) &= -Ae^{-tA}\mathbf{x}(t) + e^{-tA}(A\mathbf{x}(t) + \mathbf{r}(t)) \\
&= -Ae^{-tA}\mathbf{x}(t) + Ae^{-tA}\mathbf{x}(t) + e^{-tA}\mathbf{r}(t) \\
&= e^{-tA}\mathbf{r}(t) .
\end{aligned} \tag{4.16}$$

It follows that

$$\mathbf{z}(t) - \mathbf{z}(0) = \int_0^t e^{-sA}\mathbf{r}(s)\mathrm{d}s .$$

By (4.14), this is the same as

$$e^{-tA}\mathbf{x}(t) = \mathbf{x}_0 + \int_0^t e^{-sA}\mathbf{r}(s)\mathrm{d}s ,$$

or

$$\begin{aligned}
\mathbf{x}(t) &= e^{tA}\mathbf{x}_0 + e^{tA}\left(\int_0^t e^{-sA}\mathbf{r}(s)\mathrm{d}s\right) \\
&= e^{tA}\mathbf{x}_0 + \int_0^t e^{(t-s)A}\mathbf{r}(s)\mathrm{d}s .
\end{aligned} \tag{4.17}$$

It is now easy to differentiate and verify that the vector valued function $\mathbf{x}(t)$ defined by (4.17) does indeed satisfy (4.13) with the initial condition $\mathbf{x}(0) = \mathbf{x}_0$.

Example 3 *Solve the system differential equations*

$$x'(t) = -x(y) + y(t) + \sin(2t) \quad \text{and} \quad y(t) = -x(y) - y(t) , \tag{4.18}$$

with the initial conditions $x(0) = 3$ and $y(0) = 2$.

We let $\mathbf{x}(t) = \begin{bmatrix} x(t) \\ y(t) \end{bmatrix}$ and $\mathbf{x}_0 = \begin{bmatrix} 3 \\ 2 \end{bmatrix}$ as usual, and define

$$A = \begin{bmatrix} -1 & 1 \\ -1 & -1 \end{bmatrix} \quad \text{and} \quad \mathbf{r}(t) = \begin{bmatrix} \sin(2t) \\ 0 \end{bmatrix} .$$

with these definitions, (4.18) has the form (4.13), and hence is solved by (4.17).

We leave it as an exercise to compute that in this case

$$e^{tA} = e^{-t} \begin{bmatrix} \cos(t) & \sin(t) \\ -\sin(t) & \cos(t) \end{bmatrix} .$$

(This is a case in which the eigenvalues and eigenvectors of A are complex. However, our final expression for e^{tA} is real, as it must be.) Now, in this case

$$\int_0^t e^{(t-s)A} \mathbf{r}(s) \mathrm{d}s = \begin{bmatrix} \int_0^t e^{s-t} \cos(s-t) \sin(2s) \mathrm{d}s \\ \int_0^t e^{s-t} \sin(s-t) \sin(2s) \mathrm{d}s \end{bmatrix} .$$

The integrals may look a bit messy, but aren't so bad, and can be done on a computer anyhow. The results are:

$$\int_0^t e^{s-t} \cos(s-t) \sin(2s) \mathrm{d}s = \frac{2}{5} - \frac{4}{5} \cos^2(t) + \frac{3}{10} \sin(2t) + \frac{2}{5} e^{-t} \cos(t) - \frac{1}{5} e^{-t} \sin(t) ,$$

and

$$\int_0^t e^{s-t} \sin(s-t) \sin(2s) \mathrm{d}s = -\frac{1}{5} + \frac{2}{5} \cos^2(t) + \frac{1}{10} \sin(2t) - \frac{1}{5} e^{-t} \cos(t) - \frac{2}{5} e^{-t} \sin(t) .$$

It is now easy to add on $e^{tA} \mathbf{x}_0$, and we obtain our solutions

$$x(t) = 3e^{-t} \cos(t) + \frac{2}{5} - \frac{4}{5} \cos^2(t) + \frac{3}{10} \sin(2t) + \frac{2}{5} e^{-t} \cos(t) - \frac{1}{5} e^{-t} \sin(t) , \quad (4.19)$$

and

$$y(t) = -2e^{-t} \sin(t) - \frac{1}{5} + \frac{2}{5} \cos^2(t) + \frac{1}{10} \sin(2t) - \frac{1}{5} e^{-t} \cos(t) - \frac{2}{5} e^{-t} \sin(t) . \quad (4.20)$$

Notice that for large t, all of the terms containing a factor of e^{-t} are negligible, and so we have

$$x(t) \approx \frac{2}{5} - \frac{4}{5} \cos^2(t) + \frac{3}{10} \sin(2t) , \quad (4.21)$$

and

$$y(t) \approx -\frac{1}{5} + \frac{2}{5}\cos^2(t) + \frac{1}{10}\sin(2t) \ . \tag{4.22}$$

Here is a graph if the curve traced out by $\mathbf{x}(t)$ for $0 \leq t \leq 20$:

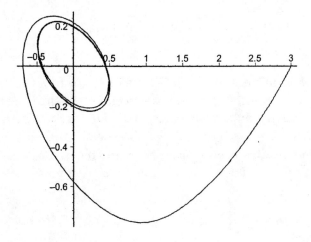

And here is a graph for $40 \leq t \leq 50$:

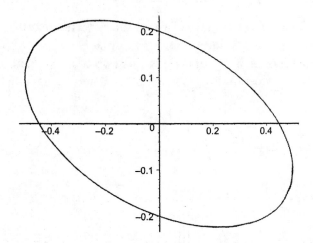

Notice that the solution seems to have settled into a "steady state" behavior of running around a fixed ellipse for large t. In fact it is easy to see that $e^{tA}\mathbf{x}_0$ tends rapidly to zero as t increases, so the dependence of the solution on the initial data is quickly washed out. The curve you see in the second graph represents the limiting "steady state" behavior *independent of the initial data.* This is a fundamental example that you should consider very carefully. The computations themselves are not too bad, especially if you use a computer to do them. But make sure you understand what computations you need to ask your computer to do if you need to solve a problem like this!

Section 5: Quadratic Forms

Definition (Quadratic Form on R^2) A function $f(x, y)$ of the two real variables x and y is called a *quadratic form* if

$$f(x, y) = ax^2 + 2bxy + cy^2$$

for some numbers a, b and c.

The term "quadratic form" is a bit archaic, and has its roots in the time before the modern mathematical notion of a "function" was well established. A modern translation of the term would be "purely quadratic function". The archaic term has the virtue of brevity, and has stuck.

Up until now we have mostly been concerned with linear functions. Quadratic forms are decidedly non–linear, but still there is a connection with matrices. Define

$$\mathbf{x} = \begin{bmatrix} x \\ y \end{bmatrix} \quad \text{and} \quad A = \begin{bmatrix} a & b \\ b & c \end{bmatrix} \ .$$

Then

$$f(x, y) = \mathbf{x} \cdot A\mathbf{x} \ .$$

We've met quadratic forms already when we discussed the maximization approach to finding eigenvalues and eigenvectors of symmetric matrices.

We have learned that since A is symmetric it has two orthonormal eigenvectors $\mathbf{u}_1$ and $\mathbf{u}_2$. Let μ_1 and μ_2 be the corresponding eigenvalues. Then we have

$$A = UDU^t$$

where

$$U = [\mathbf{u}_1, \mathbf{u}_2] \quad \text{and} \quad D = \begin{bmatrix} \mu_1 & 0 \\ 0 & \mu_2 \end{bmatrix} \ .$$

Therefore,

$$f(x, y) = \mathbf{x} \cdot A\mathbf{x} = \mathbf{x} \cdot UDU^t\mathbf{x} = (U^t\mathbf{x}) \cdot D(U^t\mathbf{x}) \ .$$

Now define $\tilde{\mathbf{x}} = \begin{bmatrix} \tilde{x} \\ \tilde{y} \end{bmatrix}$ by

$$\begin{bmatrix} \tilde{x} \\ \tilde{y} \end{bmatrix} = U^t \begin{bmatrix} x \\ y \end{bmatrix} \ .$$

Then

$$(U^t\mathbf{x}) \cdot D(U^t\mathbf{x}) = \begin{bmatrix} \tilde{x} \\ \tilde{y} \end{bmatrix} \cdot \begin{bmatrix} \mu_1 & 0 \\ 0 & \mu_2 \end{bmatrix} \begin{bmatrix} \tilde{x} \\ \tilde{y} \end{bmatrix} = \mu_1 \tilde{x}^2 + \mu_2 \tilde{y}^2 \ .$$

Define $g(\tilde{x}, \tilde{y}) = \mu_1 \tilde{x}^2 + \mu_2 \tilde{y}^2$ which is a quadratic form in the variables $\tilde{x}$ and $\tilde{y}$. As we see, when the variables $\tilde{x}$ and $\tilde{y}$ are related to x and y as above,

$$f(x, y) = g(\tilde{x}, \tilde{y}) \ .$$

The change of variables from x and y to $\tilde{x}$ and $\tilde{y}$ has eliminated the "cross term" involving both variables. Our quadratic form is now a linear combination of squares. With the cross term eliminates it is much easier to understand "what our function is doing".

Suppose that μ_1 and μ_2 are both strictly positive. Then $g(\tilde{x}, \tilde{y})$ satisfies

$$0 = g(0, 0) \leq g(\tilde{x}, \tilde{y})$$

with equality if and only if $\tilde{x} = 0$ and $\tilde{y} = 0$. This means that

$$0 = f(0, 0) \leq f(x, y)$$

with equality if and only if $x = 0$ and $y = 0$.

Definition (Positive and Negative Definite) A quadratic form $f(x, y)$ is *positive definite* in case $0 = f(0, 0) \leq f(x, y)$ with equality if and only if $x = 0$ and $y = 0$, and is *negative definite* in case $0 = f(0, 0) \geq f(x, y)$ with equality if and only if $x = 0$ and $y = 0$. A quadratic form $f(x, y)$ is *positive semidefinite* if $f(x, y) \geq 0$ for all x and y, but $f(x, y) = 0$ has solutions other than $x = y = 0$. Similarly, A quadratic form $f(x, y)$ is *positive semidefinite* if $f(x, y) \leq 0$ for all x and y, but $f(x, y) = 0$ has solutions other than $x = y = 0$.

Example 1 Consider the quadratic form

$$f(x, y) = x^2 + 4xy + y^2 = \begin{bmatrix} x \\ y \end{bmatrix} \cdot \begin{bmatrix} 1 & 2 \\ 2 & 1 \end{bmatrix} \begin{bmatrix} x \\ y \end{bmatrix} \ .$$

The characteristic polynomial of $\begin{bmatrix} 1 & 2 \\ 2 & 1 \end{bmatrix}$ is $(1 - \mu)^2 - 4 = \mu^2 - 2\mu - 3 = (\mu - 3)(\mu + 1)$.

The eigenvalues are therefore 3 and -1. One then easily finds that

$$\mathbf{u}_1 = \frac{1}{\sqrt{2}} \begin{bmatrix} 1 \\ 1 \end{bmatrix} \qquad \text{and} \qquad \mathbf{u}_2 = \frac{1}{\sqrt{2}} \begin{bmatrix} -1 \\ 1 \end{bmatrix} \qquad\qquad (5.1)$$

are an orthonormal basis of eigenvectors so that

$$A = U \begin{bmatrix} 3 & 0 \\ 0 & -1 \end{bmatrix} U^t$$

where

$$U = \frac{1}{\sqrt{2}} \begin{bmatrix} 1 & -1 \\ 1 & 1 \end{bmatrix} \ .$$

We then define

$$\begin{bmatrix} \tilde{x} \\ \tilde{y} \end{bmatrix} = \frac{1}{\sqrt{2}} \begin{bmatrix} 1 & 1 \\ -1 & 1 \end{bmatrix} \begin{bmatrix} x \\ y \end{bmatrix} ,$$

which means

$$\tilde{x} = \frac{x+y}{\sqrt{2}} \quad \text{and} \quad \tilde{y} = \frac{y-x}{\sqrt{2}} . \tag{5.2}$$

In these new coordinates, our quadratic form becomes

$$3\tilde{x}^2 - \tilde{y}^2 .$$

Indeed, we can make the substitution and check

$$
\begin{aligned}
3\tilde{x}^2 - \tilde{y}^2 &= 3\left(\frac{x+y}{\sqrt{2}}\right)^2 - \left(\frac{y-x}{\sqrt{2}}\right)^2 \\
&= \frac{3}{2}(x^2 + 2xy + y^2) - \frac{1}{2}(x^2 - 2xy + y^2) \\
&= x^2 + 4xy + y^2 .
\end{aligned}
$$

In the new variables $\tilde{x}$ and $\tilde{y}$, it is immediately clear that our quadratic form is sometimes strictly, and sometimes strictly positive, and is therefore neither positive definite nor negative definite. This is less clear in the original coordinates.

Definition (Hyperbolic) A quadratic form $f(x,y)$ is *hyperbolic* in case there are values of x and y for which $f(x,y) > 0$, and values for which $f(x,y) < 0$.

The quadratic form of Example 1 is hyperbolic. The terminology suggests that there is a geometric way of looking at quadratic forms, and indeed there is.

Let's take x and y to be the coordinates of a point in the plane as usual, and suppose that $z = f(x,y)$ is the height of a surface over the point (x,y) in the plane. That is, the surface consists of those points (x,y,z) in R^3 for which $z = f(x,y)$. we can then think of $f(x,y)$ as giving the "altitude" of the surface over the point (x,y) in the plane.

Now fix any number c, and consider the set of points (x,y) for which

$$f(x,y) = c .$$

The set of points (x,y) that solves this equation is exactly the set of points over which the altitude of the surface is c. Hence we call this set the *level set at altitude c*. It is much easier to understand what this set is if we use the $\tilde{x}$ and $\tilde{y}$ coordinates, in which the defining equation becomes

$$3\tilde{x}^2 - \tilde{y}^2 = c .$$

For each value of c, this is the equation of an hyperbola.

Here is a plot showing a number of these hyperbolas:

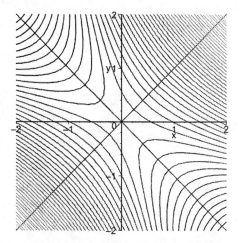

The two diagonal lines are the axes in the $\tilde{x}, \tilde{y}$ coordinate system. The $\tilde{x}$ axis, given by $\tilde{y} = 0$, is, according to (5.2), the line $y = x$. Consequently, the $\tilde{y}$ axis is the line $y = -x$. Here is a three dimensional rendering of the surface $z = f(x,y)$:

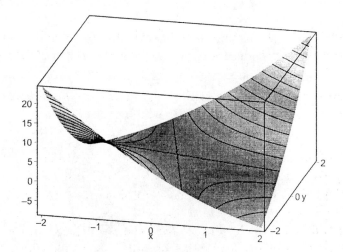

Such a surface is often described as "saddle shaped".

Example 2 *Consider*

$$f(x,y) = 2x^2 + 2xy + 2y^2 \, , \tag{5.3}$$

and determine whether its type. One easily checks that $f(x,y) = \mathbf{x} \cdot A\mathbf{x}$ for $A = \begin{bmatrix} 2 & 1 \\ 1 & 2 \end{bmatrix}$.

the eigenvalues of A are 3 and 1, and the eigenvectors are again $\mathbf{u}_1$ and $\mathbf{u}_2$ as specified in (5.1). Hence we have, using the exact same corodinate change used in Example 1, that

$$g(\tilde{x}, \tilde{y}) = 3\tilde{x}^2 + \tilde{y}^2 \, .$$

5-35

Clearly this quadratic form is positive definite. To find level sets, we have to solve

$$3\tilde{x}^2 + \tilde{y}^2 = c \ ,$$

which is evidently only possible if $c \geq 0$, in which case, the solution is an ellipse. Here is a plot showing a number of these ellipses:

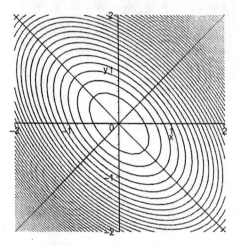

The two diagonal lines are the axes in the $\tilde{x}, \tilde{y}$ coordinate system. Here is a three dimensional rendering of the surface $z = f(x, y)$:

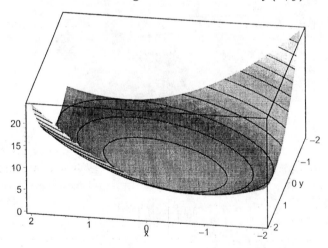

Finally, consider $f(x, y) = x^2 + 2xy + y^2$. We recognize in this case that $f(x, y) = (x+y)^2$, so that $f(x, y) \geq 0$ for all x and y, but $f(x, y) = 0$ whenever $y = -x$. hence this quadratic form is positive semidefinite.

Here is a three dimensional rendering of the surface $z = f(x, y)$ for this case:

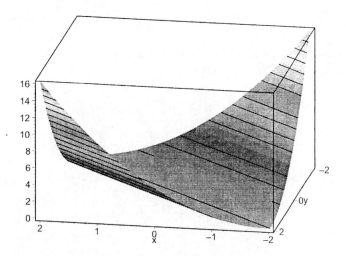

You notice that there is no saddle, just a "trough".

Determining the type of a quadratic form is important in the theory of finding minimum and maximum values of functions of two or more variables. All of the definitions made here extend easily to more than two variables – it is just that with more variables, it is not easy to draw enlightening graphs. But determining the type of a quadratic form comes down to determining the signs of the eigenvalues of the matrix giving its coefficients. we have the following theorem:

Theorem 1: *Let* $f(x_1, x_2, \ldots, x_n)$ *be a quadratic form in the* n *real variables* $x_1, x_2, \ldots, x_n$ *with*

$$f(x_1, x_2, \ldots, x_n) \begin{bmatrix} x_1 \\ x_2 \\ \vdots \\ x_n \end{bmatrix} \cdot A \begin{bmatrix} x_1 \\ x_2 \\ \vdots \\ x_n \end{bmatrix} ,$$

where A *isa symmetric* $n \times n$ *matrix. Then the quadratic form is positive definite if and only if all of the eigenvalues of* A *are strictly positive, and it is positive semidefinite if and only if all of the eigenvalues of* A *are non—negative, and at least one is zero. It is negative definite if and only if all of the eigenvalues of* A *are strictly negative, and it is negative semidefinite if and only if all of the eigenvalues of* A *are non—positive, and at least one is zero. Finally, it is hyperbolic if at least one eigenvalue is strictly positive, and one is strictly negative.*

Problems for Section 1

5.1.1 Let A be the matrix

$$A = \begin{bmatrix} 1 & 1 & 0 \\ 1 & 0 & 1 \\ 0 & 1 & 1 \end{bmatrix} .$$

(a) Find the characteristic polynomial of A, and find all of its roots.

(b) Find the eigenvalues of A, and for each eigenvalue give a basis for the corresponding eigenspace.

(c) Find an orthogonal matrix U and a diagonal matrix D so that

$$A = UDU^{-1} .$$

5.1.2 Let A be the matrix

$$A = \begin{bmatrix} -3 & 1 & 1 \\ 1 & -3 & 1 \\ 1 & 1 & -1 \end{bmatrix} .$$

(a) Find the characteristic polynomial of A, and find all of its roots.

(b) Find the eigenvalues of A, and for each eigenvalue give a basis for the corresponding eigenspace.

(c) Find an orthogonal matrix U and a diagonal matrix D so that

$$A = UDU^{-1} .$$

5.1.3 Let A be the matrix

$$A = \begin{bmatrix} 1 & 2 \\ 2 & 4 \end{bmatrix} .$$

(a) Find the characteristic polynomial of A, and find all of its roots.

(b) Find the eigenvalues of A, and for each eigenvalue give a basis for the corresponding eigenspace.

(c) Find an orthogonal matrix U and a diagonal matrix D so that

$$A = UDU^{-1} .$$

5.1.4 Consider the following matrices:

$$A = \begin{bmatrix} 1 & -2 & 0 \\ 1 & 2 & 0 \\ 0 & 2 & -1 \end{bmatrix}$$

$$B = \begin{bmatrix} 1 & 2 & 3 \\ 0 & 3 & 3 \\ 0 & 0 & -3 \end{bmatrix}$$

(a) Find all of the eigenvalues of A

(b) Find all of the eigenvalues of B

(c) Is B dianoalizable? Justify your answer.

(d) What is the largest eigenvalue of B^4, and what is the dimension of the corresponding eigenspace?

5.1.5 Consider the following matrices:

$$A = \begin{bmatrix} 0 & 1 \\ 0 & 0 \end{bmatrix} \qquad B = \begin{bmatrix} 1 & 1/2 \\ 2 & 1 \end{bmatrix}$$

$$C = \begin{bmatrix} 2 & 1 \\ 3 & 3 \end{bmatrix} \qquad D = \begin{bmatrix} 2 & 3 \\ 1 & 0 \end{bmatrix}$$

(a) Which of these matrices, if any, have only a single eigenvalue, and which have two eigenvalues?

(b) Which of these matrices, if any, can be diagonalized by a change of basis?

(c) Compute $D^2 - 2D - 3I$ where I is the two by two identity matrix.

(d) Compute B^{15}

5.1.6 Consider the following matrices:

$$A = \begin{bmatrix} 2 & 1 & 0 & 0 \\ 1 & 2 & 0 & 0 \\ 0 & 0 & 3 & 1 \\ 0 & 0 & 1 & 3 \end{bmatrix}$$

$$B = \begin{bmatrix} 1 & 0 & 0 \\ 2 & 3 & 0 \\ 1 & 3 & 3 \end{bmatrix}$$

(a) Find all of the eigenvalues and eigenvectors of A.

(b) Find all of the eigenvalues and eigenvectors of B.

(c) Is A dianoalizable? Justify your answer.

(d) Is B dianoalizable? Justify your answer.

Problems for Section 2

5.2.1 Let $\mathbf{x}(t)$ be the t–dependent vector $\mathbf{x}(t) = \begin{bmatrix} t^2 \\ t \\ t^3 \end{bmatrix}$, and let $\mathbf{y}(t)$ be the t–dependent

vector $\mathbf{y}(t) = \begin{bmatrix} t^{-1} \\ t \\ t^{-2} \end{bmatrix}$, $t \neq 0$. Let A be the matrix $A = \begin{bmatrix} 1 & 0 & 0 \\ 2 & 3 & 0 \\ 1 & 3 & 3 \end{bmatrix}$. Compute:

(a) $\mathbf{x}'(t)$

(b) $\mathbf{y}'(t)$

(c) $\frac{d}{dt}|\mathbf{x}(t)|$

(d) $\frac{d}{dt}|\mathbf{y}(t)|$

(e) $(\mathbf{x}(t) \cdot A\mathbf{y}(t))' \big|_{t=1}$

(f) For which values of t, if any, are $\mathbf{x}'(t)$ and $\mathbf{y}'(t)$ orthogonal?

5.2.2 Let $\mathbf{x}(t)$ be the t–dependent vector $\mathbf{x}(t) = \begin{bmatrix} \cos(2t) \\ \sin(2t) \\ t \end{bmatrix}$, and let $\mathbf{y}(t)$ be the t–

dependent vector $\mathbf{y}(t) = \begin{bmatrix} t \\ \cos(t) \\ \sin(t) \end{bmatrix}$, $t \neq 0$. Let A be the matrix $A = \begin{bmatrix} 1 & 0 & 0 \\ 1 & 2 & 0 \\ 1 & 1 & 1 \end{bmatrix}$. Compute:

(a) $\mathbf{x}'(t)$

(b) $\mathbf{y}'(t)$

(c) $\frac{d}{dt}|\mathbf{x}(t)|$

(d) $\frac{d}{dt}|\mathbf{y}(t)|$

(e) $(\mathbf{x}(t) \cdot A\mathbf{y}(t))' \big|_{t=1}$

(f) For which values of t, if any, are $\mathbf{x}''(t)$ and $\mathbf{y}''(t)$ orthogonal?

Problems for Section 3

5.3.1 Compute the norms of the matrices in problem 5.1.5

5.3.2 Let A be the matrix

$$A = \begin{bmatrix} 2 & 1 \\ 0 & 1 \end{bmatrix}.$$

(a) Find the the unit vectors $\mathbf{u}$ such that $|A\mathbf{u}| \geq |A\mathbf{v}|$ for any other unit vector $\mathbf{v}$. **(b)** Find the the unit vectors $\mathbf{u}$ such that $|A\mathbf{u}| \leq |A\mathbf{v}|$ for any other unit vector $\mathbf{v}$.

5.3.3 Let A be the matrix

$$A = \begin{bmatrix} 0 & 1 & 2 \\ 0 & 0 & 1 \\ 0 & 0 & 0 \end{bmatrix}.$$

(a) Find the the unit vectors $\mathbf{u}$ such that $|A\mathbf{u}| \geq |A\mathbf{v}|$ for any other unit vector $\mathbf{v}$. **(b)** Find the the unit vectors $\mathbf{u}$ such that $|A\mathbf{u}| \leq |A\mathbf{v}|$ for any other unit vector $\mathbf{v}$.

Problems for Section 4

5.4.1 Let A be the matrix

$$A = \begin{bmatrix} 1 & 1 & 0 \\ 1 & 0 & 1 \\ 0 & 1 & 1 \end{bmatrix}.$$

(See problem 5.1.1).

(a) Compute e^{tA}.

(b) Solve the system of differential equations

$$x'(t) = x(t) + y(t)$$
$$y'(t) = x(t) + z(t)$$
$$z'(t) = y(t) + z(t)$$

with the inital conditions

$$x(0) = 1 \qquad y(0) = 2 \qquad z(0) = 1.$$

5.4.2 Let A be the matrix

$$A = \begin{bmatrix} -3 & 1 & 1 \\ 1 & -3 & 1 \\ 1 & 1 & -1 \end{bmatrix}.$$

(See problem 5.1.2).

(a) Compute e^{tA}.

(b) Solve the system of differential equations

$$x'(t) = -3x(t) + y(t) + z(t)$$
$$y'(t) = x(t) - 3y(t) + z(t)$$
$$z'(t) = x(t) + y(t) - z(t)$$

with the inital conditions

$$x(0) = 1 \qquad y(0) = 2 \qquad z(0) = 1.$$

5.4.3 Let A be the matrix

$$A = \begin{bmatrix} 1 & 2 \\ 2 & 4 \end{bmatrix}.$$

(See problem 5.4.3).

(a) Compute e^{tA}.

(b) Solve the system of differential equations

$$x'(t) = x(t) + 2y(t)$$
$$y'(t) = 2x(t) + 4y(t)$$

with the initial conditions $x(0) = 1$ and $y(0) = 2$.

(c) Compute the norm $\|A\|$ of the matrix A.

Problems for Section 5

5.5.1 Which of the following quadratic forms are positive, which are negative, and which are hyperbolic?

(a) $x^2 + 2xy + 3y^2$

(b) $x^2 + 4xy + 3y^2$

(c) $2xy - x^2 - 2y^2$.

5.5.1 Which of the following quadratic forms are positive, which are negative, and which are hyperbolic?

(a) $yz + 2xy + xz$

(b) $x^2 + 2y^2 + 3y^3 - 2xy - 2xz - 2yz$

(c) $2xy + 3x^2 + 4z^2$.